MICROBIOLOGY RESEARCH ADVANCES

THE ROLE OF PHOTOSYNTHETIC MICROBES IN AGRICULTURE AND INDUSTRY

MICROBIOLOGY RESEARCH ADVANCES

Additional books and e-books in this series can be found
on Nova's website under the Series tab.

MICROBIOLOGY RESEARCH ADVANCES

THE ROLE OF PHOTOSYNTHETIC MICROBES IN AGRICULTURE AND INDUSTRY

KESHAWANAND TRIPATHI
NARENDRA KUMAR
AND
GERARD ABRAHAM
EDITORS

nova
science publishers
New York

NOTICE TO THE READER

Additional color graphics may be available in the e-book version of this book.

Library of CongressCataloging-in-Publication Data

Names: Tripathi, Keshawanand, editor.
Title: The role of photosynthetic microbes in agriculture and industry / editors, Keshawanand Tripathi, Narendra Kumar, Gerard Abraham.
Description: Hauppauge, New York : Nova Science Publisher's, Inc., [2018] | Series: Microbiology research advances | Includes bibliographical references and index.
Identifiers: LCCN 2018038730 (print) | LCCN 2018039300 (ebook) | ISBN 9781536140330 (ebook) | ISBN 9781536140323 (hardcover) | ISBN 9781536140330 (ebook)
Subjects: LCSH: Photosynthetic bacteria--Research. | Agricultural microbiology. | Industrial microbiology.
Classification: LCC QR88.5 (ebook) | LCC QR88.5 .R65 2018 (print) | DDC 579.3/8--dc23
LC record available at https://lccn.loc.gov/2018038730

Published by Nova Science Publishers, Inc. † New York

Contents

PREFACE

Application of synthetic fertilizers and pesticides coupled with excessive irrigation has resulted in the deterioration of soil fertility and pollution of the environment leading to reduced agricultural production. Therefore, the big challenge that is facing the scientists, technocrats and policy makers is to enhance the agricultural production keeping in view of the increasing population and consequent requirement for food. This is a challenging task due to the depleting land resources and challenges faced by the global climatic changes. In order to maintain the productivity and profitability of agriculture sustainable agricultural management practices are therefore needed. In this context, the exploitation of the microbial resources appears to be sustainable and viable option. The photosynthetic microbes including microalgae and cyanobacteria that thrive successfully in a wide range of ecological habitats could play a leading role in enhancing the agricultural productivity and mitigating the global climatic changes. Microalgae and cyanobacteria have tremendous potential to be exploited as promising organisms in agriculture and industry.

Biomass of the photosynthetic microbes could be used for the production of food, feed, nutraceutical and bioenergy. The simple structural organization, ease of cultivation and amenability to experimental manipulation make these microorganisms an ideal choice for the scientists. Because of their potential tremendous development and improvement have taken place in the last two decades. The major emphasis was to obtain high rate biomass production through innovations in the cultivation techniques, harvesting techniques, and techniques for conversion of biomass to various biofuels. However, the recent advancements in the field of biotechnology and engineering sciences have changed the scenario altogether. It is expected that these cutting edge technologies may lead to make the cultivation and harvest economically viable to attain the desired level of production of the product concerned.

The book entitled *The Role of Photosynthetic Microbes in Agriculture and Industry* provides an account on the exploitation of photosynthetic microorganisms such as algae and cyanobacteria and the recent innovations from an industrial perspective. It is comprised of fourteen chapters written by experts in the relevant field. The chapters basically deal with algal production and mass multiplication, algae as source of bioenergy, bioremediation, CO_2 sequestration and the industrial and commercial application of algae. This book will help the postgraduate and research students and will serve as reference for scientists, professionals and engineers working on photosynthetic microbes. We are deeply indebted to all the contributors for their kind cooperation and interest. The edited volume provides a wealth of information based on contemporary developments in algal research. Dr. Gerard Abraham and Dr. Keshawanand Tripathi thank Head, Division of Microbiology, ICAR- Indian Agricultural research Institute for support and encouragement.

Dr. Keshawanand Tripathi
Dr. Narendra Kumar
Dr. Gerard Abraham

In: The Role of Photosynthetic Microbes… ISBN: 978-1-53614-032-3
Editors: K. Tripathi, N. Kumar et al.

Chapter 1

AN INTRODUCTION TO CYANOBACTERIA: DIVERSITY AND POTENTIAL APPLICATIONS

Balu Govinda Meshram [*] ***and***
Bhupal Baburao Chaugule
Department of Botany, Savitribai Phule Pune University,
Pune, India

ABSTRACT

Cyanobacteria (blue-green algae) are an ancient group of gram-negative prokaryotic and photosynthetic microorganisms that originated before 3.5 billion years ago. During this long period of existence on earth, they have evolved incredible diversity among their forms. They are inhabitants of almost all habitats on earth including extreme environments. Some of these are living symbiotically within the tissues of host plants and fixing atmospheric nitrogen. Thus they play a crucial role as primary producers and also contribute immeasurably towards the global nitrogen budget. One more notable and significant contribution of cyanobacteria was seen in the process of endosymbiosis, through which chloroplast originated in eukaryotic algae and thus they becomes photosynthetic, autotrophic organisms. Cyanobacteria are one of the most difficult and challenging group to classify. Earlier, cyanobacteria were traditionally classified based on morphological characters. Presently, they are being studied and classified based on a polyphasic approach. Therefore, cyanobacterial taxonomy has been substantially changed. Several interesting features of cyanobacteria provoked the researchers, as many of these are found with biotechnological potentials. So, these organisms are now being used for the benefits of mankind in several ways such as a food, feed, biofertilizer, in aquaculture, water quality assessment and pollution abatement, biofuels, secondary metabolites (vitamins, enzymes, and toxins) production etc. They are also being used

[*] Corresponding Author Email: balu.meshram@gmail.com.

worldwide in the pharmaceuticals and nutraceuticals industries. This chapter is focussed on diversity among the cyanobacteria, their habitats, and their potential applications in commercial exploitation for the betterment of mankind.

Keywords: cyanobacteria, diversity, habit, habitats, potential applications

INTRODUCTION

Cyanobacteria are the most primitive, oxygenic photosynthetic prokaryotes which possess chlorophyll a, as a major photosynthetic pigment. *Acaryochloris marina* is the only species of cyanobacteria in which chlorophyll d is a major pigment (Miyashita et al., 2003). Besides this, phycobiliproteins, an accessory pigments are also present that include phycocyanin (blue) and allophycocyanin (blue) which imparts typical blue-green colour to the organisms, and species may contain phycoerythrine that colours the cells red (Zakhia et al., 2008). These pigments harvest light in the form of photon and supply it to chlorophyll a during photosynthesis. About 20–30% of a global primary photosynthetic productivity originates only from cyanobacteria (Shevela et al., 2013). These are thought to be evolved about 3.5 billion years ago (Schopf, 2002), but based on some other reports, the actual time of their evolution is thought to be closer to 2.7 billion years ago (Lee, 2008). They represents one of the most ancient, diverse and abundant group of gram-negative prokaryotic microorganisms on the planet (Komárek, 2003), because of whom oxygen level rose in the atmosphere and evolution of eukaryotic organisms began (Chapman, 2013). Based on the cellular properties and prokaryotic structure of cell constitutes, Stanier et al. (1978) proposed a name 'Cyanobacteria' to this group of organisms, and argued to be included under bacteriological code. Therefore, under the International Code of Botanical Nomenclature (now, International Code of Nomenclature for algae, fungi and plants), they are treated as 'blue-green algae', while as 'cyanobacteria' under the International Code of Nomenclature of Bacteria. Recently, Komárek and Anagnostidis (1998, 2005) and Komárek (2013) have introduced one more term 'Cyanoprokaryota' for this group. According to Guiry (2012), the estimated number of cyanobacteria is about 8000, of which described species are about 5000, with about 3000 undescribed species. Cyanobacteria are highly adaptable to varied environmental conditions as they had a long existence on the earth, and occur luxuriantly in almost all types of the natural ecosystems such as different types of soil, freshwater bodies, oceans, saline backwaters, estuaries, and hyper saline saltpans (Muthukumar et al., 2007). Many of them are also recorded from other extreme environments such as thermal springs (up to 73-74°C) (Ward and Castenholz, 2002), cold polar environments such as ice shelves, glaciers, glacial meltwater streams and ice capped lakes (Vincent, 2002), and hot deserts (Wynn-Williams, 2002). They are also found symbiotically associated with fungi in forming lichens, and other plants from the divisions: Bryophyta (hepatics, hornworts and mosses), Pteridophyta (*Azolla*), gymnosperms (cycads) and angiosperms (*Gunnera*) (Graham and Wilcox, 2000; Adams and Duggan, 2008; Sarma, 2013). In such symbiotic associations, cyanobacteria have the ability to fix atmospheric nitrogen, and then transfer it to the partner which is the key factor in the relationship (Whitton and Potts, 2012). Because of nitrogen fixing ability, their application in paddy fields was found more productive in terms of high yield in rice (De, 1939). Now more than 125 strains of cyanobacteria are known to fix nitrogen (Roger and Kulasooriya, 1980). Another great and significant input of cyanobacteria is the evolutionary event of endosymbiosis through which heterotrophic phagotrophs emerged into eukaryotic, photosynthetic organisms with membrane bounded plastid (Graham, and Wilcox, 2000; Lee, 2008; Shevela et al., 2013). Rhodophytes, chlorophytes and glaucophytes are eukaryotic algae which having plastids directly derived from cyanobacteria through primary endosymbiosis while other groups of algae have secondary or tertiary plastids (Kutschera and Niklas, 2005; Delwiche, 2007). Thus, cyanobacteria are photosynthetic ancestors of plastids in eukaryotic algae (Sharma et al., 2011). Though cyanobacteria make positive contributions to global biodiversity and environment through carbon and nitrogen fixation, they also cause severe problems by bloom formation and toxin production in aquatic bodies. Certain species of cyanobacteria such as *Microcystis aeruginosa*, *Anabaena flos-aquae*, *Aphanizomenon flos-aquae*, *Cylindrospermopsis raciborskii* and *Nodularia spumigena* form blooms in eutrophic water bodies and produce potent toxins than other bloom forming eukaryotic algae (Kulasooriya, 2011; Zaccaroni and Scaravelli, 2008). About 46 species of cyanobacteria producing toxins have been identified, of these 60% studied strains are proved to contain toxins, and thus, they deteriorate the quality of drinking water which may be hazardous to humans and animals (Zaccaroni and Scaravelli, 2008). Algae are used as a source of food in a variety of ways. Human consumption of various algae in a daily food is well known since thousands of years (Spolaore, et al., 2006). Other uses of algae include, as animal feed, chemicals and pharmaceuticals, cosmetics, biofuels, fertilizers, in waste water treatment and aquaculture (Hallmann, 2007). Besides, cyanobacteria also produce a wide variety of chemically unique secondary metabolites that having biotechnological potentials (Singh et al., 2002; Sharma et al., 2011).

HABITAT DIVERSITY

Cyanobacteria occur virtually in every habitat on earth, such as in aquatic water bodies (e.g., ponds, pools, ditches, tanks, dams, lakes, rivers, streams, and ocean) and on land (e.g., soil, rock, cemented wall, tree bark etc.).

Figure 1. Cyanobacterial habitats (A-I): A. Thick cyanobacterial mat on the rocky substratum under streaming water, B. N_2-fixing cyanobacteria in rice field, C. Running water in the river (Lotic water body), D. *Nostoc* balls abundantly growing on wet soil, E. *Nostochopsis lobatus* in a cemented tank (attached and secondarily free floating), F. *Nostoc* balls on wet stones (lithophytes), G. Epiphytic cyanobacteria on aquatic plants in a pond, H. Salt accumulating soil, I. Cyanobacteria in hot water spring.

They can also be seen growing as epiphytes (on the surface of other plants) and endophytes (inside plants), as well as in extreme environments (extremophiles). Thus, cyanobacteria form a ubiquitous group of organisms that are universally distributed (Figure 1).

AQUATIC CYANOBACTERIA

Cyanobacteria are common in aquatic environments and grow in different water bodies, where water may be stagnant (Lentic) (Figure 1E, 1G) or running (Lotic) (Figure 1A, 1C). In such habitats, they are either free floating on the surface of water (Planktic), e.g., *Merismopedia*, *Microcystis*, *Spirulina*, *Oscillatoria, Anabaena*, *Aphanizomenon*, or they are initially attached to the substratum and later becomes free floating (Tychoplanktic) e.g., *Tychonema*, *Blennothrix*, *Cylindrospermum*, *Gloeotrichia*, and *Nostochopsis*. Some cyanobacteria grow on the surface of aquatic plants (Epiphytic) (Figure 1G) or floating wooden logs, e.g., *Rivularia*, while others grow on the surface of aquatic animals (Epizoic), e.g., *Synechocystis* and *Oscillatoria*. Such cyanobacterial growths on the surface of animals like harvestman *Neosadocus* sp. (Machado and Vital, 2001) and *Prionostemma* sp. (Proud et al., 2012) have been recorded. On the contrary, some cyanobacteria grow inside the body of animals (Endozoic), and about 14 species of Oscillatoriaceae have been reported from digestive and respiratory tract of certain vertebrates (Kumar and Paliwal, 2006).

TERRESTRIAL CYANOBACTERIA

Most of the cyanobacteria are terrestrial, and grow on wet surface of the land (Figure 1D). These are classified as lithophytic (on the rock and stones) (Figure 1F) e. g. *Aphanothece*, *Chroococcus*, *Asterocapsa*, *Pleurocapsa*, *Scytonema*, *Petalonema*, *Nostoc*, *Anabaena* etc.; epipelic (on mud) e. g. *Aphanocapsa* and *Phormidium*. Some others (e.g., *Aphanothece*, *Lyngbya*, *Scytonema*, *Hapalosiphon*, and *Stigonema*) grow on the bark of trees (Corticolous) along with mosses.

CYANOBACTERIA IN EXTREME ENVIRONMENTS

Among all algae, cyanobacteria are the group of organisms that first evolved on the earth. During this long period of their existence, they have evolved many times as per the gradual environmental changes. This led in to their structural and functional modifications that assisted them to develop wide tolerance capacities in order to cope with varied environmental conditions (Mandal and Rath, 2015). Cyanobacteria had to sustain even under extreme conditions and therefore, they acclimatized and adapted accordingly to become potential extremophiles. According to cyanobacterial occurrence in these extreme conditions they are classified as thermophilic, psychrophilic and halophilic.

Thermophilic Cyanobacteria

Cyanobacteria are generally considered to be thermophilic when they grow at their optimal growth temperature range above 45°C (Castenholz, 1969). Such cyanobacteria either occur in hot waters of thermal springs (Figure 1I) or in soil crusts of deserted areas of the world. Thermal springs are scattered on most of the continents except Antarctica, and on many island groups. These resulted due to recent volcanic and/or active faulting activities (Ward et al., 2012). During such activities, surface water moves deep inside the earth, are heated and comes again to surface as hot spring water rich in sulphide. The emitted water from thermal springs provides stable temperature gradients suitable for cyanobacterial growths. Temperature of hot spring waters ranges from 51°C to 74°C (Ward et al., 2012), and sometimes may reach up to 80°C (Komárek and Anagnostidis, 1998). Species of *Synechococcus* are worldwide known from thermal springs, except Iceland, where thermophilic *Synechococcus* are completely absent (Ward et al., 2012). The maximum temperature limit tolerated by *Synechococcus* spp. during their active growth is 73-74°C, while other thermophilic cyanobacteria, e.g., *Chlorogloeopsis*, *Synechocystis*, *Phormidium*, *Oscillatoria*, *Scytonema* and *Mastigocladus laminosus* usually grow below this temperature range (Ward et al., 2012).

Psychrophilic Cyanobacteria

Among extreme habitats, cryosphere is the one where environmental conditions are very harsh for living organisms due to extreme cold and lack of liquid waters. The regions generally considered as cryosphere where temperatures remain below freezing point (0°C) and usually covered with ice during most of the year, and such localities particularly include alpines, both Polar Regions, the Arctic and Antarctic regions (Quesada and Vincent, 2012). The microorganisms growing in these habitats are called as psychrophiles (temperature optima of ≤15°C and a maximum temperature for growth ≤20°C) and psychrotolerant (temperature optima >15°C and upper limit of growth as high as 40°C) (Nadeau and Castenholz, 2000). As a result of such harsh conditions, reduced biodiversity is seen, nevertheless cyanobacterial populations predominantly represent in these regions (Zakhia et al., 2008). Such growth of cyanobacteria in cryosphere was first noted during Nordenskiölds expedition in 1870 over the Greenland Ice Cap, where cryoconite (cold rock dust) was dominated by *Calothrix parietina* (Vincent, 2002). As psychrophilic organisms, cyanobacteria have successfully defeated two major challenges needed for their survival, i.e., low temperature and low viscosity by adapting according to cryospheric conditions (D'Amico et al., 2006). About 35 unicellular and filamentous (including heterocystous N_2-fixing) species of cyanobacteria are recorded from

Antarctica, with dominant growths of *Gloeocapsa* sp., *Lyngbya attenuata*, *Oscillatoria limosa*, *O. limnetica*, *O. tenuis* and *Phormidium frigidum* (Pandey et al., 2004).

HALOPHILIC CYANOBACTERIA

Halophiles are salt loving microorganisms including cyanobacteria that inhabit hypersaline environments, such as hypersaline lakes, coastal hypersaline lagoons, saline springs, salt flats, man-made saltern evaporation ponds, and other environments (Figure 1H) where salt concentrations exceed to that of seawater (35 g/l) (Oren, 2012a). Although, the occurrence of cyanobacteria in saturated salt solutions are known from a long time, their first description from concentrated brines was probably provided by Hof and Fremy (1933). Cyanobacteria form one of the major components in these environments and contribute to primary production (Oren, 2015). With respect to salinity of environments, various terms are used to describe these habitats with inhabiting cyanobacteria (see Oren, 2012a). At high salt concentrations, they form dense communities in planktonic as well as benthic biota (Oren, 2012a). A very dense mat forming habit due to cyanobacteria can be seen prominently in hypersaline environments with salt concentrations exceeding 250 g/l, but decrease in their diversity with increasing salinity can also be noted (Chatchawan et al., 2011; Oren, 2012b). This microbial mat predominantly contains unicellular and filamentous cyanobacteria (DasSarma and Arora, 2001; Chatchawan et al., 2011). The most widely distributed species in hypersaline environments are filamentous *Coleofasciculus chthonoplastes* (=*Microcoleus chthonoplastes*) and *Halospirulina tapeticola* (=*Spirulina subsalsa*), and unicellular *Aphanothece halophytica* (Oren, 2002), whereas heterocystous forms are likely to be absent (Oren, 2014). Although many cyanobacteria with high salt tolerance ability are recently recorded from the world, their taxonomy is poorly known and extremely confusing, as the same organism has been identified with more than one name in the literatures (Oren, 2002). In order to cope with various concentrations of salts in the hypersaline habitats, cyanobacteria synthesize and accumulate considerable amount of compatible solutes. The first compatible solute found in cyanobacteria was heteroside glucosylglycerol (2-O-α-D-glucopyranosyl-(1→2)-glycerol) (Borowitzka et al. 1980), and was consecutively recorded in most of the moderately halophilic species. (e.g., *Microcoleus* sp.) (Seckbach and Oren, 2007). The less salt-tolerant cyanobacteria generally make the use of disaccharides, like sucrose and trehalose as osmotic solutes (Mackay et al., 1984; Reed et al., 1986; Erdmann and Hagemann, 2001), whereas high salt-tolerating cyanobacteria (e.g., *A. halophytica* and *Halospirulina* sp.) produce glycine betaine (Seckbach and Oren, 2007).

SYMBIOTIC CYANOBACTERIA

Besides free living aquatic and terrestrial habitats, a number of cyanobacteria live in symbiotic associations with a wide range of eukaryotic hosts. The divers host organisms include algae, fungi (forming lichen), other plants like bryophytes, pteridophytes, gymnosperms and angiosperm, and animal invertebrates (sponges, ascidians and echiuroid worms) (Peters, 1991; Usher et al., 2007; Li, 2009). Mostly the cyanobacterial symbionts are nitrogen fixers and provide fixed nitrogen to their hosts in exchange of fixed carbon by host plants (Graham and Wilcox, 2000). Among eukaryotic algae, very few organisms show symbiotic associations with cyanobacteria, particularly with diatoms and some dinoflagellates (Gordon et al., 1994; Janson, 2012). The filamentous heterocystous marine cyanobacteria namely, *Richelia intracellularis* and *Calothrix rhizosoleniae* have been found in symbiotic association with marine diatoms *Rhizosolenia*, *Hemiaulus* and *Chaetoceros*, and may make major contribution to the N budget of the oceanic areas where they are abundant (Foster et al., 2011; Adams et al., 2012). Many unicellular diazotrophic cyanobacteria (e.g., *Cyanothece*) also show such association with other diatoms (e.g., chain forming *Climacodium frauenfeldianum*) (Falco´n et al., 2002). About six genera of heterotrophic dinoflagellates viz. *Ornithocercus*, *Histioneis*, *Citharistes*, *Parahistioneis*, *Dinophysis* and *Amphiosolenia* are known to be symbiotic with a range of unicellular cyanobacteria. It is well known that nitrogen fixing cyanobacteria also coexist with symbiotic dinoflagellates (zooxanthellae) of the corals (Lesser et al., 2004; Lema et al., 2012). Cyanolichens are considered as most traditional model case of symbiosis (Mutualism) where both the partners are reciprocally benefited. Approximately 17,000 species of lichens are known, of which, about 8% species show cyanobacteria as symbionts (Oren, 2014). *Nostoc* is the most common genus occurring in lichens, but other cyanobacterial genera encountered are *Stigonema*, *Scytonema*, *Dichothrix*, *Calothrix*, *Chroococcus*, *Gloeocapsa* and *Hyella* (Rikkinen, 2013; Oren, 2014). Bryophytes are non-vascular, amphibious plants which are divided into three classes viz. Hepaticae, Anthocerotae and Musci. More than 340 genera are described under class Hepaticae (liverworts), of which two genera *Marchantia* and *Porella* show epiphytic association and other two genera *Blasia* and *Cavicularia* show endophytic association with cyanobacteria (e.g., *Nostoc*, *Stigonema* and *Calothrix*) (Adams and Duggan, 2008). Out of six genera from the class Anthocerotae (hornworts), four genera *Anthoceros*, *Phaeoceros*, *Notothylas* and *Dendroceros* have been found with endophytic cyanobacterial partners (Meeks, 2009). In a case of Musci (mosses), cyanobacterial associations are mostly epiphytic, and special symbiotic structures present in the genera of above two classes have not yet been described (Solheim and Zielke, 2002). Among pteridophytes (fern), *Azolla* is the only genus which show symbiotic association with cyanobacterium *Anabaena azollae*, and this has gained more attention due to its biofertilizer potential (Lechno-Yossef and Nierzwicki-Bauer, 2002). Interestingly, in this association, hereditary transfer of cyanobiont from one generation to the next can be seen (Adams et al., 2012). Cycads are the only gymnosperms that show cyanobacterial association. Approximately 160 species of cycads are described over 11 genera (*Cycas*, *Stangeria*, *Bowenia*, *Ceratozamia*, *Chigua*, *Dioon*, *Encephalartos*, *Lepidozamia*, *Macrozamia*, *Microcycas* and *Zamia*), and most of these having cyanobionts (e.g., *Nostoc*) in their highly specialised type of lateral roots, called coralloid roots (Lindblad et al., 1985; Lindblad, 2009; Yamada et al., 2012). The *Gunnera-Nostoc* is the only known angiosperm-cyanobacterial nitrogen fixing symbiosis till date (Osborne and Bergman, 2009).

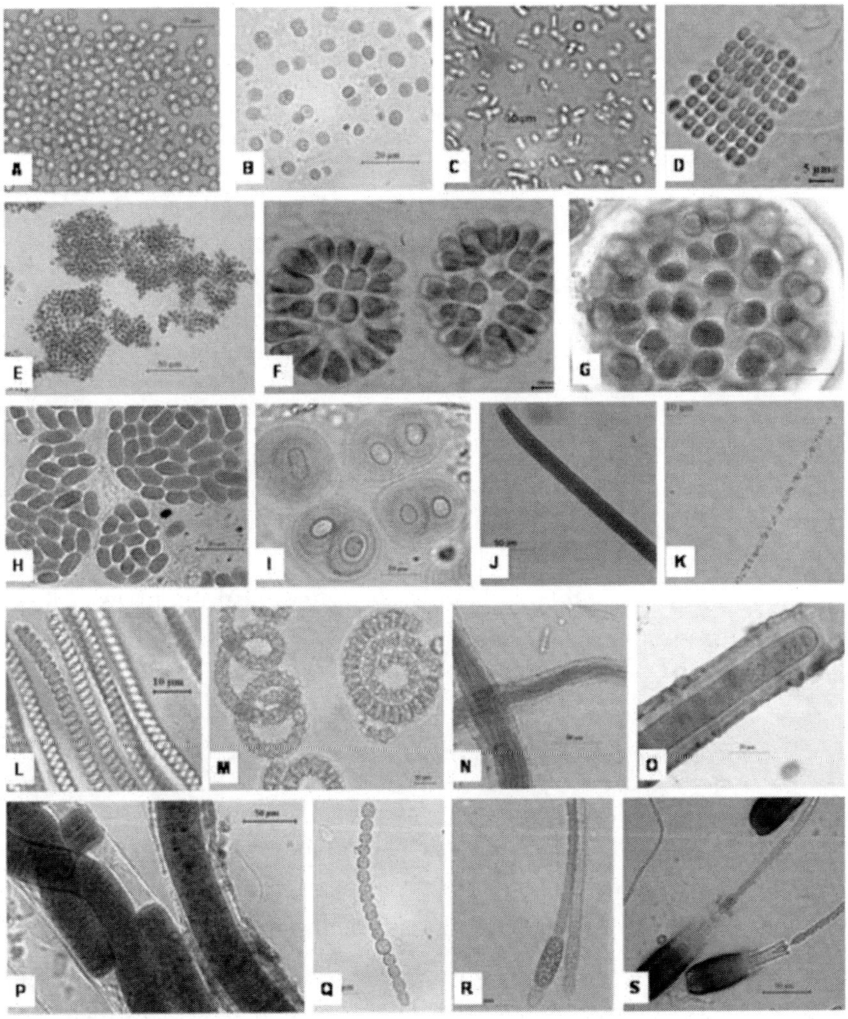

Figure 2. (Continued).

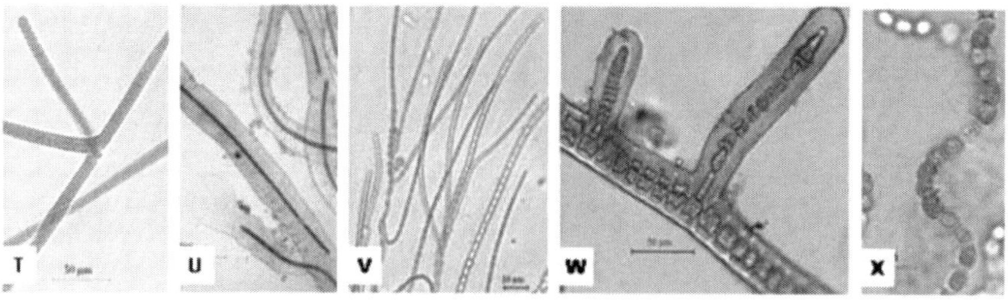

Figure 2. Thallus diversity in cyanobacteria (A-X): A. Aggregated cylindrical cells in the colony of *Aphanothece pallida,* B. Spherical cells of *Aphanocapsa muscicola,* C. Elongated cells of *Synechococcus elongates,* D. Flat plate of *Merismopedia tenuissima* with perpendicularly arranged rows of cells, E. Colony of *Microcystis flos-aquae* with densely packed numerous spherical cells, F. Colonies of *Gomphosphaeria salina* with radially arranged cells, G. Macroscopic colony of *Asterocapsa divina* with thick sheathed envelope, H. Colonies of *Rhabdoderma cavanillesiana* with long, elongated blue-green cells enveloped by thin mucilage layer, I. Two celled colonies of *Gloeothece fusco-lutea* with yellowish-brown lamellated sheath, J. Trichome of *Oscillatoria froelichii,* K. Straight trichome of *Geitlerinema amphibium,* L. Densely spirally coiled trichomes of *Spirulina subsalsa,* M. Loosely spirally coiled trichomes of *Arthrospira platensis* with distinct cross walls, N. Filaments of *Microcoleus paludosus* enclosing many twisted trichomes in bundles, O. Filamentous *Lyngbya ceylanica* var. *major* with thick, reddish brown and lamellated sheath, P. Filaments of *Blennothrix ganeshii* with 1-3 trichomes, Q. *Anabaena sphaerica* with spherical intercalary heterocyst, R. Trichomes of *Cylindrospermum michailovskoense* with large ellipsoidal akinete and basal slightly elongated heterocyst, S. *Gloeotrichia pilgeri* containing akinete and basal heterocyst, with brown coloured incomplete sheath, T. Heterocystous *Scytonema javanicum* with geminate pseudobranching, *U. Petalonema alatum* with pseudobranching, enveloped in highly thick, yellowish sheath, and also with funnel shaped pieces at the apex, V. Truly branched *Nostochopsis lobatus,* W. Truly branched *Stigonema ocellatum,* X. Unbranched filament of *Desmonostoc muscorum* with intercalary heterocyst and many akinetes in a series.

THALLUS DIVERSITY IN CYANOBACTERIA

Cyanobacteria comprise a group of highly diverse organisms in terms of their thallus structure that range from simplest unicellular to most complex branched filaments.

Coccoid Cyanobacteria (*Chroococcales*)

Coccoid thalli occur either as a single cells or in an aggregations and form variously shaped micro or macroscopic colonies within a common mucilaginous sheath (Figure 2G), wherein cells are either arranged in a perpendicular rows (Figure 2D) forming flat plate, radially arranged (Figure 2F) or randomly distributed in a spherical, hemispherical or irregular colonies (Figure 2E). Cells in a colonies may be either spherical (Figure 2B, 2E) or elongated and cylindrical (Figure 2A, 2C, 2H); these may be loosely or densely packed, and numbers may vary from few to many (Figure 2E). The thickness and colour of enveloping sheath may also vary from species to species (Figure 2H-I). The sheath is an external layer that protects inner cells from drying. Red and blue sheaths are generally found in cyanobacteria that occur in acidic and basic soils respectively, whereas yellow and brown sheaths are common in specimens that grow in a hypersaline environment (Lee, 2008).

Filamentous, Non-Heterocystous Cyanobacteria (*Oscillatoriales*)

Filamentous habit in cyanobacteria is due to repeated cell divisions occurring in a single plane at right angles to the main axis of the filament (Mur et al., 1999). Thus, a chain of cells with multicellular structure and devoid of sheath (Figure 2J) is formed, called a trichome that may be constricted or unconstricted. Trichomes may be straight (Figure 2K) or coiled (Figure 2L-M), and sometimes as aggregated or in twisted bundles, like a rope (Figure 2N). These trichomes covered with a sheath are known as filament. There may be one (Figure 2O), few (Figure 2P) or many (Figure 2N) trichomes per sheath. Such uniseriate, unbranched, filamentous structure without further cellular differentiation into heterocyst and akinetes are the characteristic features of the order Oscillatoriales. Variation in cell shape and size can also be seen among such Oscillatoriales.

Filamentous, Heterocystous Cyanobacteria (*Nostocales*)

Other filamentous cyanobacteria are characterized by the presence of three types of cells viz. vegetative cells, heterocysts and akinetes. Heterocysts (Figure 2Q-S) are larger cells and their differentiation from vegetative cells under nitrogen-limiting conditions is one of the simplest examples of cellular differentiation in multicellular organisms (Kaplan-Levy et al., 2010; Torres-Sánchez et al., 2015). Heterocysts are photysynthetically inactive cells as they neither fix CO_2 nor produce O_2, and thus, internal environment in these cells is anoxic which is ideal for nitrogen fixation by nitrogenase enzyme (Lee, 2008). They are enveloped by a thick cell wall composed of polysaccharides and glycolipids that reduce the permeation of atmospheric gases including oxygen (Zhang et al., 2006). Heterocyst in the trichome can be seen either at intercalary (Figure 2Q) or at terminal (Figure 2R-S) positions. Sometimes terminal heterocyst may born laterally, either directly on a main filament or at end cell of a short branch (1-4 celled) and are called as lateral sessile and lateral pedicelate heterocysts respectively (Figure 2V). Their differentiation and pattern formation has excellently been studied in the model cyanobacterium, *Anabaena* sp. strain PCC 7120 (Zhang et al., 2006; Torres-Sánchez et al., 2015; Muñoz-García and Ares, 2016). Another specialised cell is the akinete (Figure 2R-S) which is a resting cell and found only in certain species of Nostocales. These are dormant, spore like, non-motile cells with highly thick walls and are differentiated from vegetative cells. Akinetes are larger (up to ten folds) than vegetative cells, completely filled with food reserve and DNA, and play an important perennating role in cyanobacteria (Kaplan-Levy et al., 2010). Akinetes are formed under extreme environmental conditions, and their formation is triggered by certain environmental factors such as a low temperature, desiccation, increased level of salts and iron depletion (Tomitani et al., 2006; Olsson-Fransis et al., 2009; Carey et al., 2012). The shape, size and structure of akinetes greatly vary among the species of Nostocales. The heterocystous, filamentous cyanobacteria may be unbranched (Figure 2Q-R, 2X) or with a true branches (Figure 2V-W). Sometimes false branching (single or geminate) (Figure 2T-U) due to necridia (dead cells) formation can also be seen in some genera of this group.

Potential Applications of Cyanobacteria

Cyanobacteria are one of the most interesting groups of micro-organisms in terms of their habit and habitat diversity, and also due to their adaptive survival strategies in extreme environments. The unique and potential properties of cyanobacteria have made them more popular organisms and have fascinated the world to use them for the betterment of mankind. In this world of biotechnology, cyanobacteria has found as a most suitable resource with a number of potential applications.

Food, Feed and Nutraceuticals

Among the edible algae, cyanobacteria have a long history as human food. As told in an old text dating back to the Jin Dynasty (AD 265-316), Chinese have first used a cyanobacterium *Nostoc flagelliforme* as a source of food for about 2000 years (Spolaore et al., 2006; Barsanti and Gualtieri, 2014). Most of the cyanobacteria are relatively low in lipids but are relatively rich in carbohydrates and remarkably rich in proteins, which have made them an excellent source of nutrients not only for humans but also for domestic animals (Aaronson, 2000). Some strains of the genera *Spirulina*, *Arthrospira*, *Nostoc*, *Anabaena* and *Aphanizomenon* are utilized as food and feed in many countries such as Chile, Mexico, Peru and Philippines (Abed et al., 2009; Packer et al., 2016). In all, *Spirulina* is best known as a food supplement due its nutritional composition and digestibility, and therefore, it is being used in human nutrition as a health food supplement, as well as feed supplement in aquaculture, the aquarium and poultry industries. It is highly rich in a dietary proteins that accounted about 60-70% (Moreira et al., 2011), which is nearly threefold higher than in a beef (Kovač et al., 2013). It also contains many essential amino acids (histidine, isoleucine, leucine, lysine, methionine, phenylalanine, valine, threonine, and tryptophan) (Hudek et al., 2014), vitamins (vitamin

A, vitamins B_1, B_2, B_6, B_{12}, and C), and minerals (calcium, iron, phosphorous, magnesium, zinc, selenium, copper, manganese, chromium, potassium, and sodium) (Gutierrez-Salmean et al., 2015). The amount of β–carotene in this genus is tenfold more than any other food including carrots, and vitamin B_{12} is also accounted more as compared to any fresh plant and animal food sources (Kovač et al., 2013). It is found that, the administration of *Spirulina* at a dose of 10 g/day has significantly improved the nutritional status of malnourished children from Central Africa (Matondo et al., 2016). Like *Spirulina*, the genus *Arthrospira* also has a long history regarding its use as a health food in human nutrition. The past history of its utilization by African tribes living along Lake Chad is well described by Dangeard (1940), Brandily (1959), and Léonard and Compère (1967). These tribal peoples were making a hardened dark cake, called 'Dihe' (local name) from harvested and sun-dried *Arthrospira*, and were using it in their daily food preparations (Ciferri, 1983; Habib et al., 2008). Nearly at the same time period or even earlier, *Arthrospira* was also reported from Lake Texcoco, Mexico, where native peoples were using it as a food (Ciferri, 1983). Kanembu peoples in the Prefecture of Lac (Chad) still harvest and sell dried 'dihe' in the market for earning money and also use it as a food supplements (Habib et al., 2008). Presently, the local trading value of 'dihe' is around US$ 100,000 (Chu, 2012). Toxicological assessments by Yang et al., (2011) has proved *Arthrospira platensis* (=*Spirulina platensis*) as a safe dietary supplement for human consumption. *Arthrospira* is also rich in proteins, vitamins and some polyunsaturated fatty acids (e.g., linoleic acid, γ-linolenic acid), and therefore, it is used as a food for humans (Muhling et al., 2005; Karkos et al., 2011), and as feed for fish, shrimps and poultry in many countries like China and Japan (Habib et al., 2008). In addition, NASA (National Aeronautics and Space Administration) and ESA (European Space Agency) has also selected it as one of the primary foods for their astronauts during long-term space missions (Deng and Chow, 2010, Karkos et al., 2011). Thus, *Arthrospira* becomes most valuable and commercial microalga in terms of US$ after *Chlorella*, and in terms of total biomass produced, its market is twice or more than that of *Chlorella* (Sili et al., 2012). *Nostoc* is one of the oldest genera described under cyanobacteria. Since centauries some species of *Nostoc* such as *N. commune*, *N. flagelliforme* and *N. sphaeroides* are being used as a food in many countries including China, Japan, Thailand, Philippines, Peru, Fiji, Ecuador, Java, Mongolia, Siberia, Mexico and Nordic countries (Han et al., 2013). *N. commune* contains high amount of protein, calcium and vitamin A. As vitamin A is more in this species, it can be used for curing of nyctalopia (night blindness) (Yang et al., 2011). Toxicological studies on *N. commune* and *N. flagelliforme* showed no toxic side effects, and are found safe for human consumption as a food supplements (Takenaka et al., 1998; Yang et al., 2011). On the contrary, a neurotoxic non-protein amino acid, *β-N*-methylamino-L-alanine (BMAA) was significantly detected in 21 different *N. commune* collected from highland lakes in the mountains of Peru (Johnson et al., 2008), and in *N. flagelliforme* samples from China and Chinese markets in the United States and the United Kingdom (Roney et al., 2009). This has raised questions on safe consumptions of these cyanobacterial foods, and therefore, concern research is needed in this field. Compared to other food algae (*Nostoc*, *Spirulina* and *Arthrospira*), *Aphanizomenon flos-aquae* has a short history of consumption by humans. It was exploited in the early 1980's as a nutritious food when its bloom was first harvested from the Lake Klamath (Oregon, USA) (Carmichael et al., 2000). It also contains proteins (62%), carbohydrates (23%) and low lipids (3%) (Becker, 2007), essential amino acids (Becker, 2004), and vitamins (Becker, 1994). It was known that, the amount of vitamin B_{12} is very high in the genus *Aphanizomenon*, but it is not a true vitamin B_{12}, instead it is only a pseudovitamin B_{12} which is an inactive corrinoid, and is hardly absorbed in the mammalian intestine (Becker, 2013). Therefore, *Aphanizomenon* is not suitable for use as a vitamin B_{12} source in the humans (Miyamoto et al., 2006). *Nostochopsis lobatus* is one more cyanobacterium emerged recently as a potential source of nutritious food in China, Thailand and India (Tiwari, 1978; Chu and Tseng, 1988; Peerapornpisal et al., 2006). The local tribes in Nan Province of Northern Thailand named it as 'Lon' and are using it as their traditional food in salad dish called 'Yum Lon' (Thiamdao et al., 2012).

Chemicals and Pharmaceuticals

Cyanobacteria are emerging as an important source of a variety of potential bioactive and biotechnologically relevant chemicals. Many of these have substantial commercial value that can lead to the development of new drugs. The marine cyanobacterium, *Lyngbya majuscula* is found with such potential and show variety of chemical structures such as polyketides, lipopeptides, kalkitoxins and many others (Chu, 2012). About 424 natural products obtained from marine cyanobacteria are listed in the MarinLit database that includes 40.2% lipopeptides, 5.6% pure amino acids, 4.2% fatty acids, 4.2% macrolides, and 9.4% amides (Burja et al., 2001). Many biomedically interesting bioactive compounds are produced by cyanobacteria, and some of the potent pharmaceutically important compounds with anticancerous activities are Borophycin (*Nostoc spongiaeforme* var. *tenue*), Borophycin-8 (*Nostoc linkia*), Apratoxin A, D (*L. majuscula*), Apratoxin E-G (*L. bouillonii*), Cryptophycin 1 (*N. linkia*), Cryptophcin-8 (*N. spongiaeforme*), Largazole (*Symploca* sp.), Dolastatin-10 (*Symploca* sp.), Dolastatin-12 (*Leptolyngbya* sp.), Dolastatin-15 (*Lyngbya* sp.), Symplostatin-1 (*S. hydnoides*), Calothrixin A, B (*Calothrix* sp.), Caylobolide B (*Phormidium* spp.), Ankaraholide A (*Geitlerinema* sp.), Ethyl tumonoate A (*Oscillatoria margaritifera*), and Curacin A (*L. majuscula*) (Vijayakumar and Menakha, 2015). Some cyanobacteria have also been identified with anti-HIV compounds, for instance Cyanovirin-N, is isolated from *Nostoc ellipsosporum* (Gustafson et al., 1997) while Scytovirin from *Scytonema varium*

(Bokesch et al., 2003). In addition, five novel diacylated sulfoglycolipids and four novel acylated diglycolipids, isolated from *Scytonema* sp. (TAU strain SL-30-1-4) and *Oscillatoria raoi* (TAU strain IL-76-1-2) respectively, were also found inhibiting HIV-1 RT enzymatic activity (Reshef et al., 1997). Few more bioactive compounds with anti-HSV-1 activity include Nostoflan from *Nostoc flagelliforme* (Kanekiyo et al., 2005) and Calcium-Spirulan from *Spirulina platensis* (Mader et al., 2016). Likewise, many more diverse chemical compounds with antibacterial, antifungal, antiprotozoal and immunomodulatory activities have been identified from various cyanobacteria (Singh et al., 2011). Thus, discoveries of a wide range of novel compounds with different bioactivities have provoked the scientific communities again for the search of unexplored cyanobacteria and also for their extraordinary novel compounds with pharmaceutical and biotechnological potentials. Cyanobacteria produce certain enzymes of commercial value, among which are restriction endonucleases used in molecular biology and biotechnological studies. Many such cyanobacteria with a rich source of type-II restriction endonucleases includes, strains of *Nostoc*, *Anabaena*, *Aphanothece*, *Dactylococcopsis*, *Aphanizomenon*, *Microcystis* and *Planktothrix* (Whitehead and Brown, 1985; Lyra et al., 2000). These site-specific restriction endonucleases are A*fl*I, A*fl*II, A*fl*III from *Anabaena flos-aquae* (Whitehead and Brown, 1985), N*sp* MAC I (isoschizomer of BglII) from *Nostoc* sp. MAC PCC 8009 (Lau et al., 1985); A*sp*83/1I, A*sp*83/1II and A*sp*90I from *Anabaena* strains, A*pc*TR183I from *Aphanizomenon* strains, M*sp*199I from *Microcystis* strains, P*sc*2I, P*sc*27I and P*sc*28I from *Planktothrix* strains, and many more (A*va*II, A*va*I, A*su*II) from other cyanobacteria (e.g., *Anabaena variabilis*) (Lyra et al., 2000), and O*fo*I from *Oscillatoria foreaui* (Saravanan et al., 2003). Another class of such enzymes, called nicking endonucleases which cleave only one strand of DNA duplex has been recently reported from *Chroococcus minutus* by Sundararajan et al. (2010). The increasing levels of green house gases are continuously depleting ozone layer and therefore, unfiltered sunlight with ultraviolet radiations (UVR) is directly reaching to earth that are harmful to all living organisms. To cope with, cyanobacteria occurring in extreme environments have developed some defense mechanisms to avoid or minimize photodamage through the production of UVA absorbing sunscreen pigment i.e., Scytonemin, and UVB absorbing mycosporine-like amino acids (MAA) (Rastogi et al., 2015). Scytonemin is a lipid soluble indole alkaloid pigment molecule accumulating with yellow brown colour in a polysaccharide sheath, and is reported exclusively among cyanobacteria (e.g., *Scytonema* sp.) (Grewe and Pulz, 2012). Although, it is not universally found in all living organisms, about 300 cyanobacteria are described with scytonemin content (Soule and Garcia-Pichel, 2014). Mycosporine-like amino acids are water soluble, colourless compounds, and are mostly induced on exposure to UVA and UVB (Singh et al., 2008, Rastogi et al., 2015). These sunscreen compounds from cyanobacteria are commercially important as they can also be used for protecting humans from the effects of UV radiations that can cause skin related diseases like skin cancer.

Phycobiliproteins

All photosynthetic organisms contain pigment systems that capture light energy required for photosynthesis. Three main types of pigments in plants are chlorophylls, carotenoids and phycobiliproteins. Among cyanobacteria, phycobiliproteins are one of the most abundant proteins which serve as accessory pigments and are incorporated into a globular structure, called phycobilisomes which are associated with the outer thylakoid membrane (Glazer, 1994). These are water soluble pigments and are classified according to their absorption characteristics into three classes, as phycocyanin (PC, λ_{max} 610-620 nm), phycoerythrine (PE, λ_{max} 540-570 nm), and allophycocyanin (APC, λ_{max} 650-655 nm) (Grewe and Pulz, 2012; Griffiths et al., 2016). One of the important applications of phycobiliproteins include, their use as a natural dye in foods and natural cosmetics (e. g. lipsticks and eyeliners), replacing synthetic colourants that are toxic and unsafe to be use by humans (Grewe and Pulz, 2012; Spolaore, et al., 2006). In addition, these molecules also have certain growth promoting properties and wide range of pharmaceutical applications (Spolaore, et al., 2006). *Spirulina* is a commercially exploited cyanobacterium for the production of phycobiliproteins. 'Lina Blue', a commercial product developed from phycocyanin is marketed by Dainippon Ink and Chemicals (Sakura) for its use in chewing gum, ice sherberts, candies, popsicles, dairy products, soft drinks and wasabi (Chu, 2012). Phycobiliproteins also have fluorescence properties and therefore, they were recognized as a novel class of fluorescent tags in 1982 (Glazer, 1994). Due to fluorescent nature, phycobiliproteins are used in flow cytometry, fluorescent microscopy, immunolabeling, fluorescent activated cell-sorting, immune-histochemistry, and also as protein marker in electrophoretic techniques (Glazer, 1994; Ughy et al., 2015; Sonani et al., 2016). Nevertheless, phycobiliproteins are also known for their antioxidant, anti-inflammatory, immunomodulating and anticancer properties (Grewe and Pulz, 2012). Thus, phycobiliproteins are one of the most important natural products by cyanobacteria whose demand is continuously increasing in the market. The product cost of this native pigments vary around US$ 3 to US$ 25/mg, but it can reach to US$ 1500/mg for some cross-linked pigments (with antibodies or fluorescent molecule) (Spolaore et al., 2006).

Biofertilizer

The world's population is increasing with increased demand of food grains. On the contrary, due to increased urbanization, the land under crop cultivation is decreasing, and therefore, populated countries like China and India are facing a food crisis. To cope with, intensive crop cultivation is required to increase the productivity and yield in the crops. This can be achieved either through bringing more and more land under cultivation or by augmenting more productivity of land already under cultivation. Working on first option is not possible due to limited land resources, while second option is worthy, and can be achieved by the use of mineral rich fertile land for crop cultivation. The synthetic fertilizers used for increasing crop productivity are expensive and their disproportionate use causes deterioration of soil fertility and environmental threats. Cyanobacteria occur abundantly in rice fields (Fritsch, 1907). Many of these under nitrogen deprived condition and with nitrogenase enzyme activity perform the process of nitrogen fixation, and therefore, they can be an excellent eco-friendly alternative source of fertilizers to increase nitrogen level in the soil, to maintain its long term fertility, and thus increased crop productivity (De, 1939; Pereira et al., 2009; Kaushik, 2014). In addition, these organisms are also found to release growth promoting substances and vitamins that enhance the plant growth (Menamo and Wolde, 2013; Rana et al., 2015; Singh et al., 2016). Out of 2213 soil samples collected from Indian rice fields, about 33% were found with nitrogen fixing cyanobacteria (Venkataraman, 1975). Both free-living and symbiotically associated (e.g., *Anabaena azollae*) cyanobacteria have nitrogen fixing ability that can act as a potential biofertilizer in agriculture (Fernández-Valiente and Quesada, 2004). Such diazotrophic, free living cyanobacteria include certain coccoid (*Aphanothece*, *Chroococcidiopsis*, *Dermocarpa*, *Gloeothece*, *Myxosarcina*, *Synechococcus*, *Xenococcus*), filamentous and non-heterocystous (*Lyngbya*, *Microcoleus*, *Oscillatoria*, *Plectonema boryanum*, *Pseudanabaena*, *Schizothrix*, *Trichodesmium*), as well as filamentous and heterocystous (*Anabaena*, *Anabaenopsis*, *Aulosira*, *Cylindrospermum*, *Gloeotrichia*, *Hapalosiphon*, *Mastigocladus*, *Nodularia*, *Nostoc*, *Nostochopsis*, *Rivularia*, *Scytonema*, *Stigonema*, *Tolypothrix*, *Westiella*, *Westiellopsis*) forms (Vaishampayan et al., 2001; Kulasooriya and Magana-Arachchi, 2016). The free-living heterocytous nitrogen fixing cyanobacteria contribute an average of 20-30 Kg N per hectare, while this value goes up to 600 Kg N per hectare for the symbiotically associated *Azolla-Anabaena* (Vaishampayan et al., 2001). Many countries (e.g., India, Sri Lanka, Japan, Thailand, China, Philippines, Bangladesh, and Vietnam) are using these biofertilizers for successful increase in crop productivity (Fernández-Valiente and Quesada, 2004; Vaishampayan et al., 2001; Kaushik, 2014; Kulasooriya and Magana-Arachchi, 2016), not only for rice, but also for other plants, such as wheat (Abd-Alla et al., 1994; Rana et al., 2015), pea (Osman et al., 2010), lettuce (Menamo and Wolde, 2013) and maize (Mohan et al., 2015).

Conclusion

Cyanobacteria are amongst the most ancient groups of microorganisms which are prokaryotic and photosynthetic in nature. Due to their long period of existence on earth planet, vast phenotypic diversity is seen in these organisms. As per their morphological appearances, they are classified as coccoid (Chroococcales), filamentous non-heterocystous (Oscillatoriales), and filamentous heterocystous (Nostocales). To know the taxonomic status and also to simplify the taxonomy of these organisms, studies based on a polyphasic approach are highly essential. They have occupied a wide range of habitats including extreme environments, where other photosynthetic organisms are rarely seen. The ranges of survival strategies adapted and chemical diversity in these organisms have made them most unique and promising organisms on earth. Therefore, they have a wide array of biotechnological applications. Due to progressive urbanisation, many of such potential cyanobacteria may vanish before their scientific discovery. Therefore, systematic studies of unexplored cyanobacteria, their bioprospecting and sustainable use of available known cyanobacterial resource for betterment of mankind are major challenges in the future.

Conflict of Interest

No conflict of interest among the authors.

References

Aaronson, S. (2000). Important vegetable supplements: II C, 1 Algae. In: Kiple, K. F. and Ornelas, K. C. (eds.) *The Cambridge world history of food, part 1.* Cambridge University Press, Cambridge, pp. 231-249.

Abd-Alla, M. H., Mahmoud, A. L. E. and Issa, A. A. (1994). Cyanobacterial fertilizer improved growth of Wheat. *Phyton*, 34: 11-18.

Abed, R. M. M., Dobretsov, S. and Sudesh, K. (2009). Applications of cyanobacteria in biotechnology. *J. Appl. Microbiol*, 106(1):1-12.

Adams, D. G. and Duggan, P. S. (2008). Cyanobacteria-bryophyte symbioses. *J. Exp. Bot*, 59(5): 1047-1058.

Adams, D. G., Duggan, P. S. and Jackson, O. (2012). Cyanobacterial symbioses. In: Whitton, B. A. (ed.) *Ecology of cyanobacteria II: their diversity in space and time.* Springer, Dordrecht, pp. 593-647.

Barsanti, L. and Gualtieri, P. (2014). *Algae: anatomy, biochemistry and biotechnology.* CRC Press, Boca Raton, pp. 326.

Becker, E. W. (1994). *Microalgae: biotechnology and microbiology.* Cambridge University Press, Cambridge, pp. 293.

Becker, E. W. (2004). Microalgae in human and animal nutrition. In: Richmond, A. (ed.) *Handbook of microalgal culture: biotechnology and applied phycology.* Blackwell Science, Oxford, pp. 312-351.

Becker, E. W. (2007). Microalgae as a source of protein. *Biotechnol. Adv.*, 25: 207-210.

Becker, E. W. (2013). Microalgae for human and animal nutrition. In: Richmond, A. and Hu, Q. (eds.) *Handbook of microalgal culture: applied phycology and biotechnology.* Wiley Blackwell, Oxford, pp. 461-503.

Bokesch, H. R., O'Keefe, B. R., McKee, T. C., Pannell, L. K., Patterson, G. M. L., Gardella, R. S., Sowder II, R. C., Turpin, J., Watson, K., Buckheit Jr., R. W. and Boyd, M. R. (2003). A potent novel anti-HIV protein from the cultured cyanobacterium *Scytonema varium. Biochem.*, 42: 2578-2584.

Borowitzka, L. J., Demmerle, S., Mackay, M. A. and Norton, R. S. (1980). Carbon-13 nuclear magnetic resonance study of osmoregulation in blue-green alga. *Science*, 210: 650-651.

Brandily, M. Y. (1959). Depuis des lustres une tribu primitive du Tchad exploite la nourriture de l'an 2000. *Sci. Avenir.*, 152: 516-519. [For ages a primitive tribe of Chad exploits the food of the year 2000]

Burja, A. M., Banaigs, B., Abou-Mansour, E., Burgess, J. G. and Wright, P. C. (2001). Marine cyanobacteria-a prolific source of natural products. *Tetrahedron*, 57: 9347-9377.

Carey, C. C., Ibelings, B. W., Hoffmann, E. P., Hamilton, D. P. and Brookes, J. D. (2012). Eco-physiological adaptations that favour freshwater cyanobacteria in a changing climate. *Water Res.*, 46: 1394-1407.

Carmichael, W. W., Drapeau, C. and Anderson, D. (2000). Harvesting of *Aphanizomenon flos-aquae* Ralfs ex Born. & Flah. var. *flos-aquae* (Cyanobacteria) from Klamath Lake for human dietary use. *J. Appl. Phycol.*, 12: 585-595.

Castenholz, R. W. (1969). Thermophilic blue-green algae and the thermal environment. *Bacteriol. Rev.*, 33(4): 476-504.

Chapman, R. L. (2013). Algae: the world's most important "plants"- an introduction. *Mitig. Adapt. Strateg. Glob. Change*, 18: 5-12.

Chatchawan, T., Peerapornpisal, Y. and Komárek, J. (2011). Diversity of cyanobacteria in man-made solar saltern, Petchaburi Province, Thailand- a pilot study. *Fottea*, 11(1): 203-214.

Chu, H. J. and Tseng, C. K. (1988). Research and utilization of cyanophytes in China: a report. *Arch. Hydrobiol.*, 80 (Suppl.): 573-584.

Chu, W. L. (2012). Biotechnological applications of microalgae. *IJSME*, 6(Suppl. 1): S24-S37.

Ciferri, O. (1983). *Spirulina*, the edible microorganism. *Microbiol. Rev.*, 47(4): 551-578.

D'Amico, S., Collins, T., Marx, J. C., Feller, G. and Gerday, C. (2006). Psychrophilic microorganisms: Challenges for life. *EMBO Reports*, 7(4): pp. 385-389.

Dangeard, P. (1940). Sur une algue bleue alimentaire pour l'homme: *Arthrospira platensis* (Nordst.) Gomont. *Act. Soc. Linn. Bordx.*, 91: 39-41. [On a blue food alga for humans: *Arthrospira platensis* (Nordst.) Gomont]

DasSarma, S. and Arora, P. (2001). Halophiles. *Encyclopedia of life science*, 8: 458-466.

De, P. K. (1939). The role of blue-green algae in nitrogen fixation in rice fields. *Proc. Royal Soc. London, Ser.* B, 127(846): 121-139.

Delwiche, C. F. (2007). Algae in the warp and weave of life: bound by plastids. In: Brodie, J. and Lewis, J. (eds.) *Unravelling the algae: the past, present, and future of algal systematics*. CRC Press, Boca Raton, pp. 7-20.

Deng, R. and Chow, T. (2010). Hypolipidemic, antioxidant and anti-inflammatory activities of microalgae *Spirulina*. *Cardiovasc. Ther.*, 28(4): e33-e45.

Erdmann, N. and Hagemann, M. (2001). Salt acclimation of algae and cyanobacteria: a comparison. In: Rai, L. C. and Gaur, J. P. (eds.) *Algal adaptation to environmental stresses: physiological, biochemical and molecular mechanisms*. Springer-Verlag, Berlin, Heidelberg, pp. 323-361.

Falcon, L. I., Cipriano, F., Chistoserdov, A. Y. and Carpenter, E. J. (2002). Diversity of diazotrophic unicellular cyanobacteria in the tropical North Atlantic Ocean. *Appl. Environ. Microbiol.*, 68(11): 5760-5764.

Fernández-Valiente, E. and Quesada, A. (2004). A shallow water ecosystem: rice fields. The relevance of cyanobacteria in the ecosystem. *Limnetica*, 23(1-2): 95-108.

Foster, R. A., Kuypers, M. M. M., Vagner, M., Paerl, R. W., Musat, N. and Zehr, J. P. (2011). Nitrogen fixation and transfer in open ocean diatom-cyanobacterial symbioses. *ISME J.*, 5: 1484-1493.

Fritsch, F. E. (1907). The subaerial and freshwater algal flora of the tropics. *Ann. Bot.*, 30: 235-275.

Glazer, A. N. (1994). Phycobiliproteins – a family of valuable, widely used fluorophores. *J. Appl. Phycol.*, 6: 105-112.

Gordon, N., Angel, D. L., Neori, A., Kress, N. and Kimor, B. (1994). Heterotrophic dinoflagellates with symbiotic cyanobacteria and nitrogen limitation in the Gulf of Aqaba. *Mar. Ecol. Prog. Ser.*, 107: 83-88.

Graham, L. E. and Wilcox, L. W. (2000). *Algae*. Prentice-Hall, Inc., NJ, pp. 640.

Grewe, C. B. and Pulz, O. (2012). The biotechnology of cyanobacteria. In: Whitton, B. A. (ed.) *Ecology of cyanobacteria II: their diversity in space and time*. Springer, Dordrecht, pp. 707-739.

Griffiths, M., Harrison, S. T. L., Smit, M. and Maharajh, D. (2016). Major commercial products from micro-and macroalgae. In: Bux, F. and Chisti, Y. (eds.) *Algae biotechnology: product and processes*. Springer International Publishing, Switzerland, pp. 269-300.

Guiry, M. D. (2012). How many species of algae are there? *J. Phycol.*, 48: 1057-1063.

Gustafson, K. R., Sowder II, R. C., Henderson, L. E., Cardellina II, J. H., McMahon, J. B., Rajamani, U., Pannell, L. K. and Boyd, M. R. (1997). Isolation, primary sequence determination, and disulfide bond structure of Cyanovirin-N, an anti-HIV (Human Immunodeficiency Virus) protein from the cyanobacterium *Nostoc ellipsosporum. Biochem. Biophys. Res. Comm.*, 238: 223-228.

Gutierrez-Salmean, G., Fabila-Castillo, L. and Chamorro-Cevallos, G. (2015). Nutritional and toxicological aspects of *Spirulina* (*Arthrospira*). *Nutr. Hosp.*, 32(1): 34-40.

Habib, M. A. B., Parvin, M., Huntington, T. C. and Hasan, M. R. (2008). A review on culture, production and use of *Spirulina* as food for humans and feeds for domestic animals and fish. *FAO Fisheries and Aquaculture Circular* No. 1034, Rome-Italy, pp. 33.

Hallmann, A. (2007). Algal transgenics and biotechnology. *Transgenic Plant J.*, 1(1): 81-98.

Han, D., Deng, Z., Lu, F. and Hu, Z. (2013). Biology and biotechnology of edible *Nostoc*. In: Richmond, A. and Hu, Q. (eds.) *Handbook of microalgal culture: applied phycology and biotechnology*. Wiley Blackwell, pp. 433-444.

Hof, T. and Fremy, P. (1933). On Myxophyceae living in strong brines. *Rec. Trav. Bot. Neerl.*, 30: 140-162.

Hudek, K., Davis, L. C., Ibbini, J. and Erickson, L. (2014). Commercial products from algae. In: Bajpai, R., Prokop, A. and Zappi, M. (eds.) *Algal biorefineries, volume 1: cultivation of cells and products*. Springer Science, Dordrecht, pp. 275-295.

Janson, S. (2012). Cyanobacteria in symbiosis with diatoms. In: Rai, A. N., Bergman, B. and Rasmussen, U. (eds.) *Cyanobacteria in symbiosis*. Kluwer Academic Publishers, Netherland, pp. 1-10.

Johnson, H. E., King. S. R., Banack, S. A., Webster, C., Callanaupa, W. J. and Cox, P. A. (2008). Cyanobacteria (*Nostoc commune*) used as a dietary item in the Peruvian highlands produce neurotoxic amino acid BMAA. *J. Ethnopharmac.*, 18(1): 159-165.

Kanekiyo, K., Lee, J. B., Hayashi, K., Takenaka, H., Hayakawa, Y., Endo, S. and Hayashi, T. (2005). Isolation of an antiviral polysaccharide, Nostoflan, from a terrestrial cyanobacterium, *Nostoc flagelliforme. J. Nat. Prod.*, 68: 1037-1041.

Kaplan-Levy, R. N., Hadas, O., Summers, M. L., Rucker, J. and Sukenik, A. (2010). Akinetes: dormant cells of cyanobacteria. In: Lubzens, E., Cerda, J. and Clark, M. (eds.) *Dormancy and resistance in harsh environments*. Springer-Verlag, Berlin, Heidelberg, pp. 5-27.

Karkos, P. D., Leong, S. C., Karkos, C. D., Sivaji, N. and Assimakopoulos, D. A. (2011). *Spirulina* in clinical practice: evidence-based human applications. *Evid. Based Complement. Alternat. Med.*, doi: 10.1093/ ecam/nen058, 4 pages.

Kaushik, B. D. (2014). Developments in cyanobacterial biofertilizer. *Proc. Indian Natn. Sci. Acad.*, 80(2): 379-388.

Komárek, J. (2003). Coccoid and colonial cyanobacteria. In: Wehr, J. D. and Sheath, R. G. (eds.) *Freshwater algae of North America: ecology and classification.* Academic Press, California, pp. 59-116.

Komárek, J. (2013). *Cyanoprokaryota 3. Heterocytous genera.* In: Büdel, B., Gärtner, G., Krienitz, L. and Schagerl, M. Editors: *Süßwasserflora von Mitteleuropa* [*Freshwater flora of Central Europe*]. Springer/ Spektrum, Berlin, Heidelberg, pp. 1130.

Komárek, J. and Anagnostidis, K. (1998). *Cyanoprokaryota 1. Chroococcales.* In: Ettl, H., Gärtner, G., Heynig, H. and Mollenhauer, D. Editors: *Süßwasserflora von Mitteleuropa 19/1.* Gustav Fischer, Jena-Stuttgart- Lübeck-Ulm., pp. 548. [*Freshwater flora of Central Europe 19/1.*]

Komárek, J. and Anagnostidis, K. (2005). *Cyanoprokaryota 2. Oscillatoriales.* In: Büdel, B., Krienitz, L., Gärtner, G. and Schagerl, M. (eds.) *Süßwasserflora von Mitteleuropa* 19/2. Elsevier/Spektrum, Heidelberg, pp. 759. [*Freshwater flora of Central Europe 19/2.*]

Kovač, D. J., Simeunović, J. B., Babić, O. B., Mišan, A. C. and Milovanović, I., Lj. (2013). Algae in food and feed. *Food and Feed Res.*, 40(1): 21-32.

Kulasooriya, S. A. (2011). Cyanobacteria: pioneers of planet earth. *Ceylon J. Sci.* (Bio. Sci.), 40(2): 71-88.

Kulasooriya, S. A. and Magana-Arachchi, D. N. (2016). Nitrogen fixing cyanobacteria: their diversity, ecology and utilisation with special reference to rice cultivation. *J. Natn. Sci. Foundation Sri Lanka*, 44(2): 111-128.

Kumar, M. and Paliwal, G. S. (2006). Distributional patterns of cyanobacteria. In: Gupta, R. K., Kumar, M. and Paliwal, G. S. (eds.) *Glimpses of cyanobacteria.* Daya Publishing House, Delhi. pp. 87-106.

Kutschera, U. and Niklas, K. J. (2005). Endosymbiosis, cell evolution, and speciation. *Theory Biosci*, 124: 1-24.

Lau, R. H., Visentin, L. P., Martin, S. M., Hofman, J. D., and Doolittle, W. (1985). Site-specific restriction endonuclease from the filamentous cyanobacterium *Nostoc* sp. MAC PCC 8009. *FEBS Lett.*, 179(1): 129-132.

Lechno-Yossef, S. and Nierzwicki-Bauer, S. A. (2002). *Azolla-Anabaena* symbiosis. In: Rai, A. N., Bergman, B. and Rasmussen, U. (eds.) *Cyanobacteria in symbiosis.* Kluwer Academic Publishers, Netherland, pp. 153-178.

Lee, R. E. (2008). *Phycology.* Cambridge University Press, Cambridge, pp. 547.

Lema, K. A., Willis, B. L. and Bourne, D. G. (2012). Corals form characteristic associations with symbiotic nitrogen-fixing cyanobacteria. *Appl. Environ. Microbiol*, 78(9): 3136-3144.

Léonard, J. and Compère, P. (1967). *Spirulina platensis* (Gom.) Geitler, algue bleue de grande valeur alimentaire par sa richesse en protéines. *Bull. Jard. Bot. Nat. Belg*, 37: 1-23. [*Spirulina platensis* (Gom.) Geitler, blue seaweed of great nutritional value because of its high protein content.]

Lesser, M. P., Mazel, C. H., Gorbunov, M. Y. and Falkowski, P. G. (2004). Discovery of symbiotic nitrogen-fixing cyanobacteria in corals. *Science*, 305(5686): 997-1000.

Li, Z. (2009). Advances in marine symbiotic cyanobacteria. In: Gault, P. M. and Marler, H. J. (eds.) *Handbook on cyanobacteria: biochemistry, biotechnology and applications*. Nova Science Publishers, Inc., New York, pp. 463-483.

Lindblad, P. (2009). Cyanobacteria in symbiosis with cycads. In: Pawlowski, K. (ed.) *Prokaryotic symbionts in plants*. Springer-Verlag, Berlin, Heidelberg, pp. 225-233.

Lindblad, P., Bergman, B., Hofsten, A. V., Hallbom, L. and Nylund, J. E. (1985). The cyanobacterium-*Zamia* symbiosis: an ultrastructural study. *New Phytol*, 101(4): 707-716.

Lyra, C., Halme, T., Torsti, A. M., Tenkanen, T. and Sivonen, K. (2000). Site-specific restriction endonucleases in cyanobacteria. *J. Appl. Microbiol*, 89: 979-991.

Machado, G. and Vital, D. M. (2001). On the occurrence of epizoic cyanobacteria and liverworts on a neotropical harvestman (Arachnida: Opiliones). *Biotropica*, 33(3): 535-538.

Mackay, M. A., Norton, R. S. and Borowitzka, L. J. (1984). Organic osmoregulatory solutes in cyanobacteria. J. *Gen., Microbiol.*, 130: 2177-2191.

Mader, J., Gallo, A., Schommartz, T., Handke, W., Nagel, Gunther, P., Brune, W. and Reich, K. (2016). Calcium spirulan derived from *Spirulina platensis* inhibits herpes simplex virus 1 attachment to human keratinocytes and protects against herpes labialis. *J. Allergy. Clin. Immunol.*, 137: 197-203.

Mandal, S. and Rath, J. (2015). *Extremophilic cyanobacteria for novel drug development*. Springer International Publishing, Switzerland, pp. 92.

Matondo, F. K., Takaisi, K., Nkuadiolandu, A. B., Lukusa, A. K. and Aloni, M. N. (2016). *Spirulina* supplements improved the nutritional status of undernourished children quickly and significantly: experience from Kisantu, the Democratic republic of Congo. *Int. J. Pediatr.*, doi.org/10.1155/2016/1296414.

Meeks, J. C. (2009). Physiological adaptations in nitrogen-fixing *Nostoc*-plant symbiotic associations. In: Pawlowski, K. (eds.) *Prokaryotic symbionts in plants*. Springer, Berlin, pp. 181-205.

Menamo, M. and Wolde, Z. (2013). Effect of cyanobacteria application as biofertilizer on growth, yield and yield components of Romaine Lettuce (*Lactuca sativa* L.) on soils of Ethiopia. *Ame. Sci. Res. J. Eng. Technol. Sci.*, 4(1): 50-58.

Miyamoto, E., Tanioka, Y., Nakao, T., Barla, F., Inui, H., Fujita, T., Watanabe, F. and Nakano, Y. (2006). Purification and characterization of corrinoid-compound in edible cyanobacterium *Aphanizomenon flos-aquae* as a nutritional supplementary food. *J. Agric. Food. Chem.*, 54: 9604-9607.

Miyashita, H., Ikemoto, H., Kurano, N. and Miyachi, S. (2003). *Acaryochloris marina* gen. et sp. nov. (Cyanobacteria), an oxygenic photosynthetic prokaryote containing chl d as a major pigment. *J. Phycol.*, 39: 1247-1253.

Mohan, A., Kumar, B. and Nath, D. (2015). Cyanobacterial consortium in the improvement of Maize crop. *Int. J. Curr. Microbiol. Appl. Sci.*, 4(3): 264-274.

Moreira, L. M., Rocha, A. S. R., Ribeiro, C. L. G., Rodrigues, R. S., Soares, L. A. S. (2011). Nutritional evaluation of single cell protein produced by *Spirulina platensis*. *Afr. J. Food Sci.*, 5(15): 799-805.

Muhling, M., Belay, A. and Whitton, B. A. (2005). Variation in fatty acid composition of *Arthrospira* (Spirulina) strains. *J. Appl. Phycol.*, 17: 137-146.

Muñoz-García, J., and Ares, S. (2016). Formation and maintenance of nitrogen-fixing cell patterns in filamentous cyanobacteria. *PNAS*, 113(22): 6218-6223.

Mur, L. R., Skulberg, O. M. and Utkilen, H. (1999). Cyanobacteria in the environment. In: Chorus, I. and Bartram, J. (eds.) *Toxic cyanobacteria in water: A guide to their public health consequences, monitoring and management.* St. Edmundsbury Press, Suffolk, pp. 15-40.

Muthukumar, C., Muralitharan, G., Vijayakumar, R., Panneerselvam, A. and Thajuddin, N. (2007). Cyanobacterial biodiversity from different freshwater ponds of Thanjavur, Tamilnadu (India). *Acta. Bot. Malacitana*, 32: 17-25.

Nadeau, T. L. and Castenholz, R. W. (2000). Characterization of psychrophilic oscillatorians (Cyanobacteria) from Antarctic meltwater ponds. *J. Phycol.*, 36: 914-923.

Olsson-Francis, K., de la Torre, R., Towner, M. C. and Cockell, C. S. (2009). Survival of akinetes (resting-state cells of cyanobacteria) in low earth orbit and simulated extraterrestrial conditions. *Orig. Life Evol. Biosph.*, 39: 565-579.

Oren, A. (2002). *Halophilic microorganisms and their environments*. Kluwer Academic Publishers, Dordrecht, pp. 575.

Oren, A. (2012a). Salts and Brines. In: Whitton, B. A. (ed.) *Ecology of cyanobacteria II: their diversity in space and time*. Springer, Dordrecht, pp. 401-426.

Oren, A. (2012b). Approaches toward the study of halophilic microorganisms in their natural environments: who are they and what are they doing? In: Vreeland, R. H. (ed.) *Advances in understanding the biology of halophilic microorganisms*. Springer, Netherland, pp. 1-33.

Oren, A. (2014). Cyanobacteria: biology, ecology, and evolution. In: Sharma, N. K., Rai, A. K. and Stal, L. J. (eds.) *Cyanobacteria: an economic perspective*. Wiley Blackwell, UK, pp. 3-20.

Oren, A. (2015). Cyanobacteria in hypersaline environments: biodiversity and physiological properties. *Biodivers. Conserv.*, 24: 781-798.

Osborne, B. and Bergman, B. (2009). Why does *Gunnera* do it and other angiosperm don't? An evolutionary perspective on the *Gunnera-Nostoc* symbiosis. In: Pawlowski, K. (ed.) *Prokaryotic symbionts in plants*. Springer-Verlag, Berlin, Heidelberg, pp. 207-224.

Osman, M. E. H., El-Sheekh, M. M., El-Naggar, A. H. and Gheda, S. F. (2010). Effect of two species of cyanobacteria as biofertilizers on some metabolic activities, growth, and yield of pea plant. *Biol. Fertil. Soils*, 46: 861-875.

Packer, M. A., Harris, G. C. and Adams, S. L. (2016). Food and feed applications of algae. In: Bux, F. and Chisti, Y. (eds.) *Algae biotechnology: product and processes*. Springer International Publishing, Switzerland, pp. 217-247.

Pandey, K. D., Shukla, S. P., Shukla, P. N., Giri, D. D., Singh, J. S., Singh, P. and Kashyap, A. K. (2004). Cyanobacteria in Antarctica: Ecology, physiology and cold adaptation. *Cell. Mol. Biol.*, 50(5): 575-584.

Peerapornpisal, Y., Amomledpison, D., Rujjanawate, C., Ruangrit, K. and Kanjanapothi, D. (2006). Two endemic species of macroalgae in Nan River, Northern Thailand, as therapeutic agents. *Science Asia*, 32 (Suppl. 1): 71-76.

Pereira, I., Ortega, R., Barrientos, L., Moya, M., Reyes, G. and Kramm, V. (2009). Development of a biofertilizer based on filamentous nitrogen-fixing cyanobacteria for rice crop in Chile. *J. Appl. Phycol.*, 21: 135-144.

Peters, G. A. (1991). *Azolla* and other plant-cyanobacteria symbioses: aspects of form and function. *Pl. Soil*, 137: 25-36.

Proud, D. N., Wade, R. R., Rock, P., Townsend Jr., V. R. and Chavez, D. J. (2012). Epizoic cyanobacteria associated with a neotropical harvestman (Opiliones: Sclerosomatidae) from Costa Rica. *J. Arachnol.*, 40(2): 259-261.

Quesada, A. and Vincent, W. F. (2012). Cyanobacteria in the cryosphere: snow, ice and extreme cold. In: Whitton, B. A. (ed.) *Ecology of cyanobacteria II: their diversity in space and time*. Springer, Dordrecht, pp. 387-399.

Rana, A., Kabi, S. R., Verma, S., Adak, A., Pal, M., Shivay, Y. S., Prasanna, R. and Nain, L. (2015). Prospecting plant growth promoting bacteria and cyanobacteria as options for enrichment of macro and micronutrients in grains in rice-wheat cropping sequence. *Cogent Food Agri.*, 1(1): 1037379, 16 pages.

Rastogi, R. P., Sonani, R. R. and Madamwar, D. (2015). Cyanobacterial sunscreen scytonemin: role in photoprotection and biomedical research. *Appl. Biochem. Biotechnol.*, 176: 1551-1563.

Reed, R. H., Borowitzka, L. J., Mackay, M. A., Chudek, J. A., Foster, R., Warr, S. R. C., Moore, D. J. and Stewart, W. D. P. (1986). Organic solute accumulation in osmotically stressed cyanobacteria. *FEMS Microbiol. Rev.*, 39: 51-56.

Reshef, V., Mizrachi, E., Maretzki, T., Silberstein, C., Loya, S., Hizi, A. and Carmeli, S. (1997). New acylated sulfoglycolipids and digalactolipids and related known glycolipids from cyanobacteria with a potential to inhibit the Reverse transcriptase of HIV-1. *J. Nat. Prod.*, 60: 1251-1260.

Rikkinen, J. (2013). Molecular studies on cyanobacterial diversity in lichen symbioses. *MycoKeys*, 6: 3-32.

Roger, P. A. and Kulasooriya, S. A. (1980). *Blue-green algae and rice*. IRRI, Philippines, pp. 112.

Roney, B. R., Renhui, L., Banack, S. A., Murch, S., Honegger, R. and Cox, P. A. (2009). Consumption of *fa cai Nostoc* soup: a potential for BMAA exposure from *Nostoc* cyanobacteria in China? *Amyotroph Lateral Scler.*, 10(Suppl. 2): 44-49.

Saravanan, M., Elango, K., Chandrashekaran, S., Anand, N. and Nagaraja, V. (2003). A new type II restriction endonuclease, O*fo*I from non filamentous cyanobacterium, *Oscillatoria foreaui. Curr. Sci.*, 85: 188-191.

Sarma, T. A. (2013). *Handbook of cyanobacteria*. CRC Press, Boca Raton, pp. 802.

Schopf, J. W. (2002). The fossil record: tracing the roots of the cyanobacterial lineage. In: Whitton, B. A. and Potts, M. (eds.) *The ecology of cyanobacteria: their diversity in time and space*. Kluwer Academic Publishers, NY, pp. 13-35.

Seckbach, J. and Oren, A. (2007). Oxygenic photosynthetic microorganisms in extreme environments: possibilities and limitations. In: Seckbach, J. (ed.) *Algae and cyanobacteria in extreme environments*. Springer, Dordrecht, pp. 5-25.

Sharma, N. K., Tiwari, S. P., Tripathi, K. and Rai, A. K. (2011). Sustainability and cyanobacteria (blue-green algae): facts and challenges. *J. Appl. Phycol.*, 23: 1059-1081.

Shevela, D., Pishchalnikov, Eichacker, L. A. and Govindjee (2013). Oxygenic photosynthesis in cyanobacteria. In: Srivastava, A. K., Rai, A. N. and Neilan, B. A. (eds.) *Stress biology of cyanobacteria: molecular mechanisms to cellular responses.* CRC Press, Boca Raton, pp. 3-40.

Sili, C., Torzillo, G. and Vonshak, A. (2012). *Arthrospira* (Spirulina). In: Whitton, B. A. (ed.) *Ecology of cyanobacteria II: their diversity in space and time.* Springer, Dordrecht, pp. 677-705.

Singh, J. S., Kumar, A., Rai, A. N. and Singh, D. P. (2016). Cyanobacteria: a precious bio-resource in agriculture, ecosystem, and environmental sustainability. *Front. Microbiol.*, 7: 529 doi: 10.3389/fmicb. 2016.00529.

Singh, R. K., Tiwari, S. P., Rai, A. K. and Mohapatra, T. M. (2011). Cyanobacteria: an emerging source for drug discovery. *J. Antibiot.*, 64: 401-412.

Singh, S. C., Sinha, R. P. and Häder, D. P. (2002). Role of lipids and fatty acids in stress tolerance in cyanobacteria. *Acta. Protozool.*, 41: 297-308.

Singh, S. P., Kumari, S., Rastogi, R. P., Singh, K. L. (2008). Mycosporine-like amino acids: chemical structure, biosynthesis and significance as UV-absorbing/screening compounds. *Ind. J. Exp. Biol.*, 46: 7-17.

Solheim, B. and Zielke, M. (2002). Association between cyanobacteria and mosses. In: Rai, A. N., Bergman, B. and Rasmussen, U. (eds.) *Cyanobacteria in symbiosis*. Kluwer Academic Publishers, Netherland, pp. 137-152.

Sonani, R. R., Rastogi, R. P., Patel, R. and Madamwar, D. (2016). Recent advances in production, purification and applications of phyco-biliproteins. *World J. Biol. Chem.*, 7(1): 100-109.

Soule, T. and Garcia-Pichel, F. (2014). Ultraviolet photoprotective compounds from cyanobacteria in biomedical applications. In: Sharma, N. K., Rai, A. K. and Stal, L. J. (eds.) *Cyanobacteria: an economic perspective*. Wiley Blackwell, UK, pp. 119-143.

Spolaore, P., Joannis-Cassan, C., Duran, E. and Isambert, A. (2006). Commercial application of microalgae. *J. biosci. Bioeng.*, 101(2): 87-96.

Stanier, R. Y., Sistrom, W. R., Hansen, T. A., Whitton, B. A., Castenholz, R. W., Pfennig, N., Gorlenko, V. N., Kondratieva, E. N., Eimhjellen, K. E., Whittenbury, R., Gherna, R. L. and Truper, H. G. (1978). Proposal to place the nomenclature of the cyanobacteria (Blue-green algae) under the rules of the International Code of Nomenclature of Bacteria. *Int. J. Syst. Bacteriol.*, 28(2): 335-336.

Sundararajan, V., Bheemanaik, S. and Anand, N. (2010). Nicking endonuclease in cyanobacteria. *Rec. Res. Sci. Technol.*, 2(10): 63-65.

Takenaka, H., Yamaguchi, Y., Sakaki, S., Watarai, K., Tanaka, N., Mori, M., Seki, H., Tsuchida, M., Yamada, A., Nishimori, T. and Morinaga, T. (1998). Safety evaluation of *Nostoc flagelliforme* (*Nostocales*, *Cyanophyceae*) as a potential food. *Food Chem. Toxicol.*, 36: 1073-1077.

Thiamdao, S., Motham, M., Pekkoh, J., Mungmai, L. and Peerapornpisal, Y. (2012). *Nostochopsis lobatus* Wood em. Geitler (*Nostocales*), edible algae in Northern Thailand. *Chiang Mai J. Sci.*, 39(1): 119-127.

Tiwari, D. N. (1978). The heterocysts of the blue-green alga *Nostochopsis lobatus*: effects of cultural conditions. *New Phytol.*, 81: 653 656.

Tomitani, A., Knoll, A. H., Cavanaugh, C. M. and Ohno, T. (2006). The evolutionary diversification of cyanobacteria: molecular-phylogenetic and paleontological perspectives. *PNAS*, 103(14): 5442-5447.

Torres-Sánchez, A., Gómez-Gardeñes, J. and Falo, F. (2015). An integrative approach for modeling and simulation for heterocyst pattern formation in cyanobacteria filaments. *PLoS Comput. Biol.*, 11(3): e1004129.

Ughy, B., Nagy, C. I. and Kos, P. B. (2015). Biomedical potential of cyanobacteria and algae. *Act. Biol. Szeg.*, 59(Suppl. 2): 203-224.

Usher, K. M., Bergman, B. and Raven, J. A. (2007). Exploring cyanobacterial mutualisms. *Annu. Rev. Ecol. Evol. Syst.*, 38: 255-273.

Vaishampayan, A., Sinha, R. P., Hader, D. P., Dey, T., Gupta, A. K. Bhan, U. and Rao, A. L. (2001). Cyanobacterial biofertilizers in rice agriculture. *Bot. Rev.*, 67(4): 453-516.

Venkataraman, G. S. (1975). The role of blue-green algae in tropical rice cultivation. In: Stewart, W. D. P. (ed.) *Nitrogen fixation by free-living micro-organisms*. Cambridge University Press, Cambridge, pp. 207-2018.

Vijayakumar, S. and Menakha, M. (2015). Pharmaceutical applications of cyanobacteria-a review. *J. Acute Med.*, 5: 15-23.

Vincent, W. F. (2002). Cyanobacterial dominance in the Polar Regions. In: Whitton, B. A. and Potts, M. (eds.) *The ecology of cyanobacteria: their diversity in time and space*. Kluwer academic Publishers, NY, pp. 321-340.

Ward, D. M and Castenholz, R. W. (2002). Cyanobacteria in geothermal habitats. In: Whitton, B. A. and Potts, M. (eds.) *The ecology of cyanobacteria: their diversity in time and space*. Kluwer Academic Publisher, NY, pp. 37-59.

Ward, D. M., Castenholz, R. W. and Miller, S. R. (2012). Cyanobacteria in geothermal habitats. In: Whitton, B. A. (ed.) *Ecology of cyanobacteria II: their diversity in space and time*. Springer, Dordrecht, pp. 39-63.

Whitehead, P. R. and Brown, N. L. (1985). Three restriction endonucleases from *Anabaena flosaquae*. *J. Gen. Microbiol.*, 131: 951-958.

Whitton, B. A. and Potts, M. (2012). Introduction to the cyanobacteria. In: Whitton, B. A. (ed.) *Ecology of cyanobacteria II: their diversity in space and time*. Springer, Dordrecht, pp. 1-13.

Wynn-Williams, D. D. (2002). Cyanobacteria in deserts - life at the limit? In: Whitton, B. A. and Potts, M. (eds.) *The ecology of cyanobacteria: their diversity in time and space*. Kluwer Academic Publishers, NY, pp. 341-366.

Yamada, S., Ohkubo, S., Miyashita, H. and Setoguchi, H. (2012). Genetic diversity of symbiotic cyanobacteria in *Cycas revoluta* (Cycadaceae). *EFMS Microbiol. Ecol.*, 81: 696-706.

Yang, Y., Park, Y., Cassada, D. A., Snow, D. D., Rogers, D. G. and Lee, J. (2011). In vitro and in vivo safety assessment of edible blue-green algae, *Nostoc commune* var. *sphaeroides* Kützing and *Spirulina platensis*. *Food Chem. Toxicol.*, 49(7): 1560-1564.

Zaccaroni, A. and Scaravelli, D. (2008). Toxicity of fresh water algal toxins to humans and animals. In: Evangelista, V., Barsanti, L., Frassanito, A, M., Passarelli, V. and Gualtieri, P. (eds.) *Algal toxins: nature, occurrence, effect and detection*. Springer, Dordrecht, pp. 45-89.

Zakhia, F., Jungblut, A., Taton, A., Vincent, W. F. and Wilmotte, A. (2008). Cyanobacteria in cold ecosystems. In: Margesin, R., Schinner, F., Marx, J. C. and Gerday, C. (eds.) *Psychrophiles: from biodiversity to biotechnology*. Springer-Verlag, Berlin, pp. 121-135.

Zhang, C., Laurent, S., Sakr, S., Peng, L. and Bédu, S. (2006). Heterocyst differentiation and pattern formation in cyanobacteria: a chorus of signals. *Mol. Microbiol*, 59(2): 367-375.

In: The Role of Photosynthetic Microbes…
Editors: K. Tripathi, N. Kumar et al.
ISBN: 978-1-53614-032-3

Chapter 2

LIPID DERIVED PRODUCTS FROM MICROALGAE: DOWNSTREAM PROCESSING FOR INDUSTRIAL APPLICATION

Saumyakanti Khanra*[1]*, Kalyan Gayen*[2]*,
Gopinath Halder*[3]*, Tridib Kumar Bhowmick*[4]*,
***Gunapati Oinam*[5] *and Onkar Nath Tiwari*[6,*]**
[1,2]Department of Chemical Engineering,
NIT Agartala, West Tripura, India
[3]Department of Chemical Engineering,
NIT Durgapur, West Bengal, India
[4]Department of Bioengineering,
NIT Agartala, West Tripura, India
[5]Microbial Resources Division, DBT-IBSD, Imphal, Manipur
[6]Centre for Conservation and Utilization of Blue Green Algae,
Division of Microbiology, ICAR Indian Agricultural Research Institute (IARI), New Delhi, India

ABSTRACT

Microalgae have the ability to produce ten times oil than conventional oilseed plants and this oil can be converted into biodiesel. Moreover, microalgal based biodiesel is non-

[*] Corresponding Author Email: ontiwari1968@gmail.com.

toxic, biodegradable and sustainable. Apart from biodiesel, other lipid-derived value-added chemicals such as the eicosapentanoic acid (EPA) and decosahexaenoic acid (DHA) have a huge market. Technological advancements in terms of cultivation, harvesting, extraction, and purification are demanding for industrial scale implementation. Here, we discuss the conventional downstream processes that can scale up into industry level. Further, recent technological advancements are high that will bring down the production cost of lipid-derived products.

Keywords: extraction, lipid, microalgae, polyunsaturated fatty acids, transesterification

INTRODUCTION

Microalgae have been lauded as a potential replacement for fossil fuel over the last decade due to decrement of fossil fuel reservoirs along with its adverse environmental impacts to the society (Pragya et al., 2013., Mubarak et al., 2015; Demirbas and Fatih Demirbas, 2011; Scott et al., 2010). They are currently considered as one of the most promising candidates for biodiesel production. Since microalgal strains can be cultivated on non-arable land and in seawater, their cultivation does not compete with the food production in arable land. In addition to being a promising alternative to fossil fuel, photosynthetic activity for growing one ton of microalgae can mitigate 1.83 tons of atmospheric CO_2 (Mubarak et al., 2015; Chisti, 2008). High photosynthetic rates of microalgae serve as an effective carbon sequestration mechanism and it can accumulate lipids in their biomass up to 77% of dry cell mass (Halim et al., 2012). Biodiesel obtained from microalgae can reduce 78% CO_2 emission on a life-cycle basis, compared to conventional diesel fuel. Hence many industries prefer to use microalgae as the feedstock to produce biomass energy.

Biodiesel production from microalgal biomass consists of a number of sequential processes namely cultivation, harvesting, extraction of oil and synthesis of marketable biodiesel from extracted oil. Among these methods, lipid extraction is the key and costly step. The economics for the microalgal based biodiesel production mainly depends on the energy requirement for harvesting and extraction of lipids (Kim et al., 2013). Several technical challenges are needed to be addressed for successful implementation of industrial-scale biodiesel production from microalgae. One of the major challenges will be the maintenance of optimal large-scale outdoor conditions with high growth rate and lipids (Halim et al., 2012). This chapter focuses on different lipid extraction methods, downstream processing of microalgal lipid and lipid-derived products and the recent technological advances.

Table 1. Lipid-derived product from microalgae

Product Category	Product Name	Source	Structure	Application	Reference
Lipid	1. Biodiesel	Microalgae. Exp: *Chlorella* sp.	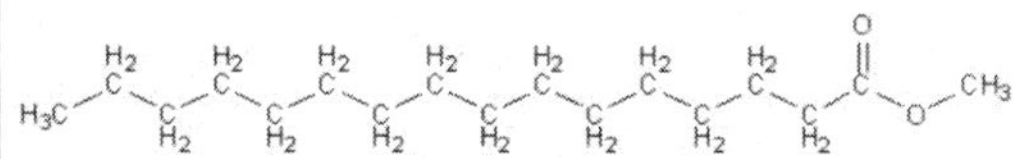	Used as an alternative fuel source for fossil fuel.	(https://www.goshen.edu/academics/chemistry/biodiesel/chemistry-of/)
	2. PUFA [Omega 3 fatty acids (EPA, DHA, etc.)]	*Phaedactylum tricornatum* (EPA) *Monodus subterraneus* (EPA) *Crypthecodinium cohnii* (DHA)	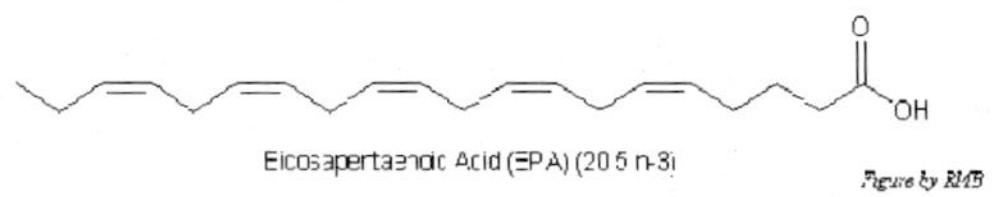 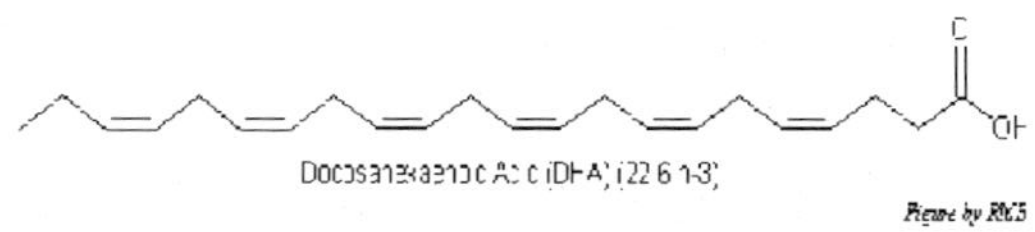	Omega-3 fatty acids as a part of diet help lower the risk of heart diseases. Omega-3 fatty acids may delay or prevent the progression of certain psychotic disorders in high-risk children and adolescents.	(http://lansbury.bwh.harvard.edu/structure_of_dha_and_epa.htm; http://www.oilgae.com/non_fuel_products/omega_3fattyacids.html)

LIPID DERIVED PRODUCTS

Recently microalgae have gained wide interest as a source of renewable biofuel as well as other valuable chemicals such as omega-3 fatty acids (eicosapentanoic acid (EPA), decosahexaenoic acid (DHA). Owing to the large variety of species (possibly in the millions), yield and composition of microalgal products can be different. Lipid-derived products from microalgae with their sources, molecular structure and application have been summarized in Table 1 and discussed in the subsequent subsections.

BIODIESEL

Neutral lipids with a lower degree of unsaturation, non-toxicity, biodegradability, and sustainability are attractive characteristics of microalgal based biodiesels (Gerpen 2005). It has been reported that algae-based biodiesel seems to be the only viable solution for replacement of petroleum-based diesel (http://www.oilgae.com/algae/oil/biod/large_scale/large_scale.html 2015). No other feedstock is able to produce a large amount of oil as per market demand. It has been estimated that around 10 million acres of land (1% of the total arable land in the United States) would be sufficient in the United States for microalgal based biodiesel production that will completely replace all the petroleum oil used currently in that country (http://www.oilgae.com/algae/oil/biod/large_scale/large_ scale.html2015).

POLYUNSATURATED FATTY ACIDS (PUFA)

Polyunsaturated fatty acids (PUFAs) are under consideration in healthcare as these compounds have properties for the prevention and treatment of several diseases like diabetes, obesity, cardiovascular diseases, and cancer (Rodriguez-Cruz et al., 2005; Wang et al., 2012). Particularly, docosahexaenoic acid (DHA) and eicosapentaeneoic acid (EPA) is under focus and these compounds fall under omega-3 fatty acids. Omega-3 fatty acids are high valued ingredients in infant formula, pharmaceuticals, and dietary supplements. Nowadays, EPA and DHA are produced from the oil of marine fish. However, it has been reported that high amount of EPA and a small amount of DHA can be producedfrom algae. Global omega-3 PUFA market is USD 9.94 billion in 2015 and market is expected to reach USD 18.95 billion by 2020 (http://www.marketsandmarkets.com/Market-Reports/omega-3-omega-6-227.html 2016; http://www.marketsandmarkets.com/Press Releases/omega-3.asp 2016). The European market analysts Frost & Sullivan reported that Omega-3 fatty acids have the greatest future (http://www.oilgae.com/non_fuel_products/omega_3fattyacids.html) .

CULTIVATION

Microalgae cultivation are employed either in open ponds and closed photobioreactors (Patil et al., 2005). Cultivation of open systems is economically more favorable where availability of land and water are cheap (Borowitzka, 1999). Open systems are widely exposed to natural sunlight where carbon dioxide is captured by microalgae from air using natural aquatic nutrients (Janssen et al., 2003). Raceway ponds are commonly used as open cultivation system (Brennan and Owende, 2010). A raceway pond costs less to build and operate and less energy intensive than photobioreactors. Though raceway pond systems are low-cost, the productivity is low as compared to photobioreactors (Zeng et al., 2011). The productivity is affected by evaporative losses, changes in temperature, photoperiod, and seasonal variations. Sometimes contamination in algal cultivation in the open pond caused serious limitation (Fon Sing et al., 2013). On the other hand, closed photobioreactors have the ability to overcome contamination problems (Gudin and Chaumont, 1991). Figure 1 portrays the schematic of the key steps involved (e.g., cultivation, harvesting, pretreatment, extraction, and purification) with downstream processing of lipid-derived products.

HARVESTING

Harvesting is a process to recover biomass from the cultivation medium. The selection of appropriate harvesting method strongly depends on the type of microalgae cultivating based on the density of cultivating media and microalgae, size of microalgae, and targeted products (Olaizola, 2000). Harvesting is one costly process accounting around (20–30%) of the total production cost (Rawat et al., 2011). Therefore, the choice of harvesting technology for mass biodiesel production is vital for efficient and economic production (Amaro et al., 2011). Various techniques (e.g., gravity sedimentation, flocculation, filtration, centrifugation, flotation, and electrophoresis) are used to harvest microalgae where flocculation is often recommended as a pretreatment owing to its energy efficiency (Mutanda et al., 2011).

PRETREATMENT

Drying

Algae are dried and ground into fine powder before extraction. Drying systems can be used to reduce the particle size of algal samples in the desired size. There are two

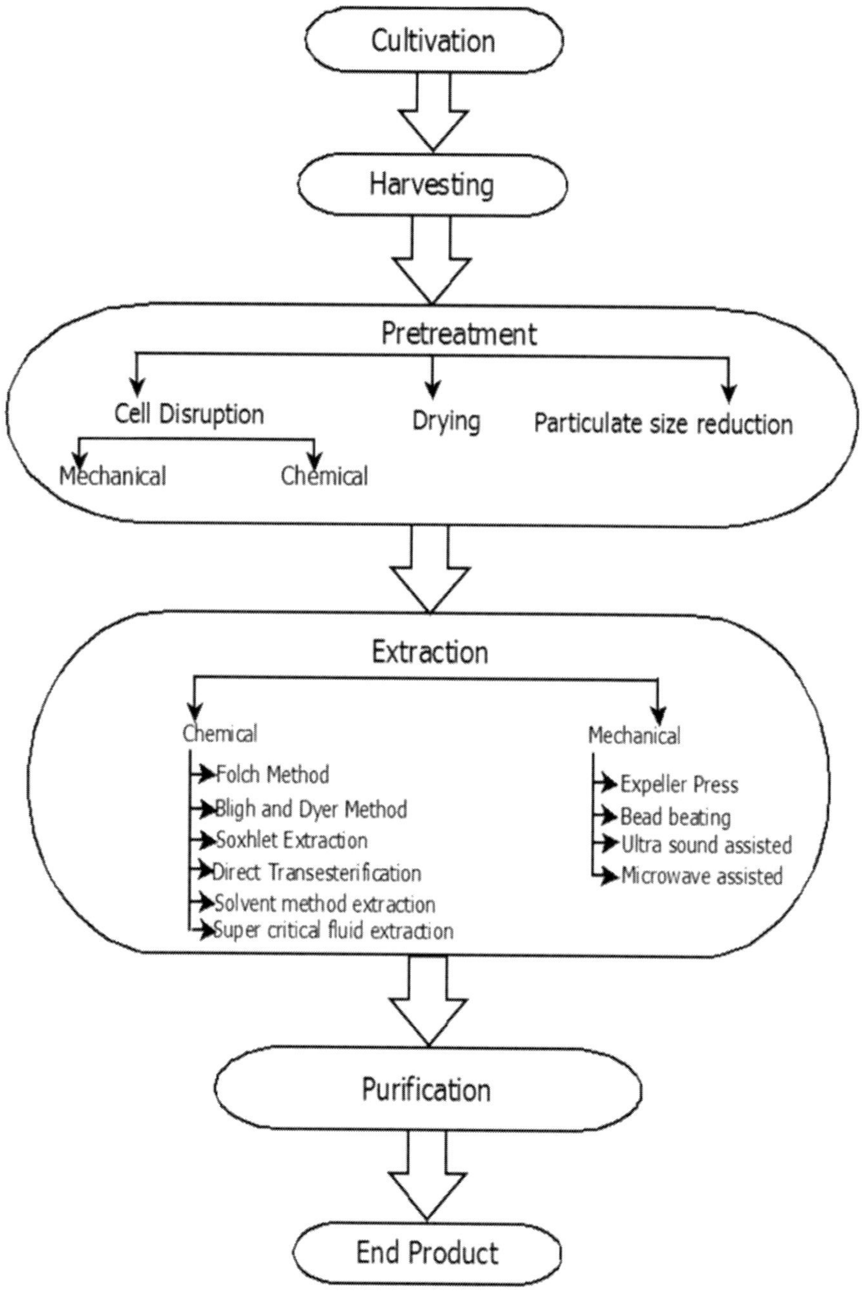

Figure 1. Schematic of cultivation, harvesting, pretreatment, extraction and purification of lipid derived microalgal products.

types of drying method: Thermal drying and Freeze drying. In case of thermal drying, a broad range of hypothesis for temperature (room temperature to 60^0C) and time (18-48 h) is employed. But generally, 60°C temperature and overnight duration are used (Rubio et al., 2010). On the other hand, in case of freeze drying, the drying duration covers up to 24-48 hour and the freezing temperatures are between -50°C and -80°C. The most important and critical part of the freeze-drying process is the freezing phase because it can be ruined the product if not properly done.

Particulate Size Reduction

After drying, microalgal biomass can be milled into powder form in various particle sizes. Reducing the size of dry algal biomass powder increases its surface area, which will improve biomass-solvent contact and shortens diffusion pathway of the extracting solvent. This is generally helpful in case of lipid extraction as it enhances the lipid recovery. Although, extremely small particle size biomass powder may lead to the higher propensity of lipid re-adsorption which in turn affects the fluid channeling in the extraction container (for Super Critical CO_2 extraction) and direct into inhomogeneous lipid extraction (Halim et al., 2012).

Cell Disruption

Cell disruption is known to increase the efficacy of lipid extraction process of microalgae. During cell disruption, cell walls of intact algal cells are broken down and intracellular lipids are released into the surrounding medium. Then, the extracting solvent can interact directly with the released lipids without having to breach the cell membrane. Hence, the process is no longer hindered by the transportation of extracting solvent across the cell structures which makes it a rapid and high yield process. Common disruption methods like ultrasonication, bead milling, and high-pressure homogenization generally require a certain amount of water, which is why cell disruption is always performed before drying.

Lipid Extraction Method

Various conventional methods are being used for lipid extraction that can be subcategorized into chemical (e.g., folch, bligh and dryer, soxhlet extraction, solvent extraction, direct transesterification and supercritical fluid extraction) and mechanical (e.g., expeller press, bead beating, ultrasound, and microwave). Recently, few emerging technologies are being reported such as single step extraction, continuous extraction, and extraction using nanotechnology that can be used for industrial application. The process and cons of the lipid extraction methods are discussed in Table 2.

Chemical Methods

Folch Method

Folch method uses the extraction solvents (chloroform and methanol in 2:1 ratio by volume) to extract the intracellular lipid from algal cells. Firstly, cells are homogenized and then mixed thoroughly with one-fourth volume of saline solution. The resultant mixture is then allowed to settle and separated into two distinct layers and lipids from the cell settle in the upper layer. This process is one of the oldest initiatives for lipid extraction and forged base of development of future extraction methods. With some modification, this method is still being used for algal lipid estimation. The main advantage of this method is that it is rapid and large number of samples can be performed at once. However, this process is less sensitive when compared to latest new methods (Ramanath et al., 2015).

Bligh and Dyer Method

The Bligh and Dyer method is one of the most broadly used techniques for extraction of lipid and also quite similar to Folch method. The fundamental difference between the methods is in solvent/solvent and solvent/tissue ratios. On the contrary to Folch method, in this process extraction solvent of chloroform and methanol in 1:2 ratio (v/v) is used. Like Folch method, chloroform layer containing lipid is separated and purified before quantified gravimetrically. This method is being used very widely by algologist all over the world for lipid estimation along with in pilot and large-scale extraction procedures (Ramanath et al., 2015).

Soxhlet Extraction

Soxhlet extraction occurs in the following steps: first, it is evaporated, and then it is recondensed and finally collected in the collection container. The efficiency of this method is very high because the sample is in recurring contact with a fresh solvent. High efficacy of this process makes this a favorite method for quantification in case of the biological samples. However, the main drawback of this process is high energy consumption and extraction time. Also, owing to the complexity of this apparatus, it is very hard to scale up and to make it a continuous process (Kim et al., 2013).

Table 2. Merits and demerits of lipid extraction methods

Method Type	Method	Merits	Demerits	Reference
Chemical	1. Folch Method	Rapid and easy processing of a large number of samples is the major advantage.	adverse effects of chloroform on the environment (EU regulation controlling chlorinated solvents), laborious (filtration etc.)	(R et al., 2015; http://www.biomedcentral.com/content/supplementary/2190-4715-24-13-S1.PDF).
	2. Bligh and Dyer Method	Simple, standard method, well established, determines total lipids, Samples can be analyzed directly with no pre-drying necessary, lipids can be used for further determinations	adverse effects of chloroform on the environment (EU regulation controlling chlorinated solvents), laborious (filtration etc.)	(http://www.biomedcentral.com/content/supplementary/2190-4715-24-13-S1.PDF)
	3.Soxhlet extraction	Simple, not very labor intensive can be operated with non-chlorinated solvents, lipids can be used for further determinations	extractable lipids are determined, not total lipids, large amounts of solvents Needed, special equipment required	(http://www.biomedcentral.com/content/supplementary/2190-4715-24-13-S1.PDF)
	4. Direct transesterifi-cation	Direct conversion of biofuel from wet algal biomass		(Ramanath et al., 2015)
	5. Solvent Extraction method	Extraction efficiency depends on the species used, a volume of the extractor, reaction time, sample volume, moisture content, types of lipids present, and in case of solvent-based methods, choice of the solvents, solvent ratio	Fire, health, and environmental hazards; regulatory issues	

Table 2. (Continued)

Method Type	Method	Merits	Demerits	Reference
	6. Supercritical Fluid Extraction	rapid process, no organic solvent or acid needed, lipids can be used for further analysis.	very expensive equipment, complex equipment, the supply of CO_2 needed	(http://www.biomedcentral.com/content/supplementary/2190-4715-24-13-S1.PDF)
Mechanical	1. Expeller press	Simple and effective, the oil recovery is in the range of 70–75%.	Heat generation and possible damage to the compounds	(Ramanath et al., 2015
	2. Bead beating	The combined effect of agitation, collision, and grinding of the beads produces a more effective disruption process. Dewatering of algal slurry is not required and this contributes to the reduction in processing costs.	Energy-intensive. The reactor should be suitably designed to reduce energy inputs, Difficult to scale up	
	3. Ultrasound-assisted extraction	Does not require the addition of beads or chemicals. Prolonged ultrasonication leads to the production of free radicals, which may be detrimental to the quality of the oil that is being extracted		
	4. Microwave-assisted process	Short reaction time, low-operating costs, and efficient extraction of algal oils.	Yet to be standardized at a commercial level, Energy demand is too high	

Organic Solvent Extraction Method

The principle behind the organic solvent extraction is that during exposure of cells to nonpolar organic solvent (e.g., chloroform or hexane), organic solvent penetrates the cell wall and bind with neutral lipids using Vander Waals forces and forms an organic solvent-lipid complex. Then these organic solvent–lipid complex permeate across the cell membrane and diffuses into the organic solvent outside the cells. As a direct result of these phenomena, neutral lipids are completely extracted out from the cells and remain dissolved in a non-polar organic solvent. Most prolifically used organic extraction solvent mixture for lipid extraction from living tissue is chloroform and methanol system in 1:2 ratio (v/v). This organic solvent mixture can totally extract the polar and neutral lipids as intracellular water acts as a ternary component. In addition, algal biomass does not need to be completely dry for this process. A mixture of Hexane and isopropanol in 3:2 ratio (v/v) is also suggested to substitute the chloroform and methanol system due to its low toxicity. This mixture works same as the chloroform-methanol system. Pure alcohol like ethanol, butanol, and isopropanol can be used in the solvent mixture for extraction because it is volatile, cheap and has a strong affinity towards lipid components. Nonetheless, polar nature of these alcohols can be a disadvantage as its limits the interaction with neutral lipids. To avoid this complexity, when alcohol is used in the solvent mixture, it is always used together with a nonpolar solvent like chloroform or hexane to ensure complete extraction of both forms neutral lipids (Halim et al., 2012).

Direct Transesterification Methods

Direct transesterification is a single step process where both extraction and esterification occurs simultaneously in the reactor. It reduces the overall production cost and consumes less time than the traditional two-step process. In this process, stainless steel reactor is submerged into the preheated isothermal fluidized sand bath in which algal biomass (wet) is kept for the desired time and removed swiftly. Simultaneously, the hydrolysis reaction is carried out in other two reactors for each condition. Subsequently, dry algal biomass is mixed with water (1:4 ratio) in a large reactor and the reaction is continued for 15, 30, 45, and 60 min at 250°C. In this process, consequent drying and dehydration convert the biomass into a solid mass which then facilitates precise loading of solids for hydrolysis reaction. Total mixture is then cooled and the solids and aqueous phase were separated using proper filter under light vacuum condition (Ramanath et al., 2015).

Supercritical Carbon Dioxide (SC-CO_2) Extraction

A general supercritical carbon dioxide (SC-CO_2) extraction component comprises of three parts: a feed pump to compress and transport liquid CO_2, extraction vessel set up inside the oven and a hot micro metering valve for depressurization of incoming SC-CO_2. After the oven is heated up sufficiently, SC-CO_2 enters into the hot oven and extraction of lipid from microalgae occurs. After decompression, CO_2 evaporates as a gas which forces the extracted lipid to precipitate and is collected in the adjoining container. One main advantage of this process is SC-CO_2 has low toxicity and high solvating power. Also, by this method extracted product come out as solvent free. The major drawback of this process is its high operational and infrastructure cost (Pragya et al., 2013).

Lipid Extraction from Wet Biomass Using Ethanol

Ethanol is used in this method to extract lipid from moist algal biomass by using ethanol at room temperature (27°C) for 30 min. The ratio of wet biomass and ethanol is 1:4 (mL/g). Lipids extracted from microalgal biomass by ethanol generally consist of various non-lipids as protein strongly bonds with carbohydrate and lipid. Crude lipid is then purified to get rid of these impurities. Hexane is used to purify and to remove these non-lipids complex because of its low toxicity. Lipid extraction yield is then calculated by quantifying the purified lipids. Finally, ethanol used in this method can be recycled using distillation column. Therefore, in this method, ethanol can be recycled to increase the product effectiveness when using at large scale. In comparison with the Bligh-Dyer method, this method is more environment-friendly because it uses ethanol. Also, when compared with the traditional Bligh-Dyer method, this method is far easy to operate at large scale. To summarize, this process has major advantages over Bligh-Dyer method, which are high extraction efficiency, shorter treatment time, and less environmental pollution. However, this process needs further large-scale evaluation with different microalgae for its applicability in the industry (Yang et al., 2014).

Mechanical Lipid Extraction Method

Expeller Press

It is one of the most simple and older techniques for oil extraction from oilseeds. Dry microalgal biomass contains oil or lipid and it can be extracted by pressing using expeller press or oil press. The rationale behind this technique is that it applies high mechanical pressure to break and crush the cells and eventually to squeeze out the oil. Extraction efficacy along with the application of pressure increases up to one point, after which it results in increased heat generation, decreased lipid recovery and choking problems (Ramesh, 2013). The major disadvantage of this process is the presence of pigments in oil. To avoid this problem, pigments have to be removed from the biomass before conversion by either solvent extraction or activated carbon adsorption, which only increase the production cost. Other disadvantages of this process include less efficiency, high maintenance cost, and requirement of skilled labor (Ramesh, 2013).

Bead Beating

Bead mill accomplishes cell disruption by using physical force. It grinds the algal cells against the surface of glass beads by agitating the beads violently. Amongst the numerous disruption techniques, it is more desirable for commercial application owing to its low operating cost and therefore also widely used in a laboratory scale (Halim et al., 2012).

Ultrasound-Assisted Extraction of Oil

The thick cell walls of microalgae block the release of intra-lipids present inside and the use of methods like solvent extraction and mechanical press yields less lipid. In ultrasound-assisted extraction method, the intensive sonication of liquid produces sound waves which transmit into the liquid media and results in alternate high-pressure and low-pressure cycles. During this cycle, the small vacuum bubbles, which are produced in the low-pressure cycle, collapse violently and result in a phenomenon called cavitation. The high pressure and high-speed liquid jets form shearing forces around the algae cells during cavitation and breaks down the cell structure in a mechanical manner and improve material transfer supporting the extraction of lipids. Hence, ultrasonic assisted extraction technique with sound waves having frequencies higher than 20 kHz is used and due to this high intensity, small vacuum bubbles are created in the liquid. This effect supports the extraction of lipids from microalgae and an oil yield enhancement by 50–500% with 10 fold reduced extraction time is achieved (Mubarak et al., 2015).

Microwave-Assisted Heating

Microwave-assisted heating is a non-contact energy source, which heats the whole sample volume concurrently in comparison with conductive heating. Microwave-assisted extraction is being utilized for efficient lipid extraction from microalgae by using traditional solvents. In traditional solvent extraction process, mass transfer happens from the direction of inside to the outside, while heat transfer happens in reverse. However, in microwave-assisted solvent extraction, both mass and heat transports happens from the inside of the extracted material to the bulk solvent. It is observed that extraction efficiency is increased along with the increase of moisture contents in the biomass that can rise up to 90% wt. Currently, microwave energy is being used in the fast production of biodiesel from vegetable oil via transesterification process where the commercially available microwave is being used in continuous or batch (Iqbal and Theegala, 2013).

EMERGING AND INNOVATIVE TECHNOLOGIES FOR EXTRACTION

Origin Oil Single-Step Extraction of Microalgal Lipids

Origin Oil Inc., a USA establishment has devised a novel lipid extraction method from microalgae. This method devised by Origin Oil performs (a patent pending technology) three sequential steps (dewatering, cell disruption, and lipid extraction) in a single downstream step instead of following the conventional sequence-based downstream processing. This process is being mentioned as OriginOil Single-Step Extraction™. This extraction procedure has various advantages. One major benefit is that it significantly reduces the energy cost as it combines three steps (dewatering, cell disruption, and lipid extraction) in a single downstream step. This extraction method also does not utilize any toxic solvent and, hence, there is no need for solvent recovery or purification (Halim et al., 2012).

Process for Extracting Lipids from Microalgae

This procedure involves the step of pre-treating the non-homogenized microalgal sample. Pre-treatment was to be performed by aliphatic alcohol for a previously determined time span. This pre-treatment process released a substantial amount of lipid from the cells. These released lipids were then processed with a transesterification agent to form fatty acid methyl esters (FAME). The FAMEs are then separated from the resultant mixture for further purification to remove any impurities or other solvents. FAME produced by this method is suited for green biodiesel product. Also, pre-treatment process used in this method does not require energy input which makes it less expensive from energy requiring cell membrane disruptive technologies (Hutton and Lehr, 2013).

Oil Extraction from Microalgae

This investigation relates to recovery of lipid and biomass from microalgae using aqueous alcohol processing methods. In particular, the extraction of lipids and precipitation of protein components of biomass are obtained from oleaginous microalgae. According to the inventor, technology is capable of extraction and separation of lipid and protein from algal biomass towards biofuel production (Wang, 2012).

METHODS FOR EXTRACTION OF LIPIDS FROM WET ALGAL BIOMASS

This invention provides a unique procedure for lipid extraction from intact or lysed microalgae in aqueous culture using a partially water-soluble co-solvent with, or without a second organic solvent, and/or pressurized CO_2 in the extraction methods. Such a process can also be implemented at industrial scale to improve production rates and lower costs (Donohue and Williams, 2014).

Algal Cell Lysis and Lipid Extraction Using Electromagnetic Radiation-Excitable Metallic Nanoparticles

This current disclosure relates to systems and methods of extracting lipids from algae by providing metallic nanoparticles to the algal cells and then exciting the nanoparticles using electromagnetic radiation, particularly radio frequency (RF) or microwave radiation. These lipids may then be further processed to produce useful products, such as plastics, jet fuel, cosmetics, and biodiesel (Costas and Eck, 2014).

Procedure for Extracting of Lipids from Algae without Cell Sacrifice

In this procedure, lipids are extracted from the cells by subjecting it to an electric field in an aqueous medium which is enough to release the intracellular lipids from the algal cells. This electric field is configured to extract lipid from algal cells in an aqueous medium which passes between the two electrodes forming the above-mentioned field (Reep and Green, 2012).

Lipid Extraction from Microalgae Using a Single Ionic Liquid

This is a one-step process, which includes the lyses of the microalgal cell wall and intracellular lipid separation towards biofuel production using 1-butyl-3-methylimidazolium, a hydrophilic ionic liquid. This ionic hydrophilic liquid lyses the microalgal cell walls which form two un-mixable layers. One layer contains the lipid contents of the cell. From the mixture, gravity causes the hydrophobic part to go up. It is then separated from the mixture and purified. This ionic hydrophilic liquid can be recycled for another algal sample (Di Salvo et al., 2013).

PURIFICATION PROCESS FOR HIGH-VALUE PRODUCTS (DHA AND EPA)

Docosahexaenoic acid (DHA) and eicosapentaenoic acid (EPA) in its crude form are not suitable for consumption due to its turbid appearance, odor, and taste. Hence, there is a need for purification and refining. This can be accomplished by simply employing the standard vegetable oil refining steps which are degumming, caustic refining, bleaching, and deodorization. During this purification and refining process, special care should be taken in terms of operation speed and process control as the oil is sensitive to oxidation. Martek Biosciences Corporation is known to have recovered and purified DHA from algal oil (Cuellar-Bermudez et al., 2015). They have formulated an optimized process for purification based on the parameters like taste, oxidative stability, and odor. Guil-Guerrero et al., (2000) have purified EPA from algal biomass of *Porphyridium cruentum* by simultaneous oil saponification and fatty acid extraction. The saponification process is performed by potassium hydroxide and ethanol for 1 hour at 60° C. The urea method is performed to concentrate the polyunsaturated fatty acid (PUFA). In the end, silica gel column chromatography is used to separate the EPA methyl esters from PUFA concentrate. Guil-Guerrero et al., (2000) recorded a 50.8% of EPA recovery with a purity level of 97%.

CONCLUSION

Currently, substantial research efforts are being made on the extraction of oil from microalgae to develop technology at industrial scale from both government and industrial sector. Still, no clear-cut front-runner technologies are available for lipid extraction for commercial purpose. Industries like OriginOil Inc. and other emerging inventors have come up with better solutions than traditional extraction techniques. These inventors are being patented for commercial production. To extract lipid from a specific species of algae, extraction process should be chosen based on the performance of the particular technique and the end product application must also be considered. Present techniques offered a lot of promise for commercial extraction technology and we can hope that with all these recent research developments, a breakthrough is on the verge.

Conflict of Interest

No conflict of interest among the authors.

References

Amaro, H. M., Guedes, A. C., and Malcata, F. X. (2011). Advances and perspectives in using microalgae to produce biodiesel. *Appl. Energy,* 88 (10):3402-3410.

Amaro, H. M., Guedes, A. C., and Malcata, F. X. (2011). Advances and perspectives in using microalgae to produce biodiesel. *Applied. Energy*, 88 (10):3402-3410.

Borowitzka, M. A. (1999). Commercial production of microalgae: ponds, tanks, tubes and fermenters. *J. Biotechnol*, 70 (1–3):313-321.

Borowitzka, M. A. (1999). Commercial production of microalgae: ponds, tanks, tubes and fermenters. *J. Biotechnol*, 70 (1–3):313-321.

Brennan, L., and Owende, P. (2010). Biofuels from microalgae-a review of technologies for production, processing, and extractions of biofuels and co-products. *Renew. Sust. Energ. Rev,* 14 (2):557-577.

Brennan, L., and Owende, P. (2010). Biofuels from microalgae—A review of technologies for production, processing, and extractions of biofuels and co-products. *Renew. Sust. Energ. Rev* 14 (2):557-577.

Chisti, Y. (2008). Biodiesel from microalgae beats bioethanol. *Tren. Biotech*, 26 (3):126-131.

Chisti, Y. (2008). Biodiesel from microalgae beats bioethanol. *Tren. Biotechnol*, 26 (3):126-131. doi: 10.1016/j.tibtech.2007.12.002.

Costas, C. R., and Eck, C. R. (2014). "Algal cell lysis and lipid extraction using electromagnetic radiation-excitable metallic Nanoparticles." *US Patent App US8722375B2*.

Costas, C. R., and Eck, C. R. (2014). "Algal cell lysis and lipid extraction using electromagnetic radiation-excitable metallic nanoparticles." US8722375B2.

Cuellar-Bermudez, S. P., Aguilar-Hernandez, I., Cardenas-Chavez, D. L., Ornelas-Soto, N., Romero-Ogawa, M. A., and Parra-Saldivar, R. (2015). Extraction and purification of high-value metabolites from microalgae: essential lipids, astaxanthin and phycobiliproteins. *Micro. Biotechnol*, 8 (2):190-209.

Cuellar-Bermudez, S. P., Aguilar-Hernandez, I., Cardenas-Chavez, D. L., Ornelas-Soto, N., Romero-Ogawa, M. A., and Parra-Saldivar, R. (2015). Extraction and purification of high-value metabolites from microalgae: essential lipids, astaxanthin and phycobiliproteins. *Microbiol. Biotechnol,* 8 (2):190-209.

Demirbas, A., and Fatih, D. M. (2011). Importance of algae oil as a source of biodiesel. *Energy. Convers. Manag*, 52 (1):163-170.

Demirbas, A., and Fatih, D. M. (2011). Importance of algae oil as a source of biodiesel. *Energy. Convers. Manag* 52 (1):163-170.

Di Salvo, R., Reich, A., Dykes, H. W. H., and Teixeira, R. (2013). "Lipid extraction from microalgae using a single ionic liquid." *US 20110130551A1.*

Di-Salvo, R., Reich, A., Dykes, H. W. H., and Teixeira, R. (2013). Lipid extraction from microalgae using a single ionic liquid. US Patent App.

Donohue, M., and Williams, S. (2014). "Methods for extraction of lipids from wet algal biomass." *US W02014052357A1.*

Donohue, M., and Williams, S. (2015). *Methods for extraction of lipids from wet algal biomass. US Patent App. 14/430,922.*

Fon Sing, S., Isdepsky, A., Borowitzka, M., and Moheimani, N. (2013). Production of biofuels from microalgae. *Mitigation and Adaptation Strategies for Global Change* 18 (1):47-72.

Fon-Sing, S., Isdepsky, A., Borowitzka, M., and Moheimani, N. (2013). Production of biofuels from microalgae. *Mitigation and adaptation strategies for global change* 18 (1):47-72.

Gerpen, J. V. (2005). Biodiesel processing and production. *Fuel. Process. Technol*, 86 (10):1097-1107.

Gerpen, J. V. (2005). Biodiesel processing and production. *Fullerene. Sci. Technol,* 86 (10):1097-1107.

Gudin, C., and Chaumont, D. (1991). Cell fragility-The key problem of microalgae mass production in closed photobioreactors. *Biores. Technol*, 38 (2–3):145-151.

Gudin, C., and Chaumont, D. (1991). Cell fragility-The key problem of microalgae mass production in closed photobioreactors. *Biores. Technol*, 38 (2–3):145-151.

Guil-Guerrero, J., Belarbi, E. H., and Rebolloso-Fuentes, M. (2000). Eicosapentaenoic and arachidonic acids purification from the red microalga. *Porphyridium cruentum. Bioseparation,* 9 (5):299-306.

Guil-Guerrero, J., Belarbi, E. H., Rebolloso-Fuentes, M. (2000). Eicosapentaenoic and arachidonic acids purification from the red microalga *Porphyridium cruentum. Bioseparation* 9 (5):299-306.

Halim, R., Danquah, M. K., and Webley P. A. (2012). Extraction of oil from microalgae for biodiesel production: A review. *Biotechnol. Adv*, 30 (3):709-732.

Halim, R., Danquah, M. K., and Webley, P. A. (2012). Extraction of oil from microalgae for biodiesel production: A review. *Biotechnol. Adv*, 30 (3):709-732.

Hutton, M. W., and Lehr, C. R. (2013). *Process for extracting lipids from microalgae.*

Iqbal, J., and Theegala, C. (2013). Microwave-assisted lipid extraction from microalgae using biodiesel as co-solvent. *Algal. Res*, 2 (1):34-42.

Janssen, M., Tramper, J., Mur, L. R., and Wijffels, R. H. (2003) Enclosed outdoor photobioreactors: Light regime, photosynthetic efficiency, scale-up, and future prospects. *Biotechnol. Bioeng,* 81 (2):193-210.

Kim, J., Yoo, G., Lee, H., Lim, J., Kim, K., Kim, C. W., Park, M. S., and Yang J-W. (2013) Methods of downstream processing for the production of biodiesel from microalgae. *Biotechnol. Adv*, 31 (6):862-876.

Kumar, R. R., Rao, P. H., and Arumugam, M. (2015). Lipid extraction methods from microalgae: A comprehensive review. *Front. Energy. Res,* (2), 1-9.

Mubarak, M., Shaija, A., and Suchithra, T. V. (2015). A review on the extraction of lipid from microalgae for biodiesel production. *Algal. Res*, 7 (0):117-123.

Mutanda, T., Ramesh, D., Karthikeyan, S., Kumari, S., Anandraj, A., and Bux, F. (2011). Bioprospecting for hyper-lipid producing microalgal strains for sustainable biofuel production. *Biores. Technol,* 102 (1):57-70.

Olaizola, M, (2000). Commercial production of astaxanthin from Haematococcus pluvialis using 25,000-liter outdoor photobioreactors. *Appl. Phycol,* 12 (3-5):499-506.

Patil, V., Reitan, and K. I., Knutsen, G., Mortensen, L. M., Källqvist, T., Olsen, E., Vogt, G., and Gislerød, H. R. (2007). Microalgae as source of polyunsaturated fatty acids for aquaculture. *Plant. Biol*, 15(1):1–9.

Pragya, N., Pandey, K. K., and Sahoo, P. K. (2013). A review on harvesting, oil extraction and biofuels production technologies from microalgae. *Renew. Sust. Energ. Rev*, (24), 159–171.

Ramesh, D. (2013). Lipid Identification and Extraction Techniques. In: Bux F (ed) *Biotechnological Applications of Microalgae: Biodiesel and Value-Added Products.* CRC Press, pp. 89-97.

Rawat, I., Ranjith K. R., Mutanda, T., and Bux, F. (2011). Dual role of microalgae: Phycoremediation of domestic wastewater and biomass production for sustainable biofuels production. *Appl. Energy*, 88 (10):3411-3424.

Reep, P., and Green M. P. (2012). "Procedure for extracting of lipids from algae without cell sacrifice." *US 2012/0040428A1.*

Rodriguez-Cruz, M., Tovar, A. R., del-Prado, M., and Torres, N. (2005). Molecular mechanisms of action and health benefits of polyunsaturated fatty acids. *Rev. Invest. Clin,* 57 (3):457-472.

Scott, S. A., Davey, M. P., Dennis, J. S., Horst, I., Howe, C. J., Lea-Smith, D. J., and Smith, A. G (2010). Biodiesel from algae: challenges and prospects. *Curr. Opin. Biotechnol,* 21 (3):277-286.

Wang T. (2012). "Oil extraction from microalgae." *US 2012/0238732A1*.

Wang, J., Luo, T., Li S., Zhao, J. (2012). The powerful applications of polyunsaturated fatty acids in improving the therapeutic efficacy of anticancer drugs. *Expert. Opin. Drug. Deliv*, 9 (1):1-7.

Yang, F., Xiang, W., Sun, X., Wu, H., Li, T., Long, L. (2014). A Novel Lipid Extraction Method from Wet Microalga Picochlorum sp. at Room Temperature. *Marine Drugs* 12 (3):1258-1270.

Zeng, X., Danquah, M. K., Chen, X. D., Lu, Y. (2011). Microalgae bioengineering: From CO_2 fixation to biofuel production. *Renew. Sust. Energ. Rev*, 15 (6):3252-3260.

Online References

http://lansbury.bwh.harvard.edu/structure_of_dha_and_epa.htm. Accessed 30-Jun-2015 2015.

http://www.biomedcentral.com/content/supplementary/2190-4715-24-13-S1.PDF. Accessed 30-Jun-2015 2015.

https://www.marketsandmarkets.com/Market-Reports/omega-3-omega-6-227.html (2016) Omega-3 PUFA Market by Type (DHA, EPA, ALA), Application (Dietary Supplements, Functional Foods & Beverages, Pharmaceuticals, Infant Formula), Source (Marine, Plant), Sub-source), and Region-Global Forecasts to 2020. Accessed 18-05-2016.

https://www.marketsandmarkets.com/PressReleases/omega-3.asp (2016). 2016.

http://www.oilgae.com/algae/oil/biod/large_scale/large_scale.html (2015). Accessed 03-Sep-2015.

http://www.oilgae.com/non_fuel_products/omega_3fattyacids.html. Accessed 07-Aug-2015 2015.

https://en.wikipedia.org/wiki/Freeze-drying. Accessed 16-Jul-2015 2015.

https://www.goshen.edu/academics/chemistry/biodiesel/chemistry-of/. Accessed 13 - Jul – 2015.

In: The Role of Photosynthetic Microbes…
Editors: K. Tripathi, N. Kumar et al.
ISBN: 978-1-53614-032-3

Chapter 3

MICRO-ALGAE AS AN EFFECTIVE TOOL FOR WASTEWATER TREATMENT AND MANAGEMENT

Jaspal Singh Chauhan[1,*] and Vineet Kumar Maurya[2]
[1]Department of Himalayan Aquatic Biodiversity
School of Life Sciences, HNB Garhwal University,
Srinagar Garhwal, India
[2]School of Life Sciences, HNB Garhwal University,
Srinagar Garhwal, India

ABSTRACT

With the rise in developmental activities, a load of pollution in the environment has increased. Most of the water bodies remain polluted due to the wastewater released by agricultural, domestic and industrial activities. Treatment of wastewater by physical, chemical and biological approaches is implied in order to cope with this problem and discharge an apparently clean effluent into natural water bodies. Recently, for the biological treatment of wastewater micro-algae have gained importance since they are able to accumulate plant nutrients, heavy metals, pesticides, organic and inorganic toxic substances and radioactive matters in their bodies. Further, the biomass left after the treatment is used for biogas generation and bio-compost. Biological wastewater treatment systems with micro-algae are now widely accepted as effective and low-cost natural treatment systems for purification of wastewater. In this chapter, we highlight the role of micro-algae in the treatment and management of wastewater.

---

[*]Corresponding Author Email: jaspal.env@gmail.com.

Keywords: micro-algae, water-purification, wastewater, water pollution

INTRODUCTION

Un-managed industrial development has deteriorated the quality of freshwater resources around the globe and hence caused the shortage of freshwater. Discharge of wastewater directly from industries without treatment has adversely affected the receiving water body. The water quality norms say that before discharging wastewater in any aquatic system it is mandatory to treat the wastewater and bring it up to a standard permissible limit. The initial cost, as well as operating cost of wastewater treatment plant including primary, secondary or advanced stages, is highly expensive and only a few industries follow the norms strictly. Therefore current research is focused to find out treatment methods which are ecofriendly, efficient and low cost. In recent years, microalgae have received substantial attention in wastewater treatment since they are able to accumulate heavy metals, pesticides, organic and inorganic toxic substances and radioactive matter in their bodies (Oswald, 1988; Sakaguchi et al., 1981). Microalgae require nutrients for growth which are enormously found in the wastewater. Thus, wastewater can be used as a food source for microalgae, which would achieve the removal of nutrients and other pollutants from wastewater. Microalgae also utilize organic carbonaceous matter and convert it into cellular constituents such as lipids and carbohydrates (Wang et al., 2010). Further, the algal biomass harvested after treatment is used for different purposes such as biofuel, biogas, biofertilizer generation, etc. Microalgae based treatment supersedes conventional biological treatment systems because besides removing dissolved organic matter it also efficiently removes contaminants such as nutrients (C, N and P), pathogens and heavy metals (Lau et al., 1996; Pahad and Rao, 1962). Various technologies have been used in microalgae based wastewater treatment, starting from simple stabilization oxidation ponds to more advance high rate oxidation ponds (HROP), high rate algae ponds (HIRAP) and activated algae ponds. This chapter gives an overview of algal applications in wastewater treatment with reference to the removal of nutrient, pathogen and heavy metals.

CONVENTIONAL WASTEWATER TREATMENT

Wastewater treatment involves physical, chemical and biological processes in primary, secondary or tertiary stages to improve wastewater quality up to a permissible limit. Figure 1 shows the flow chart of wastewater treatment process.

Primary treatment removes large and heavy materials that float or readily settle out by the physical processes of screening, commination, grit removal, and sedimentation.

The secondary treatment is the biological process that removes the soluble organic matter and suspended solids left from primary treatment with the help of micro-organisms. Tertiary or advanced treatment is the process in which remaining nitrates and phosphates are removed. The major disadvantages with conventional wastewater treatment are:

- High operational cost and maintenance requirements
- Continuously generate large amounts of sludge that create the problem of disposal. Handling and disposal of sludge is typically very expensive operation of a wastewater treatment plant.
- It requires a constant energy supply and high energy consumption
- Skilled and trained operators are required who can monitor the system and react to changes immediately
- This treatment method can only reduce nitrogen and phosphorus concentrations from wastewater to a limited quantity. Further, discharge of such wastewater having nitrates and phosphates impairs environmental quality.
- Wastewater contains undesirable organic matter that is subjected to biodegradation by micro-organisms and converted to smaller molecules (CO_2, NH_3, PO_4^{3-} etc.) in presence of oxygen. The oxygen is supplied externally to the system and requires a lot of expertise, money and manpower.

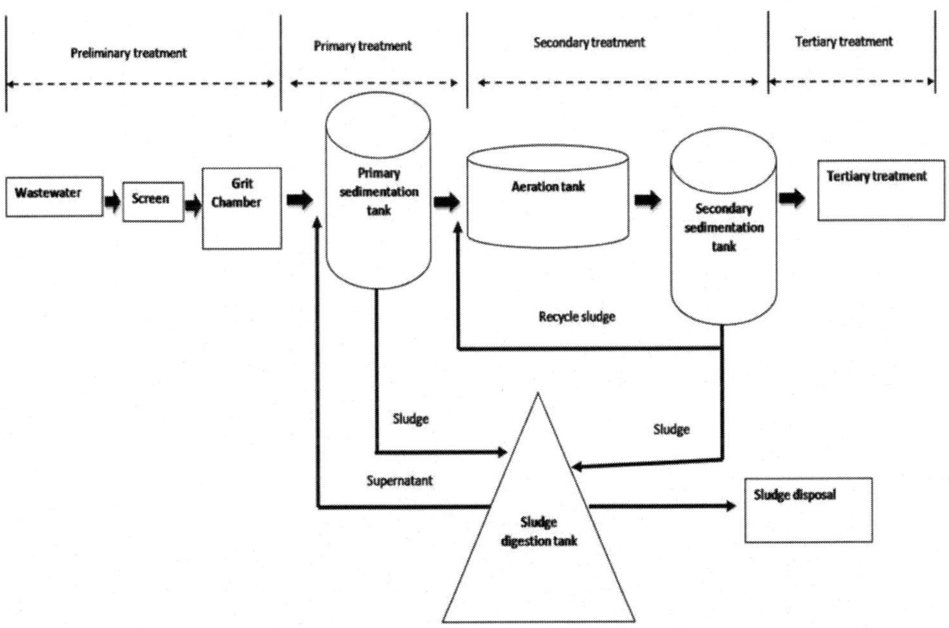

Figure 1. The flow chart of wastewater treatment process (Abdel-Raouf et al., 2012).

MICROALGAE BASED WASTEWATER TREATMENT

Microalgae based wastewater treatment was first introduced by William Oswald who described the basic role of microalgae in wastewater treatment such as nutrients uptake and oxygen generation (Oswald, 1988; Oswald and Gotaas, 1955). Figure 2 explains the process of wastewater treatment in presence of microalgae. Microalgae are microscopic photosynthetic organisms found in aquatic ecosystems and have a photosynthetic mechanism very similar to plants. During photosynthesis, microalgae use solar light as an energy source and CO_2 as carbon source and uptake nitrogen and phosphorus for their growth. Thus, during wastewater treatment this process of photosynthesis reduces the load of nutrients (N, P) in wastewater and also contributes to reduce CO_2 produced by bacteria present in wastewater. In addition, microalgae produce oxygen, as a by-product of photosynthesis, which is used by aerobic bacteria to oxidize organic matter present in the wastewater. In fact, microalgae can help to reduce the need for mechanical aeration during wastewater treatment. Microalgae can also work in the absence of light as heterotrophic organism and use oxygen to assimilate organic carbon. In addition, microalgal treatment provides an environment that increases the death of pathogenic organisms due to elevated pH and antibacterial substances that may be excreted by microalgae (Polprasert et al., 1983). Various technologies have been used in microalgae based wastewater treatment, starting from simple stabilization oxidation ponds to more advance high rate algae ponds. There are primarily two designs that incorporate algae based treatments (EPA, 2002; Larsdotter et al., 2007).

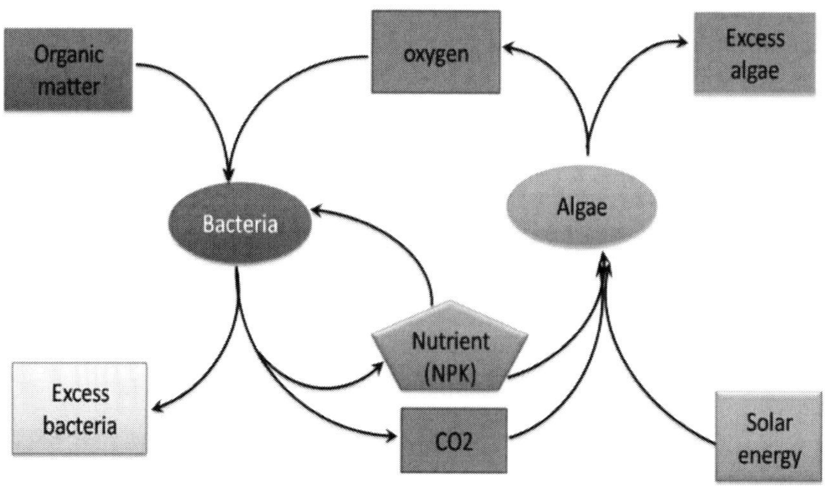

Source: Oswald and Gotass, 1957.

Figure 2. Process of wastewater treatment by microalgae in a pond.

Facultative Waste Stabilization Ponds (Also Called Facultative Lagoons or Ponds)

The facultative lagoons are shallow ponds (4-8 feet deep) and are the most common form of oxidation ponds that exhibit aerobic conditions on the surface due to photosynthesis by algae and anaerobic conditions in the bottom layers. The oxygen produced by algae is then utilized by bacteria to assimilate carbon and remove BOD from wastewater. Further, in facultative ponds pH increases due to algal growth that volatilizes ammonia, removing N from wastewater (Larsdotter et al., 2007). Due to less operational costs, facultative ponds are economically beneficial in the long run, but requires large land area and are therefore beneficial for the treatment of wastewater from small scale industries or small rural communities (EPA, 2002).

High Rate Algae Ponds (HRAP)

High-rate ponds are also known as raceway ponds (EPA, 2002; Larsdotter et al., 2007) and were developed as an effective medium for removal of organic matter, suspended solids and pathogen. They are shallow and therefore light reaches up to the bottom part making it completely oxygenated (Oswald, 1978). The term 'high rate' is used because the algal growth rate in these ponds is very high as compared to conventional waste stabilization ponds. HRAPs are one of the most cost-effective reactors available for wastewater management and for efficient capture of solar energy (Oswald, 1995). Well designed and operated HRAPs can efficiently remove more than 90% of the BOD and up to 80% of the nitrogen and phosphorus from wastewater (Oswald, 1988). HRAPs are designed to be shallow (30–100 cm) with a raceway shape and have a paddle wheel for gentle mixing of wastewater with algae (Figure 3). Continuous mixing by paddlewheel exposes the algal cells periodically to light and thereby keeps them active on the substrate. As a result, treatment is more efficient and thereby less land intense. The removal of N in HRAPs is due to volatilization of ammonia at high pH and metabolic uptake of N by algae for its growth.

Role of Microalgae in Wastewater Treatment

Algae are used in wastewater treatment for the removal of coliform bacteria, organic pollutant, nutrient (N and P) and also for the removal of heavy metals. The following sections highlights the role of the algae with respect to these processes.

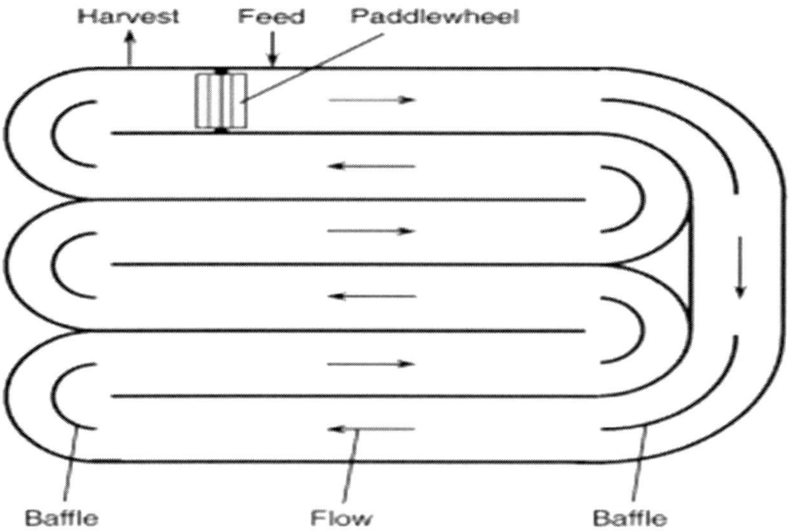

Figure 3. Design of Raceway pond (Sompech K. et al., 2012).

Removal of Pathogen

Removal of pathogens by algae-based treatment ponds is a complicated phenomenon that may result due to physical, chemical, biological reactions or environmental factors such as predation, temperature, sunlight, dissolved oxygen, pH, sedimentation and starvation (Fallowfield et al., 1996, Polprasert et al., 1983). Several hypotheses were made to explain the causes of bacterial reduction among which few are mentioned below:

- A high level of pH is common in algae-based treatment ponds. Algal photosynthesis causes an increase in the pH due to the simultaneous removal of CO_2 and H^+ ions and the uptake of bicarbonate when the algae are carbon limited (Craggs et al., 1997; Fallowfield et al., 1996). According to Rose et al., (1996) a pH of 9.2 for 24 hours kills 100% of *E. coli* and most pathogenic bacteria and viruses. Similarly, Pahad and Rao, (1962) also observed that growth of *E. coli* in wastewater is inhibited at a pH higher than 9.2.
- Presence of antibacterial substances excreted by algae and the production of toxic extracellular chemicals secreted by certain algal species also kills the bacteria.

Heavy Metal Removal

Microalgae also require metals for their metabolic functions and selected microalgae have the potential to accumulate high concentrations of metals from the contaminated aquatic ecosystems. Microalgae produce peptide molecules which bind the heavy metals forming organometallic complexes, which are then contained in vacuoles to maintain the cytoplasmic concentration of metal ions thereby neutralizing the potential toxic effect of the heavy metals (Perales-Vela et al., 2006). Metal accumulation in algae involves two processes: a passive uptake (also known as biosorption) followed by an active uptake or chemisorption (Bates et al., 1982). During the passive uptake, the metal ions are quickly

Table 1. List of metal specific hyper accumulator algal species

S.No	Metal	Hyper accumulator algae	References
1.	Cadmium (Cd^{2+})	*Chaetoceros calcitrans* *Scenedesmus abundans* *Desmodesmus pleiomorphus* *Tetraselmis chuii*	Sjahrul and Arifin, (2012) Monteiroet al., (2009) Monteiroet al., (2010) Sjahrul and Arifin, (2012)
2.	Lead (Pb^{2+})	*Oscillatorialaete-virens* *Pseudochlorococcum typicum* *Spirulina (Arthrospira) platensis* *Sargassum vulgare* *Sargassum natans*	Miranda et al., (2012) Shanab et al., (2012) Arunakumaraet al., (2008) Holan and Volesky, (1994) Holan and Volesky,(1994)
3.	Zinc (Zn^{2+})	*Isochrysis galbana* *Planothidium lanceolatum* *Scenedesmus subspicatus* *Sargassum fluitans*	Sbihiet al., 2012 Sbihiet al., (2012) Schmitt et al.,(2001) Figueira et al. (1997)
4.	Mercury (Hg^{2+})	*Pseudochlorococcum typicum* *Dunaliella* *Chlorella pyrenoidosa* *Spirulina platensis*	Shanab et al., (2012) Imani et al,(2011) Yao et al., (2011) Gannikar (2002)
5.	Nickel (Ni^{2+})	*Chlorella miniata* *Chlorella vulgaris* *Spirulina* *Fucus vesiculosus*	Wong et al., (2000) Al-Rub et al., (2004) Doshiet al., (2007) Holan and Volesky,(1994)
6.	Chromium (Cr^{3+})	*Spirulinaplatensis* *Nostoc spongiaeforme* *Nostoc linckia* *Chlorella marina*	Shashirekhaet al., (2008) Kaushiket al., (2008) Kaushik et al., (2008) Kumar et al., (2015)
7.	Arsenic (As^{2+})	*Tetraselmis chuil* *Spirogyra hyalina*	Irgolic et al.,(1977) Nirmal Kumar and Cini, (2012)
8.	Gold (Au^{2+})	*Ascophyllum nodosum*	Kuyucak and Volesky, (1989)
9.	Copper (Cu^{2+})	*Chlorella vulgaris,* *Spirulina maxima* *Synechocystis* sps. *Sargassum sinicola*	Chan et al, (2014) Prado et al, (2010)

adsorbed over the cell surface and this process is metabolism-independent. Then chemisorption, a metabolism-dependent process starts in which the metal ions are slowly transferred across the cell membrane into the cytoplasm. After metal uptake, recovery of valuable elements such as gold and silver is also possible.

Other benefits of metal removal by micro-algae includes low cost, rapid metal uptake capability, time and energy saving, high efficiency, and applicability of water containing high metal concentrations or relatively low contaminant levels (Monteiro et al., 2012). The removal efficiency of heavy metals by algae varies with the type of metal, metal concentration and type of algal species. Some studies showed that hyper accumulation of chromium by *Oscillatoria*, cadmium, copper and zinc by *Chlorella vulgaris*, lead by *Chlamydomonas* and molybdenum by *Scenedes muschlorelloides* is successful (Laliberteetal., 1994, Filip et al., 1979, Nakajima et al., 1981, Ting et al., 1989, Hassett et al., 1981, Sakaguchi et al., 1981). In a study by El-Sheekh et al., (2005) monocultures of *Nostoc muscorum* showed the removal of Cu, Co, Pb and Mn by 64.4, 2.20, 84.6 and 64.10% while *Anabaena subcylindrica* showed 33.3, 33.3, 86.2 and 40%, respectively from sterilized sewage wastewater. In mix culture of *Nostoc muscorum* and *Anabaena subcylindrica* 75, 11.8, 100 and 61.5% removal of Cu, Co, Pb and Mn was observed from sewage wastewater. Ajayan et al., (2011) observed the Cu, Zn and Co removal of 60, 42.9 and 29.6%, with *Oscillatoria quadripunctutata,* while Pb removal of 34.6% was observed with *Scenedes musbijuga* in sewage wastewater. A list of metal specific hyper accumulator algal species is mentioned in Table 1.

Removal of N and/or P

Organic nitrogen is one of the key elements of important biological substances like enzymes, peptides, protein, chlorophyll and adenosine diphosphate (ADP) and adenosine triphosphate (ATP). Organic nitrogen is derived from inorganic sources mostly nitrite (NO_2^-), nitrate (NO_3^-), ammonia (NH_3), ammonium (NH_4^+), nitric acid (HNO_3), and nitrogen gas (N_2). In the conventional treatment of wastewater, after secondary treatment, the treated water still contains inorganic nutrients (N, P) in theform of NH_4 (ammonia), NO_2^-(nitrite), NO_3^-(nitrate) and PO_4^{3-}(orthophosphate). The adverse effects of nutrient enrichment in receiving aquatic bodies are eutrophication which stimulates the growth of unwanted plants, algae and aquatic macrophytes. Increase in concentration of nitrogenous compounds in wastewater effluents leads to toxicity to aquatic organisms. The nitrate concentrations above 45 g/m^3 in drinking water results in methemoglobinemia disorder in infants (Lincolin and Earle, 1990). Microalgae can use most forms of nitrogen such as NH_4^-, NO_3^-, NO_2^- and N_2, therefore, it is a better alternative for wastewater treatment. A eukaryotic alga only has an ability to convert inorganic nitrogen in the forms of nitrite, nitrate and ammonium to organic nitrogen through a process called assimilation (Lovaie and Noüe, 1985). Most of the nitrogen in algal cell bound to proteins which comprise 45-

60% of dry weight. Phosphorus also plays a crucial role in cell growth, metabolism and is found in biologically important structures like proteins, phospholipids and nucleic acids. During algal metabolism, phosphorus mostly in the form of $H_2PO_4^-$ and HPO_4^{2-} and is incorporated into organic compounds through a process called phosphorylation. During phosphorylation formation of ATP from ADP takes place and microalgae are able to store phosphorus in excess within the cell in the form of polyphosphate (volutin) granules (Fogg, 1975). Microalgae using nitrogen and phosphorus in growth may remove nutrients load of wastewater within a few hours to a few days. Lau et al., (1996) studied *Chlorella vulgaris* for the removal of nutrients from wastewater and found that the nutrient removal efficiency of 86% (inorganic N) and 70% (inorganic P) respectively. In a similar study, Colak and Kaya (1988) reported removal of nitrogen (50.2%) and phosphorus (85.7%) from industrial wastewater and (97.8%) of phosphorus from domestic wastewater treated by microalgae.

Important Factors Affecting Algal Wastewater Treatment

Temperature

Generally, the optimum temperatures for algal growth range between 15–25°C (Goldman and Carpenter, 1974). However, different species of algae have different effects of temperature. The rate of metabolism and growth is directly proportional to temperature until it reaches the optimum temperature (Borowitzka, 1998). Munoz et al., (2004), observed that an increase in the temperature from 25 to 30°C double the removal efficiency of a symbiotic microcosm formed by *C. sorokiniana* and *R. basilensis* strains.

Light

The light energy is directly responsible for growth of microalgae. The synthesis of food in the algal cell by photosynthesis depends on the efficiency of converting the light energy to chemical energy. Through photosynthesis, microalgae need light to produce adenosine triphosphate (ATP) that is used as an energy source for cell synthesis, growth and maintenance. However, according to Oswald (1988) only about 10% of the light is converted to chemical energy and the rest escapes as heat. Not only light duration but also light intensity, and wavelength are important parameters that affect the algal growth (Carvalho et al., 2011).

pH

The pH of the wastewater influences the metabolism and growth of microalgae. Every microalgal species has its own optimum range of pH at which they can grow and function properly. Mostly the algal species shows optimal growth response at pH range of 7.5-8.5 (Chisti, 2008; Marcel et al., 2003; Molina et al., 2001). In a study, Fontes et al. (1987) observed that optimal growth of the cyanobacterium *Anabaena variabilis* at pH 8.2–8.4, decreased growth at pH range 7.4–7.8 and no growth above pH 9. Most microalgal species can ideally grow under neutral pH or slightly higher. Also, there are some species that prefer acidic conditions. Hodaifa et al., (2009) found that pH has great influence on the growth and biomass composition of microalgae *Scenedesmus obliquus*. They found that high reduction of BOD occurs at elevated pH and the highest protein and chlorophyll content in the cell occur when alga is cultivated under pH 7.

Carbon and Nutrients

Through photosynthesis, microalgae uptake nitrogen and phosphorus in the presence of light and CO_2 to make their own organic molecules. For optimum growth, microalgae need carbon, nitrogen, and phosphorus in a ratio of about 106: 10: 1. High ratio of N: P (i.e., about 30:1) results in P limitation, whereas low ratios of N: P about 5:1 results N limitation (Darley, 1982). Based on the typical wastewater concentrations of these nutrients, phosphorus was found to rarely affect microalgal growth, while nitrogen was found to have an effect. Overall, wastewater has abundant concentrations of nitrogen and phosphorus and they do not seem to be an issue for microalgae growth. Microalgal growth in wastewater is more likely limited by carbon and light (Noue et al., 1992).

The inorganic carbon normally used by microalgae is CO_2 and HCO_3^-. Besides these, some algal species also use organic carbon sources such as sugars, organic acids, acetate or glycerol (Borowitzka, 1998, Fogg, 1975, Kawaguchi, 1980, Ogbonna et al., 2000). The amount of dissolved CO_2 in wastewater varies with pH. At lower pH the concentration of CO_2 is abundant but at higher pH values (i.e., greater than 9), there is a reduction in CO_2 concentration. Furthermore, at higher pH most of the carbon will be in the form of carbonate (CO_3^{2-}) that cannot be utilised by microalgae unless it is converted to CO_2, a reaction favoured by the carbonic anhydrase enzyme (Borowitzka, 1998). Low CO_2 concentration lowers algal growth but this is not an issue in wastewater treatment as CO_2 is continuously released by bacterial respiration.

IMPORTANCE OF ALGAE BASED WASTEWATER TREATMENT

Algae based wastewater treatment has the following advantages over conventional wastewater treatment.

- fficiently remove ammonia, total nitrogen and phosphorus from wastewater effluent (Ruiz et al., 2011)
- Reduce the aeration requirements,
- CO_2 which is a big reason for global warming is sequester by algae (Hende et al., 2011).
- The overall cost of treatment decreases (Oswald, 2003).
- A simpler and sustainable treatment option.
- Decrease the pathogen content in the final effluent without the use of chemicals.
- Decrease sludge quantities thereby sludge disposal cost.
- Produce algal biomass that is a source for many useful products including biogas and biofuel, bioethanol, biobutanol and vegetable oil (Sheehan et al., 1998).
- Algae like *Chlamydomonas reinhardtii* (a green-algae), would sometimes switch from the production of oxygen to the production of hydrogen (Melis et al., 2000).
- Algal biomass can be burnt to produce heat and electricity

CONCLUSION

Water is essential for human consumption, economic growth and sustainable development of the environment. Due to urbanization and population growth freshwater resources are getting limited and the problem of water shortage is becoming more serious. In addition, urban population growth combined with rapid agricultural and industrial development has not only enhanced the pressure on fresh water but also generated a huge amount of wastewater. This wastewater is mostly disposed into the river and ponds thereby destroying the environment. There is an urgent need of wastewater treatment methods that can efficiently treat effluent with minimal energy and money consumption and low cost. Using microalgae in wastewater treatment has significantly improved the treatment process through reducing the pollutant concentrations efficiently. Furthermore, it is ecofriendly, low cost, low energy, natural and user-friendly. Microalgae play a significant role in wastewater treatment process by taking up nutrients, reducing BOD and CO_2 and inactivating pathogens. A microalgae based wastewater treatment system like high rate algal ponds gives efficient results when provided appropriate environmental and operational conditions such as temperature, light, carbon, nutrients and mixing, etc. The major constraints in an algal based treatment system are contamination, land requirement, grazing, diseases, premature collapse and lack of species control. These problems can be minimized by modifying the operational conditions and controlling the microbial population.

Acknowledgments

JSC thankful to acknowledges Head, Department of Himalayan Aquatic Biodiversity School of Life Sciences, Srinagar (UK) for the providing facility.

Conflict of Interest

No conflict of interest among the authors.

References

Ajayan, K.V., Selvaraju, M., and Thirugnamoorthy, K. (2011). Growth and heavy metals accumulation potential of microalgae grown in sewage wastewater and petrochemical effluent. *Pak. J. Biol. Sci*, 14:805–811.

Al-Rub, F. A., El-Naas, M. H., Benyahia, F., and Ashour, I. (2004). Biosorption of nickel on blank alginate beads, free and immobilized algal cells. *Process. Biochem.* 39, 1767–1773.

Abdel-Raouf, N., Al-Homaidan A. A., Ibraheem I. B. M (2012). Microalgae and wastewater treatment. *Saudi. J. Biol. Sci*, 19, 257–275.

Bala, K., Rani, N., and Kaushik, A. (2008). Chromium (VI) tolerance in two halotolerant strains of *Nostoc*. *J. Environ. Biol*, 29 (2): 155-158.

Bates, S.S., Tessier, A., Campbell, P. G. C., and Buffle, J. (1982). Zinc adsorption and transport by *Chlamydomonas variabilis* and *Scenedes mussubspicatus* (*Chlorophyceae*) grown in semi continuous culture. *J. Phycol*, 18: 521–529.

Borowitzka, M. A. (1998). *Limits to growth, in wastewater treatment with algae*. Wong Y. S and Tam N. F. Y., (eds). Springer Verlag. pp. 203–226.

Carvalho, A. P., Silva, S. O., Baptista, J. M., and Malcata, F. X. (2011). Light requirements in microalgal photobioreactors: an overview of bio photonic aspects. *Appl. Microbiol. Biotechnol*, 1275-1288.

Chan, A., Hamidreza, S., and Ed, Mc Bean. (2014). Heavy metal removal (copper and zinc) in secondary effluents from waste water treatment plants by microalgae. *ACS Sustain. Chem. Eng*, 2(2): 130- 137.

Chisti, Y. (2008). Biodiesel from microalgae beats bioethanol. *Trends. Biotechnol*, 26: 126-131.

Colak, O., and Kaya, Z., (1988). A study on the possibilities of biological wastewater treatment using algae. *Doga. Biyolji. Serisi*, 12 (1): 18–29.

Craggs, R. J., McAuley, P. J., and Smith, V. J. (1997). Wastewater nutrient removal by marine Microalgae Grown on a Corrugated Raceway. *Water. Res*, 31(7): 1701-1707.

Darley, W. M. (1982). *Algal biology: A physiological approach.* Oxford and Boston and St. Louis, Mo.: Blackwell Scientific Publications.

Doshi, H., Ray, A., and Kothari, I. L. (2007). Bioremediation potential of live and dead Spirulina: spectroscopic, kinetics and SEM studies. *Biotechnol. Bioeng.* 96 (6):1051–1063.

El-Sheekh, M.M., El-Shouny, W. A., Osman, M. E. H., and El-Gammal, E. W. E. (2005). Growth and heavy metals removal efficiency of *Nostoc muscorum* and *Anabaena cylindrica* in sewage and industrial wastewater effluents. *Environ. Toxicol. Pharmacol*, 19:357–365.

EPA, U.S. (2002). *Wastewater technology fact sheet-facultative lagoons.*

Fallowfield, H. J., Cromar, N. J., and Evison, L. M. (1996). Coliform die-off rate constants in a high rate algal pond and the effect of operational and environmental variables. *Water Sci. Technol*, 34(11): 141-147.

Figueira, M. M., Volesky, B., and Ciminelli, V. S. T. (1997). Assessment of interference in biosorption of a heavy metal. *Biotechnol. Bioeng*, 54:344.

Filip, D. S., Peters, T., Adams, V. D., and Middlebrooks. E. J. (1979). Residual heavy metal removal by an algae-intermittent sand filtration system. *Water. Res.* 13:305-313.

Fogg, G. E. (1975). *Algal cultures and phytoplankton ecology.* Second edition 1975 ed. Wisconsin: The university of Wisconsin press.

Fontes, A. G., Vargas, M. A., Moreno, J., Guerrero, M. G., and Losada, M. (1987). Factors affecting the production of biomass by a nitrogen-fixing blue-green alga in outdoor culture. *Biomass,* 13: 33–43.

Garnikar, D. (2002). Accumulation of copper, mercury and lead in *Spirulina platensis* studied in Zarrouk's medium. *The Journal of KMITNB* 12(4): 333-35.

Goldman, J. C., and Carpenter, E. J. (1974). A kinetic approach to the effect of temperature on algal growth. *Limnol. Oceanogr*, 19(5): 756-766.

Hassett, J. M., Jennett, J. C., and Smith, J. E. (1981). Microplate technique for determining accumulation of metals by algae. *Appli. Environ. Microbiol*, 41:1097-106.

Hodaifa, G, Martínez, M. E., and Sánchez, S. (2009). Daily doses of light in relation to the growth of *Scenedesmus obliquus* in diluted three-phase olive mill wastewater. *J. Chem. Technol. Biotechnol.* 84: 1550-1558.

Holan, Z. R., and Volesky, B. (1994). Biosorption of lead and nickel by biomass of marine. *Algal. Biotechnol. Bioeng*, 43: 1001 -1009

Imani S., Rezaei-Zarchi S., Hashemi M., Borna H., Javid A., Ali, M., and Hossein, B.A. (2011) Hg, Cd and Pb heavy metal bioremediation by *Dunaliella* algae. *J. Med. Plants. Res.* 5(13): 2775-2780.

Irgolic, K. J., Woolson, E. A., Stockton, R. A., Newman, R. D., Bottino, N. R., Zingaro, R. A., Kearney, P. C., Pyles, R. A., Maeda, S., McShane, W. J., and Cox, E. R. (1977). Characterization of arsenic compounds formed by daphnia magna and *Tetraselmis chuii* from inorganic arsenate. *Environ. Health. Perspect*, 19:61–66.

Kawaguchi, K. (1980). *Microalgae production systems in Asia, in algae biomass*, G. Shelef and C. J. Soeder, (eds). Elsevier, Amsterdam, 25–33.

Kumar, D., Santhanam, S. P., Jayalakshmi, T., Nandakumar, R., Ananth, S., Shenbaga, D. A., and Prasath, B. (2015). Excessive nutrients and heavy metals removal from diverse waste waters using marine microalga *Chlorella marina* (Butcher). *Indian. J. Geo. Mar. Sci*. 44 (1): 97-103.

Kuyucak, N., and Volesky, B. (1989). Desorption of cobalt-laden algal biosorbent. *Biotechnol Bioeng*. 33: 815.

Laliberte, G., Proulx, D., De, P. N., and De La, Noüe J. (1994). Algal technology in waste water treatment. In: H. Kausch and W. Lampert (eds.), *Advances in Limnology*. E. Schweizerbartsche Verlagsbuch-handlung, Stuttgart, 283-382.

Larsdotter, K., Jansen, J. L. C., and Dalhammar, G. (2007). Biologically mediated phosphorus precipitation in wastewater treatment with microalgae. *Environ. Technol.*, 28: 953 - 960.

Lau, P. S., Tam, N. F. Y., and Wong, Y. S. (1996) Wastewater nutrients removal by *Chlorella vulgaris*: optimization through acclimation. *Environ. Technol.* 17(2): 183–189.

Lincolin, E. P, and Earle, J. F. K. (1990). Wastewater treatment with microalgae. In: Akatsuka, I. (eds.), *Introduction to Applied Phycology*. SPB Academic Publishing by the Hague, The Netherlands, pp. 429–446.

Lovaie, A., and De, La. Noüe, J. (1985). Hyper concentrated cultures of *Scenedesmus obliquus*: A new approach for wastewater biological tertiary treatment, *Water. Res*, 19:1437-42.

Marcel, J., Johannes, T., Luuc, R. M., and René, H. W. (2003). Enclosed outdoor photobioreactors: Light regime, photosynthetic efficiency, scale-up, and future prospects. *Biotechnol. Bioengin*, 81: 193-210.

Miranda, J., Krishnakumar, G., D, and Silva, A., (2012). Removal of Pb from aqueous system by live *Oscillatorialaete-virens* (Crouan and Crouan) gomont isolated from industrial effluents. *World. J. Microbiol. Biotechnol*, 28:3053–3065.

Molina E, Fernández J, Acién F. G., and Chisti Y, (2001). Tubular photobioreactor design for algal cultures. *J. Biotechnol*, 92: 113-131.

Monteiro, C. M., Castro, P. M. L., and Malcata, F. X. (2010). Cadmium removal by two strains of *Desmodesmus pleio morphus* cells. *Water. Air. Soil. Pollut*, 208, 17–27.

Monteiro, C. M., Castro, P. M. L., and Malcata, F. X. (2009). Use of the microalga *Scenedesmus obliquus* to remove cadmium cations from aqueous solutions. *World. J. Microbiol. Biotechnol,* 25: 1573–1578.

Monteiro, C. M., Castro, P. M. L., and Malcata, F. X. (2012). Metal uptake by microalgae: underlying mechanisms and practical applications. *Biotechnol. Prog.* 28 (2): 299–311.

Munoz, R., Köllner, C., Guieysse, B., and Mattiasson, B. (2004). Photo-synthetically oxygenated salicylate biodegradation in a continuous stirred tank photobioreactor. *Biotechnol. Bioengin*, 87(6): 797–803.

Nakajima, A., Horikoshi, T., and Sakaguchi, T. (1981). Studies on the accumulation heavy metal elements in biological system XVII. Selective accumulation of heavy metal ions by *Chlorella vulgaris. Eur. J. App. Microbiol. Biotechnol,* 12:76-83.

Nirmal, K. J. I., Cini, O. (2012). Removal of heavy metals by biosorption using freshwater alga *Spirogyra hyalina. J. Environ. Biol*, 33: 27-31.

Noue, J. D., Laliberte, G., and Proulx, D. (1992). Algae and waste water. *J. Appl. Phycol*, 247-254.

Ogbonna, J. C., Yoshizawa, H., and Tanaka, H. (2000). Treatment of high strength organic wastewater by a mixed culture of photosynthetic microorganisms. *J. Appl. Phycol.* 12: 277–284.

Oswald, W. J. (1978). Engineering aspects of micro algae. *Handbook of Microbiology.* Florida: CRC Press, Boca Raton.

Oswald, W. J. (1988). Micro-algae and waste-water treatment. In: *Micro-algal Biotechnology*. Cambridge University Press, Cambridge. pp. 305-328.

Oswald, W. J. (1995). Ponds in the twenty-first century. *Water. Sci. Technol*, 31:1–8.

Oswald, W. J., and Gotaas, H. B. (1955). Photosynthesis in sewage treatment. Proceedings separate 686, *Proc. Am. Soc. Civil. Eng*, pp 73–105.

Parhad, N. M., and Rao, N. U. (1976). Decrease of bacterial content in different types of stabilization ponds. *Ind. J. Environ. Health,* 18:33–46.

Perales-Vela, H. V., Pena-Castro, J. M., and Canizares-Villanueva, R. O. (2006). Heavy metal detoxification in eukaryotic microalgae. *Chemosphere*, 64: 1–10.

Polprasert, C., Dissanayake, M., and Thanh, N.C. (1983). Bacterial die-off kinetics in waste Stabilization Ponds. *J. Water. Pollut. Control. Fed.* 55(3): 285-296.

Prado, M. P., Baudilio, A. V., Serviere-Zaragoza. E., and Lia C. Mendez-Rudriguez. (2010). Copper and Cadmium biosorption by dried seaweed *Sargassum sinicola* in saline wastewater. *Water. Air. Soil. Pollut*, 210: 197-202.

Rose, J. B., Dickson, L. J., Farrah, S. R., and Carnahan R. P. (1996). Removal of pathogenic and indicator micro-organisms by a full scale water reclamation facility. *Water. Res.* 30: 2785–2797.

Sakaguchi, T., Nakajima, A., and Horikoshi, T. (1981). Studies on the accumulation heavy metal elements in biological system XVIII. Accumulation of molybdenum by green microalgae. *Eur. J. App. Microbiol. Biotechnol*, 12:84-89.

Sbihi, K., Cherifi, O., El, Gharmali. A., Oudra, B., and Aziz, F. (2012). Accumulation and toxicological effects of cadmium, copper and zinc on the growth and photosynthesis of the freshwater diatom *Planothidium lanceolatum* (Brébisson) Lange-Bertalot: A laboratory study. *J. Mater. Environ. Sci.* 3 (3): 497–506.

Schmitt, D., Müller, A., Csögör, Z., Frimmel, F. H. and Posten, C. (2001). The adsorption kinetics of metal ions onto different microalgae and siliceous earth. *Water. Res*, 35(3): 779–785.

Shanab, S., Essa, A., and Shalaby, E. (2012). Bioremoval capacity of three heavy metals by some microalgae species (Egyptian Isolates).*Plant. Signal. Behav*, 7(3): 392–399.

Shashirekha, V., Sridharan, M. R., and Swamy, M. (2008). Biosorption of trivalent chromium by free and immobilized blue green algae: Kinetics and equilibrium studies. *J. Environ. Sci. Health A*. 43(4), 390-401.

Sjahrul, M., and Arifin, Dr. (2012). Phytoremediation of Cd by marine phytoplanktons, *Tetracelmis chuii* and *Chaetoceros calcitrans*. *Int. J. Chem*, 4(1): 69–74.

Sompech, K., Chisti, Y., and Srinophakun, T. (2012). *Design of raceway ponds for producing microalgae*. 3 (4):387-397.

Ting, Y. P., Lawson, E., and Prince, I. G. (1989). Uptake of cadmium and zinc by alga *Chlorella vulgaris*: Part I. İndividual ion species. *Biotechnol. Bioeng*. 34:990-99.

Wang, X. S., Chen, L. F., Li, F. Y., Chen, K. L., Wan, W. Y., and Tang, Y. J. (2010). Removal of Cr (VI) with wheat-residue derived black carbon: reaction mechanism and adsorption performance. *J. Hazard. Mater*. 175: 816–822.

Wong, J. P. K., Wong, Y. S., and Tam, N. F. Y. (2000). Nickel biosorption by two *chlorella* species, *C. Vulgaris* (a commercial species) and *C. Miniata* (a local isolate). *Bioresour. Technol*. 73: 133–137.

Yao, J., Li, W., Xia, F et al., (2011). Heavy metals and PCDD/Fs in solid waste incinerator fly ash in Zhejiang province, China: Chemical and bioanalytical characterization. *Environ. Monit. Assess*, 184 (6): 3711–3720.

In: The Role of Photosynthetic Microbes…
Editors: K. Tripathi, N. Kumar et al.
ISBN: 978-1-53614-032-3

Chapter 4

MICROALGAE AS A SUSTAINABLE SOURCE OF BIOENERGY: PRESENT STATUS AND FUTURE PROSPECTS

Surendra Singh*[*]*, Rishibha Dixit and Ankita Kachhwaha
Algal Biotechnology Laboratory, Department of P. G. Studies and Research in Biological Science, Rani Durgavati University,
Jabalpur, Madhya Pradesh, India

ABSTRACT

An enormous amount of interest has been raised on the use of microalgae-based technologies for the production of a sustainable source of bioenergy and high-value co-products. Biotechnological exploitation of microalgae for human welfare is a recent phenomenon although these wonderful organisms exist on this planet since archeological era. Microalgae are eukaryotic photosynthetic microorganism known for their rapid growth. The main microalgae are *Scenedesmus, Chlorella*, *Spirulina, Dunaliella* and *Haematococcus* are currently cultivated photo-synthetically for the production of verity of bioenergy and valuable products. Micro-algal biomass have high biotechnological potential and it is being use as a source of drugs in pharmaceutical industries, bio-chemicals, biofuels (bio-diesel, bio-gas, bio-ethanol and bio-butanol), bio-fertilizer, bio-pigments and dye, renewable food, Polyunsaturated fatty acid (DHA, EPA, GLA), feed, cosmetic, sink for greenhouse gases, soil amelioration, bioremediation and other applications such as treatment of wastewater. A major bottleneck in the application of microalgae to such processes is low productivity of the culture, both in terms of biomass and product. Comparison of productivity between economically important microalgae

[*]Corresponding Author Email: singhbiosci@yahoo.co.in.

and biomass production system are not easy because of geographical variations in climatic conditions, culture strategies, type of algal species, harvesting procedure etc., In brief, this technology requires considerable R&D, with the major challenge to obtain high production to justify the unavoidably huge capital and operating costs.

Keywords: biofuels, bioenergy, microalgae, micro-algal biomass

INTRODUCTION

Microalgae are the photosynthetic microorganism that rapidly grows and can live in rough conditions due to their simple structure. For example, green algae (Chlorophyta) and diatoms (Bacillariophyta) (Li et al., 2008). The thalli of algae display a wide range of organization, ranging from single cells (*Chlorella* sp.), through motile (*Chlamydomonas* sp., *Dunaliella* sp.), colonial (*Volvox* sp., *Botryococcus* sp.), filamentous (*Spirulina* sp., *Spirogyra* sp.) and plant-like (*Chara* sp.) to giant seaweeds (*Postelsia* sp., *Fucus* sp.) (Fritsch, 1935). Microalgae are present in the aquatic and terrestrial environment, representing a big variety of species living in a wide range of environmental condition, receiving sufficient solar radiation for photosynthesis. They use solar energy and converts CO_2, other simple inorganic compound to myriad molecules. Out of 50,000 species, only a limited number of around 30,000 have been studied and analyzed (Richmond et al., 2004).

Microalgae can produce a variety of high-value biological derivatives with many possible commercial applications including synthesis of a large range of fine chemicals and bulk products such as fat, natural dye, pigments, polyunsaturated fatty acid, sugars, antioxidants and bioactive compounds. Microalgae explores a large number of biotechnology areas including biofertilizer, bioremediation, heavy metal absorption, secondary sewage removal, carbon dioxide sequestration, cosmetics, pharmaceuticals (Phycocyanin, Phycoerythrin, β-carotene, Astaxanthin, Leutinm), nutrition and food additives (Polyunsaturated fatty acids), Vitamins, Arachidonic acid (AA), Biotin, α-tocopherol (Vitamin E), B-1,3-glucan), aquaculture, pollution prevention and biofuels (bio-diesel, bio-ethanol, bio-butanol, bio-electricity, bio-hydrogen and bio-methane).

Screening and evaluation of naturally occurring microalgal strains are done that exhibit high growth rate and high oil content. Selected strains can be improved genetically for robustness in performance under diverse environmental and cultural conditions. Large-scale photo-bioreactors are designed and strain is grown in outdoors and harvested by downstream processing (i.e., harvesting, dewatering and drying).

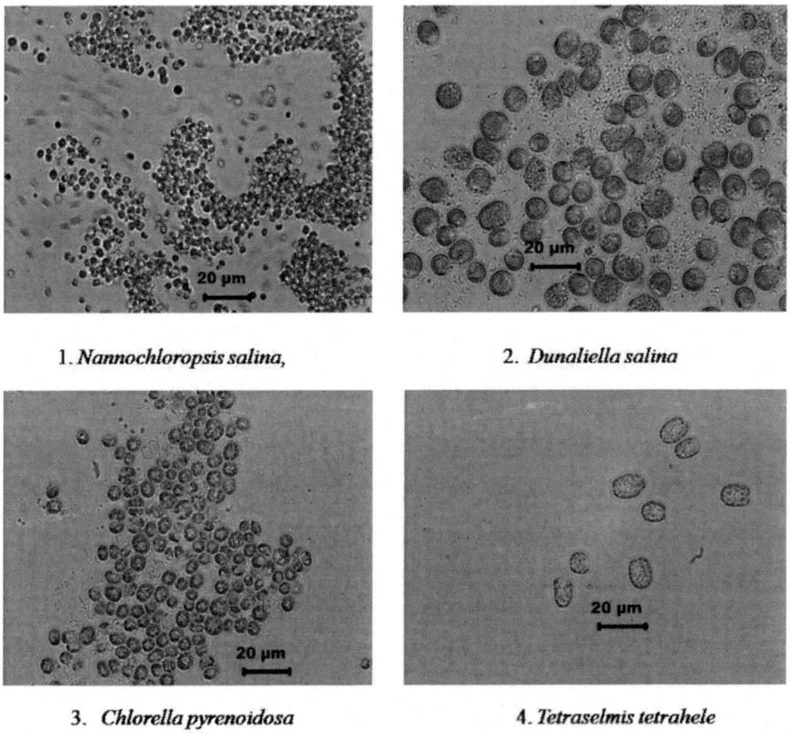

Figure 1. Shows some of the promising microalgal strains with potential for biofuel production.

Microalgal biomass is converted into biofuels and high-value co-products (Figure 1). One of the key reasons more algae-based technologies have not yet proven to be commercially successful is that, there are several fundamental technical problems that remain to be solved before algae-based technologies can become common. Refinement of the cultivation process, downstream processing of biomass and development of an economically viable model for commercialization of microalgae-based biofuels and biomaterials are the main challenges.

BIODIESEL

Biodiesel is a mixture of monoalkyl esters of long chain fatty acids (FAME) derived from a renewable lipid feedstock such as *Jatropha*, microalgal oil etc., (Demirbas, 2009). Micro-algal lipid are mostly neutral lipid with the lower degree of unsaturation making them a potential replacement for fossil fuel. Microalgae are the most efficient biological producer of oil and a versatile biomass source (Li et al., 2008).

Table 1. Richest lipid-containing microalgae species considered for biodiesel production

Microalgal strains	Lipid % in dry weight	References
Botryococcus braunii	25-75	Dayananda et al. (2005); Chisti, (2007)
Chlorella emersonii	63	Gouveia and Oliveira, (2009)
Chlorella minutissima	57	Gouveia and Oliveira, (2009)
Chlorella protothecoides	55	Xiong et al., (2008)
Chlorella sorokiana	22	Gouveia and Oliveira, (2009)
Chlorella vulgaris	14 - 56	Gouveia and Oliveira, (2009)
Dunaliella salina	28.1	Mohapatra, (2006)
Dunaliella tertiolecta	36 - 42	Tsukahara and Sawayama, 2005
Isochrysis galbana	21.2	Mohapatra, (2006)
Nannochloropsis species	28.7	Gouveia and Oliveira, (2009)
Scenedesmus dimorphus	16 - 40	Becker, (1994)
Scenedesmus obliquus	12 - 14	Becker, (1994)
Scenedesmus quadricauda	19.9	Mohapatra, (2006)
Tetraselmis suecia	15 – 23	Chisti, (2007); Huntley and Redalje (2007)

Oil content found in microalgae is generally much higher than the other vegetable crops (Mata et al., 2010). Table 1, listed some typical lipid containing microalgae sp. that are used as feed-stock for biodiesel (Chisti, 2007). Production of biodiesel requires selection of high-oil content strains and cost-effective methods of harvesting, oil extraction and conversion of oil into biodiesel (Parmar et al., 2011).

Methods for Extraction of Lipid

For the extraction of lipid from microalgae different methods are used such as expeller/oil press, liquid-liquid extraction (solvent extraction), supercritical fluid extraction (SFE) and ultrasound techniques (Harun et al., 2010). Algal oil can be converted into biodiesel through a process called transesterification. It is a process in which triglycerides are transformed into fatty acid alkyl esters, in the presence of an alcohol and a catalyst along with glycerol as a byproduct (Vasudevan and Briggs, 2008; Sharma and Singh, 2009).

Conversion of Microalgal Lipidinto Biodiesel by Transesterification

The most common industrial process for the biodiesel production from microalgae is base-catalyzed transesterification with alcohol (Figure 2). The transesterification is the reversible reaction of fat or oil with an alcohol to form fatty acid alkyl ester. The reaction requires a 3:1 molar alcohol to oil ratio, but excess alcohol is (usually methyl alcohol is used) added to drive the equilibrium toward the product side (Fukuda et al., 2001). Excess of methyl alcohol ensures that the reaction is driven in the direction of biodiesel production. The yield of methyl esters exceeds 98% on a dry weight basis. Triglycerides are first converted to diglycerides, monoglycerides and finally glycerol. The alkali-catalyzed trans-esterification is about 4000 times faster than the acid catalyzed reaction. Alkalis such as sodium and potassium hydroxide are commonly used as commercial catalysts at a concentration of about 1% by weight of oil. Alkoxides such as sodium methoxide are even better catalysts than sodium hydroxide and are being increasingly used. However, the use of lipases offers important advantages (Ma and Hanna, 1999; Banerjee et al., 2002).

Alkali-catalyzed transesterification is carried out at approximately at 60°C under atmospheric pressure, as methanol boils off at 65°C at atmospheric pressure. Under these conditions, the reaction takes about 90 minutes to complete and high temperature can be used in combination with higher pressure. Methanol and oil do not mix; hence the reaction mixture contains two liquid phases. Other alcohols can also be used, but methanol is the least expensive. To prevent yield loss due to saponification reactions (i.e., soap formation), the oil and alcohol must be dry and the oil should have a minimum of free fatty acids. Biodiesel is recovered by repeated washing with water to remove glycerol and methanol (Banerjee et al., 2002). This process of biodiesel production is found to be the most efficient and least corrosive of all the processes as the reaction rate is reasonably high even at a low temperature of 60°C.

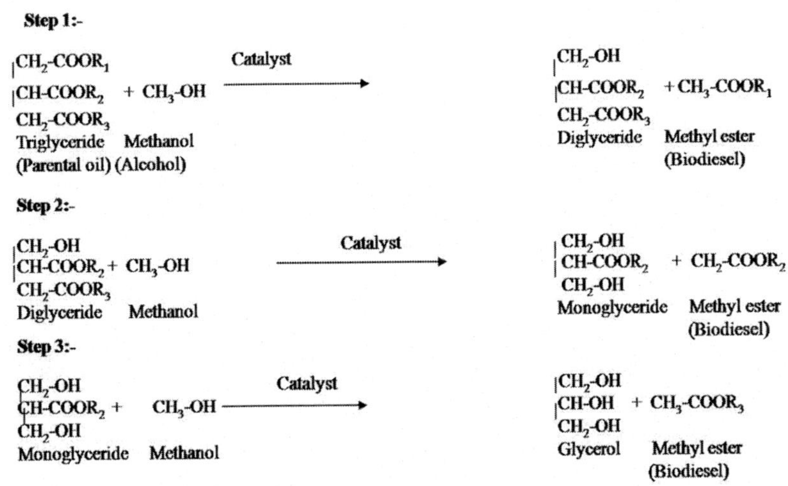

Figure 2. Transesterification of lipid to biodiesel, Where R1, R2 and R3 are long chains of hydrocarbon groups (Bajhaiya et al., 2010).

BIOETHANOL

Bioethanol production from microalgae like *Chlorella vulgaris*, *Chlorococcum reinhardtii* and *Chlorococcum infusionum* result in better conversion rates than that of macroalgae because of a number of factors such as cell disruption efficiency, fermentable sugar content and cell wall structure. Microalgae are less robust than vascular macroalgae due to a thinner cell wall and lack of lignin resulting in the more efficient release of fermentable sugars. Moreover, microalgae are rich in carbohydrates (in the form of glucose, starch/cellulose, glycogen, hexoses, pentose and other polysaccharides) that can be converted into fermentable sugars for bioethanol production via fermentation and proteins that can be used as carbon sources for fermentation by bacteria, yeast or fungi (Wayman, 1996). Hence, it is expected that the overall bioethanol production process can be simplified due to the non-requirement of chemical and enzymatic pre-treatment (Table 2) of the biomass, hydrolysis, fermentation and product recovery. Simultaneous occurrences or a combination of these process steps could hugely impact on the process economics of bioethanol production from microalgae (Figure 3) (Harun et al., 2009, 2010a, 2010b).

The best-reported yield thus far was 0.4 g of ethanol from 1g of dried *C. vulgaris* in 24 h fermentation with genetically engineered E. coli SJL2526 (Lee et al., 2011). Interestingly, usage of waste biomass from the supercritical CO_2 extraction of *Chlorococcum* sp. lipid for biodiesel production generated similar yield albeit at a longer fermentation time (Harun et al., 2010).

On the other hand, simultaneous biodiesel and bioethanol production from microalgae is also possible, in which micro-algal lipid is extracted prior to the fermentation process. This concept has been proven viable in a recent study in which lipid from *Chlorococum* sp. was extracted with supercritical CO_2 at 60°C and subsequently subjected to fermentation by the yeast Saccharomyces bayanus (Harun et al., 2010a). From the report, micro-algal biomass with pre-extracted lipid gave 60% higher ethanol concentration for all samples than the dried microalgae biomass without lipid extraction. This is because supercritical CO_2 can act as a superior pre-treatment method to breakdown micro-algal cell wall causing the simultaneous release of lipid and carbohydrate embedded within the cell wall.

Maximum bioethanol yield of 3.83 g/L was achieved from 10 g/L of lipid extracted microalgae residue. In other words, lipid extraction from microalgae biomass for biodiesel production and pre-treatment step to release carbohydrates for bioethanol production can occur in just one single step which greatly enhanced the viability of microalgae biofuel production in commercial scale. Apart from supercritical CO_2, other lipid extraction methods such as ultrasonication, chemical solvent, microwave and bead-beater have been studied in order to get a comprehensive comparison between the methods.

Bacteria, yeast or fungi are microorganisms used to ferment carbohydrates to produce ethanol under anaerobic conditions. In addition to ethanol as main products, CO_2 and water are also formed as by-products (Harun et al., 2010b; 2010a). According to following simplified reaction equation below, stoichiometric yields are 0.51 kg ethanol and 0.49 kg CO_2, per kg of carbon sugar, i.e., glucose.

$$C_6H_{12}O_6 \rightarrow 2CH_3CH_2OH + 2CO_2$$

Table 2. Bioethanol production from different microalgae strains through various pretreatment methods (Lam and Lee, 2012)

Feedstock	Pre-treatment	Ethanol yield (g ethanol/g substrate)	Reference
Chlorococum sp.	Supercritical CO_2	0.38	Harun et al., (2010a)
Chlorococum humicolo	Acid	0.52	Harun and Danquah, 2011
Chlorococum infusionum	Alkaline	0.26	Harun et al., (2010b)
Chlamydomonas reinhardtii	Enzymatic	0.24	Choi et al.., (2010)

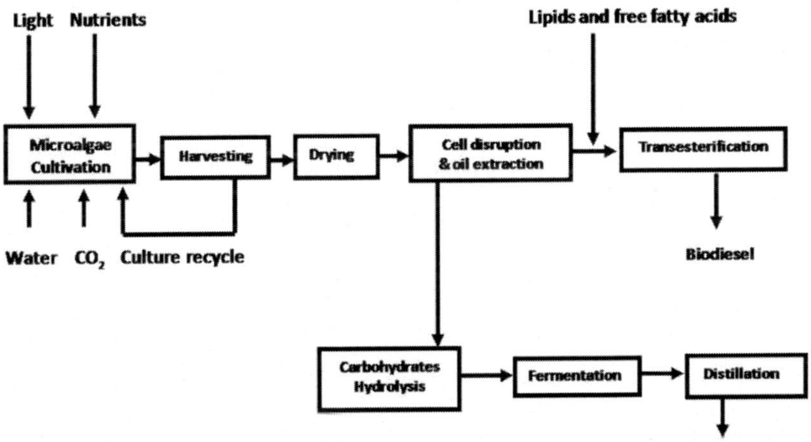

Figure 3. Overall process of biodiesel and bioethanol production from microalgal biomass.

The prospective feed-stocks and feasibility of the various processes for ethanol production have been reported by various authors. However, bioethanol production by fermentation has not been studied extensively. The most recent work on bioethanol production by fermentation has been reported by Harun et al., (2010). They have carried out experiments, for studying the suitability of microalgae (*Chlorococum sp.*) as a substrate, using yeast for fermentation. A productivity level of around 38% dry biomass has been reported, which supports the suitability of microalgae as a promising substrate for bioethanol production. In the research work, Moen (2008), has demonstrated that brown seaweed produces higher bioethanol compared to other algal species. Another study by Hirayama et al., (1998) proposed a self-fermentation of algae to obtain ethanol. The reported advantages of this technique over conventional fermentation are comparatively simple processes with shorter fermentation time. Ueda et al., (1996) have patented a two-stage process for microalgae fermentation. In the first stage, microalgae undergo fermentation in the anaerobic and dark environment and ethanol is produced. The ethanol this produced can be purified and used as fuel. The CO_2 produced in the fermentation process was recycled to algal cultivation ponds as a nutrient to grow microalgae. The second stage involved the utilization of remaining algal biomass slurry left after fermentation, which may be used in anaerobic digestion processes while keeping the pH in the range of 6-9. This process produced methane which can further be converted to produce electricity.

BIOMETHANE

The application of methane fermentation technology to algae has received considerable attention because it produces valuable by-products such as biogas. Biogas mainly consists of a mixture of methane (55-75%) and carbon dioxide (25-45%) produced during anaerobic digestion by anaerobic microorganisms. Methane from anaerobic digestion can be used as fuel gas and also be converted to generate electricity (Holm et al., 2009). Organic material like biomass can be used to produce biogas via anaerobic digestion and fermentation (Hankamer et al., 2007). Organic biopolymers (e.g., carbohydrate, lipid and protein) are hydrolyzed and broken down into monomers, which are then converted into a methane-rich gas via fermentation. Carbon dioxide is the second main component found in biogas (approximately 25-50%) and, like other interfering impurities, has to be removed before the methane is used (Hankamer et al., 2007). Residual biomass from anaerobic digestion can be further reprocessing to make fertilizers. In addition to being renewable and sustainable, this would encourage sustainable agricultural practices in providing greater efficiencies and reduce algal production costs.

Microalgae contain almost no lignin and lower cellulose; therefore demonstrate good process stability and high conversion efficiency for anaerobic digestion (Vergara et al., 2008). The biogas production from this anaerobic digestion process is primarily affected by its organic loading, temperatures, pH and retention time in reactors. Basically, long solid retention time and high organic loading rate give significant results in high methane yield (Chynoweth, 2005). In addition, anaerobic digestion can operate in either mesophilic (35°C) or thermophilic (55°C) conditions. Methane in the form of compressed natural gas is used as a vehicle fuel and is claimed to be more environmentally friendly than fossil fuels such as gasoline/petrol and diesel.

Converti et al., (2009) showed biogas production and purification by a two-step bench-scale biological system, consisting of fed-batch pulse-feeding anaerobic digestion of mixed sludge, followed by methane enrichment of biogas by the use of the cyanobacteria *Arthrospira platensis*. The composition of biogas was nearly constant, and methane and carbon dioxide percentages ranged between 70.5-76.0% and 13.2-19.5%, respectively. The data of carbon dioxide removal from biogas revealed the existence of a linear relationship between the rates of *Arthrospira platensis* growth and carbon dioxide removal from biogas and allowed calculating carbon utilization efficiency for biomass production of almost 95% (Converti et al., 2009). Microalgae offer good potential for biogas production but commercial productions have still not been implemented.

Advantages of Using Microalgae for Bioenergy Production over the Other Feed-Stock

Many research reports and articles described several advantages of using microalgae for bio-energy production in comparison with other available feed-stock (Tsukahara and Sawayama, 2005; Chisti, 2007; Li et al., 2008(a); Li et al., 2008(b). From a practical point of view, microalgae are easy to cultivate, can grow with little or even no attention, using water unsuitable for human consumption. There are several aspects of microalgal bio-energy production that is combined to capture the interest of researchers and entrepreneurs around the world (Figure 4). These include:

- Microalgae reproduce themselves using photosynthesis to convert sun energy into chemical energy, completing an entire growth cycle every few days (approximately 1-10 days) (Sheehan et al., 1998). Its high photosynthetic yield that is, about 3-8% of solar energy can be converted to biomass as compared to terrestrial plants where it is approximately 0.5%.
- Ability to synthesize and accumulate large quantities of lipid per dry weight biomass (DCW) (40-86%). Potential to produce 10-20 times higher yield of oil compared to other oleaginous seeds or vegetable oils (Verma et al., 2010).
- Therefore, the competition for arable soil with other crops, in particular for human consumption, is greatly reduced. They are nonfood-based feed-stock resources for biofuel. They use otherwise non-productive, non-arable land (Verma et al., 2010).
- They use a wide variety of water sources (fresh, brackish, seawater and wastewater) and have potential to grow in saline water and harsh conditions. Moreover, they can grow almost anywhere, requiring sunlight and some simple

nutrients, although the growth rates can be accelerated by the addition of specific nutrients and sufficient aeration (Aslan and Kapdan, 2006).

- Wastewater treatment by removal of NH^{4+}, NO^{3-}, PO_4^{3-}, make algae to grow using these water contaminants as nutrients (Wang et al., 2008). They provide a pathway for the removal of chemical and organic contaminants, heavy metals and pathogens from wastewater while producing biomass for biofuel production, Savings on requirements for chemical remediation (Wang et al., 2008) and possible minimization of freshwater use for biomass production (Li et al., 2008).
- From an environmental perspective, their use help to reduce the pollutants and greenhouse gas emissions and micro-algal biofuels are not toxic and environmentally friendly as it produces substantially less carbon monoxide and 100% less sulfur dioxide emissions with no unburned hydrocarbons and accordingly it is an ideal fuel for heavily polluted cities. Bioenergy represents a carbon dioxide-cycle in combustion. They are biodegradable and contribute to sustainability and are renewable by nature. There are so many benefits to the environment, economy and consumers in using microalgal bio-energy.
- From the energy point of view, bioenergy is a renewable and clean energy source. Furthermore, their use helps to reduce energy dependence on petroleum fuel.
- Economics of biofuel production from microalgae can be improved by capturing additional revenue from co-production of food, feed and high-value products, wastewater treatment and net fertilizer value in case of nitrogen-fixing algae.
- After oil extraction the resulting algal biomass can be processed into ethanol, methane, livestock feed, used as an organic fertilizer due to its high N:P ratio or simply burned for energy cogeneration (electricity and heat) (Wang et al., 2008).
- Combined with their ability to grow under harsher conditions, and their reduced needs for nutrients, they can be grown in areas unsuitable for agricultural purposes independently of the seasonal weather changes, thus not competing for arable land use, and can use wastewaters as the culture medium, not requiring the use of freshwater (Mata et al., 2010).
- Depending on the micro-algal species other compounds may also be extracted, with valuable applications in different industrial sectors, including a large range of fine chemicals and bulk products, such as fats, polyunsaturated fatty acids, oil, natural dye, sugars, pigments, antioxidants, high-value bioactive compounds and other fine chemicals and biomass (Li et al., 2008(a); Li et al., 2008(b); Raja et al., 2008).
- Microalgae can be produced all year round to them and therefore, quantity of oil production exceeds the yield of the best oilseed crops, e.g., biodiesel yield of 58,700 l ha^{-1} for microalgae containing only 30% oil by wt., compared with 1190

l $ha^{-1}$for rapeseed or Canola (Schenk et al., 2008), 1892 l ha^{-1} for *Jatropha* (Chisti., 2007), and 2590 l ha^{-1} for Karanj (Pongamiapinnata).

- Because of this variety of high-value biological derivatives, with many possible commercial applications, microalgae can potentially revolutionize a large number of biotechnology areas including bioenergy, cosmetics, pharmaceuticals, nutrition and food additives, aquaculture, and pollution prevention (Raja et al., 2008; Rosenberg et al., 2008).

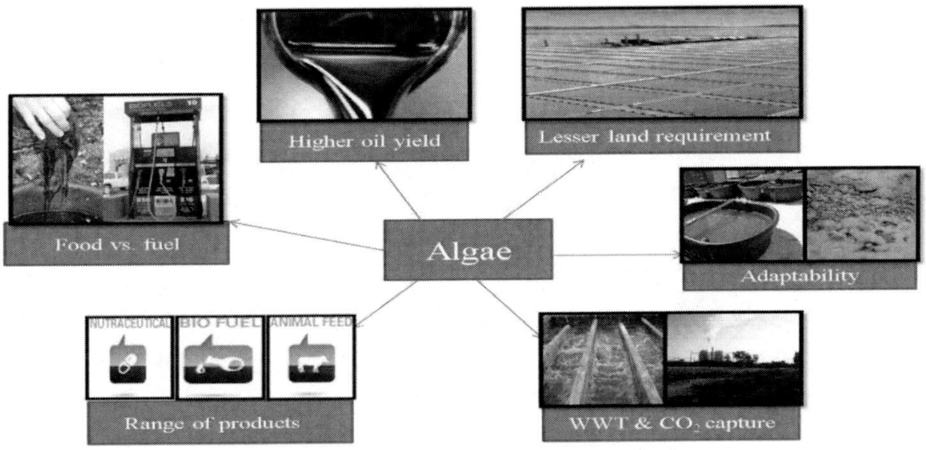

Figure 4. Advantages of microalgae for bioenergy production.

GLOBAL SCENARIO OF BIOFUELS PRODUCTION

The world has been confronted with an energy crisis due to depletion of finite resources of fossil fuel. Continued use of petroleum-based fuels is now widely recognized as unsustainable because of depleting supplies and contribution of these fuels to pollute the environment. India meets nearly 75-80% of its total petroleum requirements through imports. The hydrocarbons sector plays a vital role in the economic growth of the country. With the increase in the share of hydrocarbons in the energy supply, this shares of imported energy is expected to exceed 90% by 2030 (Report of the Working Group, 2006). In the U.S, the federal government passed the energy independence and security act (EISA) in 2007 which requires a gradual increase in the production of renewable fuels to reach 36 billion gallons per year by 2022 (Yaug et al., 2010). The International Energy Agency (IEA) has reported in that the world's primary energy need is projected to grow by 55% between 2005 and 2030, at an average annual rate of 1.8% per year. Fossil fuels remain the dominant source of primary energy, accounting for 84% of the overall increase in demand between 2005 and 2030 (World Energy Outlook, 2007). Energy and capital have reported that, by 2025, the world's demand for oil will shoot up to 60%, while production capacity would be thrown back to 1985 levels.

According to the Energy Information Agency (EIA) report, petroleum consumption fell by 90,000 barrels per day in 2008-2009 (World Energy Outlook, 2007). China's annual oil consumption growth rates are around 7.5% and while that of India's is 5.5% and both are expected to take a quantum leap over the next decade (The Truth about oil).

INDIAN SCENARIO FOR BIOFUEL PRODUCTION

The increasing import of fuel has necessitated the search for other liquid fuels as an alternative to diesel, which is being used in large quantities in the transport of industrial and agricultural sector (Meher et al., 2006). According to an estimate, automobiles alone contribute to 70% of the total petroleum consumptions. The country faces problems with regard to the fuel requirement for increased transportation demand (Grow diesel News, 2008). The consumption of petroleum products in India has been increasing at an annual growth rate of 5-6% in the year 2005-2006, it was about 112 million ton (Grow Diesel News, 2008). According to the annual report (2006-2007) of Ministry of Petroleum and Natural Gas (MoPNG), India has imported about 99 million ton of crude oil during the year 2005-2006, causing a heavy burden of Rs. 171,702 crore on foreign exchange (Annual Report, 2007). For the 2007-2008, the crude oil import bill was Rs. 272,699 crore ($68 billion), having more than 75% of oil import dependency. India's transportation fuel requirements are unique in the world. India consumes almost five times more diesel fuel than gasoline, whereas, almost all other countries in the world use more gasoline than diesel fuel. Diesel burns roughly 64 million ton, or 450 million barrels, a year, as opposed to about 84 million barrels of gasoline (Annual report, 2007). During the last two decades, diesel consumption has increased enormously. The estimated annual consumption of petroleum product in India is nearly about 120 million ton per year, and no other feedstock except microalgae has the capacity to replace this large volume of oil. To elaborate, it has been calculated that, in order for a crop such as a soybean or palm to yield enough oil capable of replacing petro-diesel completely, a very large percentage of current land available need to be utilized only for biodiesel crop production, which is quite infeasible (Banerjee et al., 2002). For small countries, in fact, it implies that all land available in the country be dedicated to biodiesel crop production. However, if the feedstock were to be algae, owing to its very high yield of oil per acre of cultivation, it has been estimated that less than 2-3 percent of total Indian cropping land is sufficient to produce enough biodiesel to replace all petro-diesel currently used in the country (Bajhaiya et al., 2010).

Currently, a litre of gasoline normally costs 2.5 times more than a litre of diesel fuel. Taking advantage of this cost differential, Indian car manufacturers have been investing heavily in the production of diesel vehicles. As such, there are substantial numbers of vehicles on the road that demand diesel and would not require the relatively expensive retrofits needed to use Compressed Natural Gas (CNG). The final factor making biodiesel production in India attractive is the potential to cultivate cheap feed-stocks. India's tropical climate is conducive to grow various species of microalgae, which serves as a natural benefit over other countries for the production of algal biodiesel. According to current research, microalgae seem to be the promising renewable source of energy for India. Petroleum diesel fuel has been sold at government-subsidized rates in India to keep the transport costs low and increase GDP.

Future Prospect of Microalgal Biofuels

Biomass is focused as an alternative energy source, as it's a renewable resource and it can fix atmospheric CO_2 through photosynthesis. Among biomass, algae (cyanobacteria and microalgae) usually have a higher photosynthetic efficiency than other biomass producing plants (Banerjee et al., 2002). The future of algae biofuels is quite bright in many ways. Some market analysts feel that in spite of all the positive factors, e.g., fast growth potential, non-interference with food crops, CO_2 sequestration and the biggest advantage of algal biodiesel is that, it's a sustainable source of liquid transportation fuel and derives energy from the sun. According to current research, microalgae seem to be the promising renewable source of energy for India. Biodiesel from microalgae appears to be a feasible solution to India, for replacing petro-diesel. The microalgae produced sufficient quantity of biodiesel to completely replace petroleum (Laura et al., 2009). While traditional high oil crops, such palm can produce 2000 to 2500 litre of biodiesel per acre, algae can yield 19,000-57,000 litre per acre (Banerjee et al., 2002). So the adoption of large-scale biodiesel production and consumption can potentially lower India's dependence on foreign countries for oil and helps to improve air quality in major cities like Delhi, Kolkata, Bangalore, Chennai, reclaims unusable wastelands, employs unemployed Indians, and keeps the country's economy on track for its planned 8 to 10% annual GDP growth according to 11th five year plan of India (Bajhaiya et al., 2010).

Conclusion

The enormous amount of burning of fossil fuel has increased the CO_2 level in the atmosphere, causing global warming, ozone layer depletion, smog, and acid rain that is health hazards to all living creatures. Thus, in India, search for alternatives to energy produced from petroleum, natural gas, coal, hydro and nuclear energy is of special importance. Thus, the use of micro-algal bioenergy is comparatively much more important for us than for the rest of the developed countries. Microalgal biomass is focused as an alternative energy source, as it is a renewable resource and it can fix atmospheric CO_2 through photosynthesis.

Microalgae usually have a higher photosynthetic efficiency than other biomass producing plants. They are capable of generating widest range of renewable bioenergy like biodiesel, bioethanol, bio-butanol, bioelectricity, bio-hydrogen and bio-methane etc., According to the current research, microalgae seem to be the promising renewable source of bio-energy in India, for replacing energy obtained from fossil fuel. Microalgae also produce valuable co-products like Phycocyanin, β-carotene, Astaxanthin, Leutinm, Eicosapentaenoic acid (EPA), Docosahexaenoic acid (DHA), Ascorbic acid (Vitamin C), Arachidonic acid (AA), Biotin, a-tocopherol (Vitamin E), B-1,3-glucan which are likely to reduce the cost of bioenergy production.

Acknowledgments

The authors thankful to the Head Department of Post Graduate Studies and Research in Biological Science, Rani Durgavati University, Jabalpur-482001, (M. P.) India, for providing necessary facilities.

Conflict of Interest

No conflict of interest among the authors.

References

Aslan, S., and Kapdan, I. K. (2006). Batch kinetics of nitrogen and phosphorus removal from synthetic wastewater by algae. *Ecol. Eng,* 28:64-70.

Bajhaiya, A. K., Mandotra, S. K., Suseela, M. R., Toppo, Kiran and Ranade, S. (2010). Algal Biodiesel: The next generation biofuel for India. *Asian J. Exp. Biol. Sci,* 1(4):728-739.

Banerjee, A., Sharma, R., Chisti, Y. and Banerjee, U. C. (2002). *Botryo-coccus braunii*: a renewable source of hydrocarbons and other chemicals. *Cri. Rev. Biotech,* 22:245-279.

Becker, E. W. (1994). *Microalgae: Biotechnology and Microbiology*. Cambridge University Press, Cambridge, UK.

Chisti, Y. (2007). Biodiesel from microalgae. *Biotech. Adv.*, 25:294-306.

Choi, S. P., Nguyen, M. T. and Sim, S. J. (2010). Enzymatic pretreatment of *Chlamydomonas reinhardtii* biomass for ethanol production. *Bioresour. Technol,* 101:5330-6.

Converti, A., Oliveira, R. P., Torres, B. R., Lodi, A. and Zilli, M. (2009). Biogas production and valorization by means of a two-step biological process. *Bioresour. Technol*, 100:5771-5776.

Dayananda, C., Sarada, R., Bhattacharya, S. and Ravishankar, G. A. (2005). Effect of media and culture conditions on growth and hydrocarbon production by *Botryococcus braunii*. *Pro. Biochem*, 40:3125-3131.

Demirbas, A. (2009). Progress and recent trends in biodiesel fuels. *Energy. Convers. Manag,* 50(1):14-34.

Fritsch, F. E. (1935). *The structure and reproduction of Algae.* Vols I and II. Cambridge University Press, Cambridge, UK.

Fukuda, H., Kondo, A. and Noda, H. (2001). Biodiesel fuel production by transesterification of oils. *J. Biosci. Bioeng,* 92:405-16.

Gouveia, L. and Oliveira, A. C. (2009). Microalgae as raw material for biofuels production. *J. Ind. Microbiol. Biotechnol,* 36:269-274.

Hankamer, B., Lehr, F., Rupprecht, J., Mssgnug, J. H., Posten, C., and Kruse, O. (2007). Photosynthetic biomass and H_2 production by green algae: from bioengineering to bioreactor scale-up. *Plant. Physiol.* 131:10-21.

Harun, R., Danquah, M. K., and Forde, G. M. (2010a). Microalgal biomass as a fermentation feedstock for bioethanol production. *J. Chem. Technol. Biotechnol,* 85:199-203.

Harun, R., Jason, W. S. Y., Cherrington, T., Danquah, M. K. (2010b). Exploring alkaline pre-treatment of microalgal biomass for bioethanol production. *Appl. Energy,* 88:3464-7.

Harun, R., and Danquah, M. K. (2011). Influence of acid pre-treatment on microalgal biomass for bioethanol production. *Proc. Biochem.* 46:304-309.

Harun, R., Danquah, M. K., and Forde, G. M. (2009). Microalgal biomass as a fermentation feedstock for bioethanol production. *J. Chem. Technol. Biotechnol*, 85: 199-203.

Hirayama, S., Ueda, R., Ogushi, Y., Hirano, A., Samejima, Y. and Hon-Nami K. (1998). Ethanol production from carbon dioxide by fermentative microalgae. *Stud. Surf. Sci. Catal.*, 114:657-60.

Holm-Nielsen, J. B., Al Seadi, T. and Oleskowicz-Popiel, P. (2009). The future of anaerobic digestion and biogas utilization. *Bioresour. Technol.*, 100:5478-84.

Huntley, M. E., and Redalje, D. G. (2007). CO_2 mitigation and renewable oil from photosynthetic microbes: a new appraisal. *Mitigat. Adapt. Strat. Global Change,* 12:573-608.

Laura, B. L., Eric, B. S., Peters, J. W., Matthew and Posewitz, C. (2009). Engineering algae for biohydrogen and biofuels production. *Curr. Opin. Biotechnol,* 20:264-271.

Li, Q., Du, W. and Liu, D. (2008a). Perspectives of microbial oils for biodiesel production. *Appl. Microbiol. Biotechnol*, 80:749-756.

Li, Y., Horsman, M., Wu, N., Lan, C. Q., and Dubois-Calero, N. (2008b). Biofuels from microalgae, *Biotechnol. Prog,* 24: 815-820.

Ma, F. R. and Hanna, M.A. (1999). Biodiesel production: A review. *Bioresour. Technol*, 70:1-15.

Mata Teresa, M., Martins Antonio, A. and Caetano Nidia, S. (2010). Microalgae for biodiesel production and other applications: A review. *Renew. Sust. Energ Rev,* 14:217-32.

Meher, L. C. Vidya, S. D. and Naik, S. N. (2006). Technical aspects of biodiesel production by transesterification: A review. *Renew. Sust. Energ. Rev,* 10:248-68.

Mohapatra, P. K. (2006). Biotechnological approaches to microalga culture. In: *Textbook of environmental biotechnology.* IK International Publishing, New Delhi, India.

Molina, G., Acien-Fernandez, F. G., Camacho, G., and Chisti, Y. (1999). Photobioreactors: light regime, mass transfer and scale up. *J. Biotechnol,* 70:231-247.

Parmar, A., Singh, N. K., Pandey, A., Gnansounou, E. and Datta, M. (2011). Cyanobacteria and microalgae: A positive prospect for biofuels. *Bioresour. Technol*, 102:10163-10172.

Raja, R., Hemaiswarya, S., Kumar, N. A., Sridhar, S. and Rengasamy, R. (2008). A perspective on the biotechnological potential of microalgae. *Crit. Rev. Microbiol,* 34(2):77-88.

Richmond, A. (2004). *Handbook of microalgal culture: biotechnology and applied phycology.* Blackwell Science Ltd.

Rosenberg, J. N., Oyler, G. A., Wilkinson, L. and Betenbaugh, M. J. (2008). A green light for engineered algae redirecting metabolism to fuel a biotechnology revolution. *Curr. Opin. Biotechnol,* 19(5):430-6.

Sharma, Y. C., Singh, B. and Korstad, J. (2011). Latest developments on application of heterogenous basic catalysts for an efficient and eco-friendly synthesis of biodiesel: A review. *Fuel*, 90:1309-1324.

Sheehan, J., Dunahay, T., Benemann, J. and Roessler, P. (1998). *A look back at the U.S. department of energy's aquatic species program-biodiesel from algae.* National Renewable Energy Laboratory, Golden, CO., Report NREL/TP-580-24190.

Tsukahara, K. and Sawayama, S. (2005). Liquid fuel production using microalgae. *Journal of the Japan Petroleum Institute,* 48(5):251-9.

Ueda, R., Hirayama, S., Sugata, K. and Nakayama, H. (1996). *Process for the production of ethanol from microalgae.*

Vasudevan, P. and Briggs, M. (2008). Biodiesel production-current state of the art and challenges. *J. Ind. Microbiol. Biotechnol*, 35:421-430.

Vergara-Fernandez, A., Vargas, G., Alarcon, N. and Velasco, A. (2008). Evaluation of marine algae as a source of biogas in a two-stage anaerobic reactor system. *Biomass. Bioenergy*, 32: 338-44.

Verma, N. M., Mehoratra, S., Shukla, A. and Mishra, N. B. (2010). Prospective of biodiesel production utilizing microalgae as the cell factories: A comprehensive discussion, *Afr. J. Biotechnol,* 9(10):1402-1411.

Wang, B., Li, Y., Wu, N. and Lan, C. Q. (2008). CO_2 bio-mitigation using microalgae. *Appl. Microbiol. Biotechnol,* 79(5):707-18.

Wayman, C. (1996). *Handbook of Bioethanol: Production and Utilization*, Taylor & Francis, ISBN 1-56032- 553-4, Washington, USA.

Xiong, W., Li, X., Xiang, J. and Wu, Q. (2008). High-density fermentation of microalga *Chlorella protothecoides* in bioreactor for micro biodiesel production. *Appl. Microbiol. Biotechnol,* 78:29-36.

Yang, Y. F., Feng, C. P., Inamori, Y. and Maekawa, T. (2004). Analysis of energy conversion characteristics in liquefaction of algae. *Resour. Conserv. Recycl,* 43(1):21-33.

In: The Role of Photosynthetic Microbes…
Editors: K. Tripathi, N. Kumar et al.
ISBN: 978-1-53614-032-3

Chapter 5

ALGAL BASED CO_2 SEQUESTRATION: A SUSTAINABLE ROUTE FOR CO_2 MITIGATION

Shailendra Kumar Singh[1,*], Sushil Kumar Singh[1], Vinod Kumar Kannaujiya[2], Md. Akhlaqur Rahman[3], Kritika Dixit[4], Adinath[5], Suman Kapur[1] and Shanthy Sundaram[4]

[1]Department of Biological Science, Birla Institute of Technology and Science, Pilani-Hyderabad Campus, Hyderabad-Telangana, India
[2]Department of Botany, MMV, Banaras Hindu University, Varanasi, India
[3]Department of Biotechnology, S. S. Khanna Girl's Degree College, Allahabad, India
[4]Centre of Biotechnology, University of Allahabad, Allahabad, U. P. India
[5]Nehru Gram Bharati, Allahabad

ABSTRACT

Our world is no different from the single cell of an organism. As the cells require a constant supply of energy to keeps them alive, our society also demands energy for uninterrupted socio-economic development. However, over the past 150 years, to keep pace with demand, humankind has consumed enormous amount of carbon-based fuels in the development of our society. Interestingly, community has now become more concerned with the impact of harmful consequences of use of carbon-based fuels and is eager to explore alternate sustainable routes to minimize the harmful impact. Among all possible routes, CO_2 sequestration via group of unicellular or simple, multicellular, fast growing photosynthetic microorganisms has received significant attention. Studies and investigation on algal based CO_2 sequestration technologies are steadily growing. But the technology is still in its infancy and much remains to be done at the level of optimizing

[*] Corresponding Author Email: shailbiochem@gmail.com

the biology, engineering, life cycle analysis, as well as in testing, demonstration pilot-scale systems business models. In this article, we attempt to elucidate various CO_2 fixation pathways to understand the mechanism at biochemical level. Article also discusses major challenges associated with the development of economically viable algae-based CO_2 sequestration technology. It may help researchers to explore opportunities for mitigating significant quantities of CO_2 emission at larger scale.

Keywords: algae, biofuel, CO_2 mitigation, global warming, sustainability

INTRODUCTION

Reducing greenhouse gases (GHGs) emissions to acceptable levels is a major challenge in the 21st century. Apart from global warming, GHGs also have other harmful effects on the environment such as, smog pollution, depletion in ozone layer, acidification of ocean and changes on plant growth. Due to these facts, 192 countries have adopted 2°C or 1.5°C (relative to pre-industrial levels), through Kyoto Protocol, as global warming limit that would prevent dangerous anthropogenic interference with the climate system (Naam, 2009). Among GHGs, carbon dioxide (CO_2) is considered as core gas of greenhouse effect in terms of total contribution (72% of emitted GHGs) and its global climate change potential (retain in atmosphere for 100 - 200 years) (Singh et al., 2016). To meet GHG emissions reduction target, nowadays most research focuses on carbon capture at fossil fuel-powered energy plants. A complimentary range of technological approaches, such as adopting clean fuels and clean coal technologies and CO_2 capture and storage (CCS), have been proposed by researchers to capture CO_2, effectively and economically(Leung et al., 2014). This integrated carbon capturing system includes carbon capture, transport, storage and monitoring. The capture and compression steps, which are expensive and could escalate the cost of producing electricity by about 20 to 50% (Mathieu and Bolland, 2013). Thus, CCS technologies are not commercially viable. To achieve commercial viability, they need to be accomplished in cost-effective and sustainable way. In this context, atmospheric CO_2 fixation by photosynthetic microorganisms such as microalgae and cyanobacteria seem to be a promising way to recycle CO_2 through photosynthesis. In the recent years, algal technology has received widespread attention to protect our environment along with the production of carbon neutral fuel (Hannon et al., 2010, Singh et al., 2016).

These tiny organisms can not only consume CO_2 but can also be cultivated in wide variety of water sources (fresh, brackish, seawater and wastewater) and provide non-food-based feedstock alternative for biofuel production throughout the year (Singh et al., 2013, Rahman et al., 2014). Due to their rapid growth potential, massive availability and remarkable metabolic plasticity, they have gained attention in bioindustrial applications such as biofuel (Singh et al., 2013), bioremediants (Singh et al., 2014), CO_2 sequester

(Chiu et al., 2008, Singh et al., 2016), nutritional supplements, pigments, secondary metabolites, etc.

With this in mind, intent of present article is to anticipate the major sustainability concerns of CO_2 emission and beneficial environmental and societal consequences if commercial-scale algal biofuel production is widely deployed. The present chapter will also provide a brief overview of mechanism and bioenergy application of algal based CO_2 sequestration.

WHY CO_2 MITIGATION IS CRUCIAL FOR ENVIRONMENT

First, let's understand why reducing CO_2 emissions must be a priority in mitigating the climate change. Life requires energy to survive. Since beginning of industrialization, burning of fossil fuels for energy (first coal, then oil and gas) have grown emission of anthropogenic green-house gases (GHGs) such as carbon dioxide (CO_2), sulfur dioxide (SO_2), and nitrogen oxides (NO_2), exponentially (McCarthy et al., 2016). The primary green house gas, CO_2 is naturally present in the Earth's atmosphere as a component of carbon cycle. Conversely adding more CO_2 has the long-lived impact on the natural CO_2 sinks like forests and ocean. However, human activities are altering constantly the dynamic balance of normal carbon cycle. Nearly 36% of CO_2 emissions are attributable to manufacturing industries such as chemicals, petrochemicals, iron and steel, cement, paper and pulp, and other minerals and metals, which account for more than two-thirds of the carbon dioxide emissions (Kuramochi et al., 2013). CO_2 gas has natural heat-trapping property. Addition of CO_2 could enhances the ability of Earth's atmosphere to entrap heat from sun and pushing the world into hazardous high temperature territory. Adverse CO_2effects on climate could impact till thousands of years after CO_2 emissions cease (Muradov, 2014). Oceans, the major sink for atmospheric CO_2, are also no longer capable to take up the increase in CO_2 concentration. This has led to an awareness to adopt new and sustainable solution for CO_2 management with lower pollution rate. The statistical data of Scripps Institute of Oceanography and Earth Policy Institute (Figure 1) show that worldwide atmospheric CO_2 concentrations elevated gradually year by year; 396.48 $mg.L^{-1}$ in 2013, up from 389.86 $mg.L^{-1}$ in 2010 and from 290 $mg.L^{-1}$ in 1880 (https://scripps.ucsd.edu/news/7992).

Concomitant effect of rise in atmospheric CO_2 on climate change underlies the urgent need for implementation of carbon mitigation approaches to alleviate the serious global climate repercussions. Thus, the present CO_2 management projects of industries concentrate on emission reductions from current operations through energy efficiency improvements, operating practice changes, or process changes. The Kyoto Protocol of 1997 called for a 5.2% decrease in GHGs level globally from 1990 values (Parry et al., 1998). To search effective CO_2 reduction strategy, numerous efforts have been explored including chemical reaction-based approaches, direct injection to underground or into the ocean and biological CO_2 mitigation. The first two approaches, chemical and direct injection are not only costly (use so much inputs and energy) but also unsustainable (store enormous amount of flue gases) making mitigation benefits marginal (Muradov, 2014). These procedures may only be considered as short-term CO_2 mitigation solution (Mathieu and Bolland, 2013). Therefore, development of CO_2-neutral energy resources for mitigation of CO_2 related emissions is one of the much-needed challenges for global environmental sustainability. In this scenario, biological CO_2 mitigation in which CO_2 is transformed into important commercial products appears to be economically sustainable approach (Singh et al., 2014).

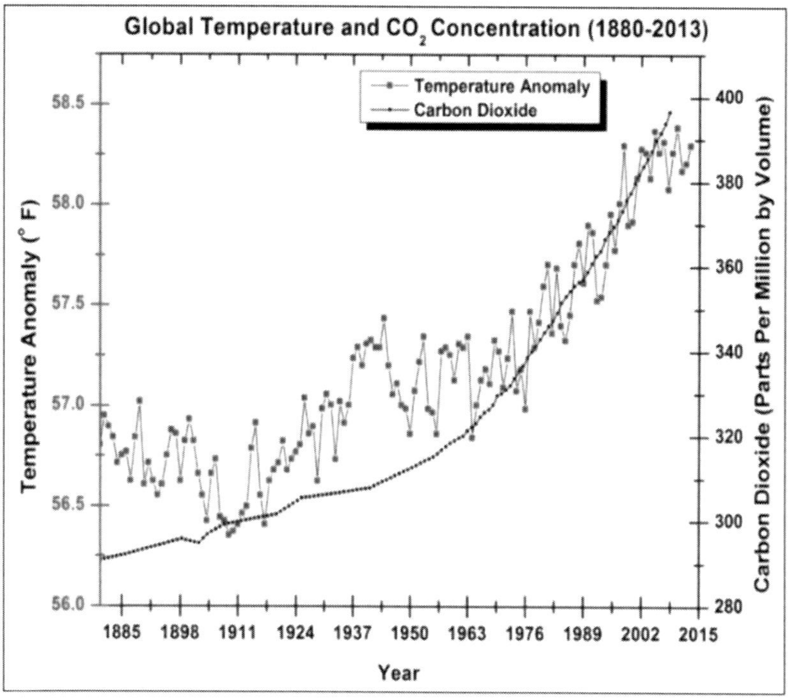

Figure 1. Atmospheric CO_2 concentration and temperature anomaly (Data from https://scripps.ucsd.edu/news/7992).

NATURE'S WAY TO RECYCLE CO_2

Recycling is an important way to conserve our natural resources and greatly contribute towards improving the surrounding environment. Nature has evolved sophisticated mechanisms for recycling CO_2 through photosynthetic and chemolithoautotrophic organisms (Jajesniak et al., 2014). These organisms exhibit superb capability in assimilating CO_2 and transform them into various biomolecules. Figure 2 illustrates the CO_2-utilizing microorganisms that have been studied extensively and have potential to potential to be genetically modified for industrial level bioprocesses. Table 1 summarizes their industrial relevance and challenges for exploitation in biological CO_2 mitigation. Among, CO_2-utilizing organisms, in the current scenario, research endeavor on photosynthetic microorganisms such as microalgae and cyanobacteria have started to gain momentum. It is noteworthy that photosynthetic microorganisms are not developed by nature for industrial level production of valuable biochemicals. Many of their inherent properties such as e.g., growth characteristics, various metabolites produced, thermal stability, and tolerance towards inhibitors are not fit for industrial bioprocesses. Developments in genetic engineering tools have made it possible to improve their phenotypic characters and to expand their repertoire for biochemical production.

Among these CO_2 fixing photosynthetic microorganisms, algae and cyanobacteria are perhaps the most significant in their industrial relevance and researched for carbon fixation. Annually, around 500 billion tonnes of CO_2 is fixed by terrestrial vegetation which is 20 times more than the total CO_2 released from fossil fuel consumption (Skjanes et al., 2007). In comparison to terrestrial plants, algal species are able to grow much faster with 10–50 times superior CO_2-fixation efficiency (Singh et al., 2014). Their potential is attributed to their wide habitat, high biomass ability (Davis et al., 2016), fast CO_2 uptake (Chen et al., 2016) and utilization (to produce products of high commercial value (Markou and Nerantzis, 2013). These useful combinations provide the main rationale for algal biotechnology and propel them to the forefront for sustainable development options.

BIOCHEMICAL MECHANISM OF CO_2 BIOTRANSFORMATION IN PHOTOSYNTHETIC MICROORGANISMS

Over billions of years ago, algae have originated on earth and been used by humans for centuries. These tiny photosynthetic microorganisms are capable of reducing CO_2 emissions from the atmosphere, through biotransformation into biomass in the presence of light energy. The overall reaction for photosynthesis is given by:

$$CO_2 + H_2O + LIGHT \rightarrow (CH_2O)_n + O_2$$

Figure 3 depicts three different inorganic carbon assimilation routes in algae: (i) CO_2 assimilation via the plasmatic membrane; (ii) the use of bicarbonate by inducing specific enzyme isoforms 1 through 9 of carbonic anhydrase, which converts the H into CO_2; and (iii) direct transport of H via the plasmatic membrane. Incorporated CO_2 is carboxylated by ribulose 1, 5 bisphosphate carboxylase/dehydrogenase (RuBisCO) which uses it as

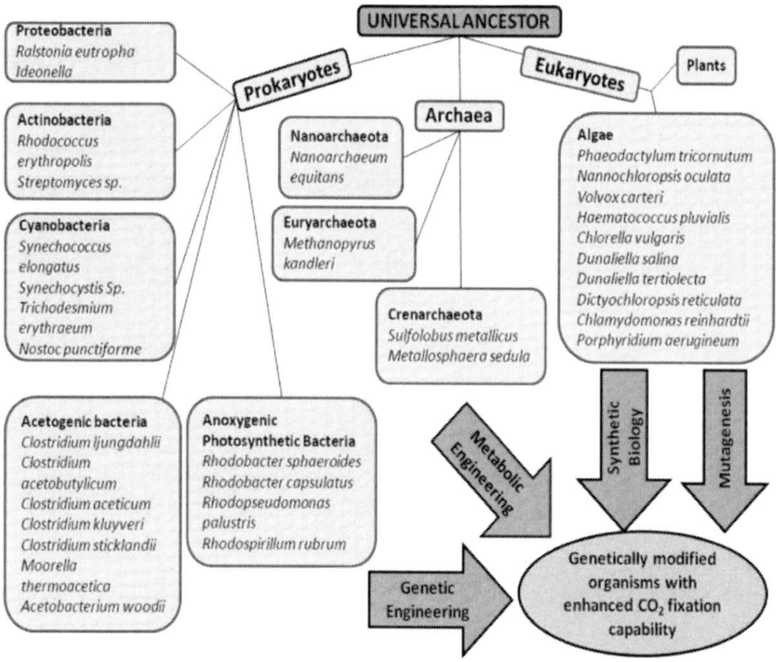

Figure 2. Phylogenetic representation of nature's organism for CO_2 fixation and technologies for enhancement of CO_2 fixing ability.

a substrate to produce 3 carbon phosphoglycerate (PGA) as the end product. Further reduction of PGA caused by the electron transporter nicotinamide adenine dinucleotide phosphate (NADPH) leads to the production of a series of intermediary phosphorylated sugars and finally to glucose which is known as the Calvin-Benson cycle and occurs in intra cytoplasmatic membrane (Morris, 1980).

Thus, there are mainly two rate determinant reactions in algal CO_2 fixation pathways (Singh et al., 2014a) as shown in Figure 3 (i) conversion of the HCO_3^- into CO_2 using enzyme carbonic anhydrase and (ii) CO_2 carboxylation by ribulose 1, 5 bisphosphate carboxylase (RuBisCO). The rate of HCO_3^- conversion into CO_2 reaction may be slow due to limited CO_2 production. Elevated efficiency of enzyme carbonic anhydrase may increase the intracellular levels of CO_2 to 1000 times higher concentrations than the external fluid, resulting in efficient carbon fixation. Similarly, at the normal atmospheric level of CO_2 (390 ppm), RuBisCO has low affinity for CO_2. Normally, it is only half saturated with the atmospheric CO_2. RuBisCO also performs oxygenase activity which uses incorporated CO_2 as a substrate to produce 3 carbon phosphoglycerate (PGA) as the end product (Beardall and Raven, 2016). The oxygenase activity of RuBisCO inhibits biomass formation by around 50% (Giordano et al., 2005). It has no use to cell and its synthesis consumes significant amount of cellular energy and also releases previously fixed CO_2 by the carboxylase activity of RuBisCO.

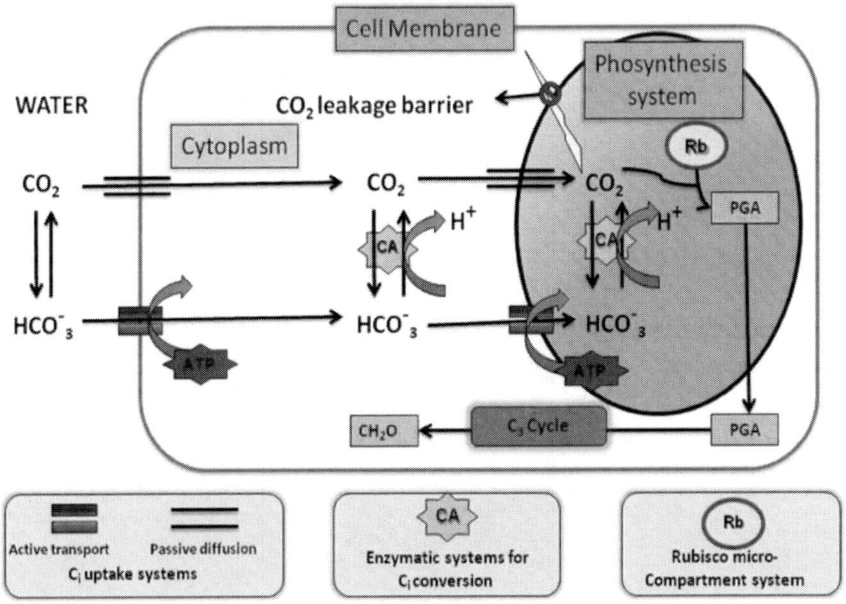

Figure 3. Schematic representation CO_2 concentrating mechanisms in algae (Singh et al., 2014a).

Fixation Rate Regulation via CO_2 Concentrating Mechanisms

To overcome the slow conversion of HCO_3^- and low affinity of RuBisCO for CO_2, most algae and cyanobacteria have different CO_2 concentrating mechanisms (CCMs). CCMs are activated only at low dissolved carbon concentration. The function of the CCMs is to raise the intracellular inorganic carbon levels, compensating for limitations in the CO_2 supply that could reduce the photosynthetic rates (Price et al., 2008). This mechanism is responsible for pumping CO_2 to the carboxylation sites. The expression of the enzyme carbonic anhydrase (CA) has been associated with induction of the CCMs. CA catalyzes the inter-conversion of CO_2 and HCO_3^- and is an important component in the intracellular mobilization of the HCO_3^- pool, by catalyzing the production of CO_2 for RuBisCO (Singh et al., 2014a).

$$CO_2 + H_2O \rightarrow HCO_3^- + H^-$$

The maximum value of dissolved inorganic carbon till it is active depends upon the strain, pH, light availability, preadaptation of cells, etc. The understanding in CCMs system of algae may act as an enhancer for higher growth rates in algae and hence can be used for improvement in photobioreactor productivity.

Table 1. Industrial importance and technological challenges of CO_2-utilizing organisms

Group	Industrial importance	Challenges
Archaea	• Many anaerobic methanogens naturally produce methane (CH_4) from CO_2. • Good thermo stable enzyme source (e.g., Esterases, carbonic anhydrase, Polymerarses). • Some archaeal species can naturally accumulate PHA.	• Gas fermentation is not yet developed. • Hard to imitate growth conditions. • Genetic engineering tools are not available for most of archaeal strains.
β-Proteobacteria	• Aerobic microorganisms, metabolically diverse and contain chemolithoautotrophs, photoautotrophs, and generalist heterotrophs. Various carbon sources and their utilization pathways. • Some strains have natural ability to store and convert intracellularly CO_2 to PHA. • Genetic engineering tools are available.	• Gas fermentation is not yet developed. • Easier in cultivation compared to archaea.
Clostridia	• Diverse range of carbon substrates e.g., simple or complex carbohydrates, CO_2/H_2 and CO. • Diverse pathways for production of useful metabolites/industrial products e.g., ethanol, acetate, acetone, lactate, butanol, 2, 3-butanediol, valeroate, caproate, carpylate and closthioamide. • Tolerance to toxic metabolites and substrates. • Genetic modification tools available.	• Anaerobic cultivation is an expensive process. • Gas fermentation is not yet developed.
Algae	• Broad habitat mostly in moist surroundings. • Able to cultivate in open ponds or closed photobioreactors. • Fastest growing photosynthetic organism with fast CO_2 uptake. • High cell density, High lipid content, high-value by products (e.g., proteins, nutrients, biofertilizers). • Genetic engineering tools are available.	• Light quality affects growth (e.g., intensity, wavelength). • Huge water amount required. • Nitrogen and phosphorous required as a fertilizer. • Biodiesel produced have poor cold flow character and less oxidative stability.
Cyanobacteria	• Cultivation is easy with simple nutrition and low-cost. • Photosynthetic activity and growth are higher than algae and plants. • Considerable amount of lipid content in thylakoid membranes. • Broad range of fuels could produce (e.g., H_2, ethanol, biodiesel, methane, etc.). • Genetic modification tools are available.	• Physical factors like temperature 20 - 30°C, pH 7 - 9 and light intensity can affect productivity. • After carbon, nitrogen is also a limiting nutrient for high biomass production. • Agitation, aeration, and mechanical stirring are essential which could increase the cultivation cost.

ALGAE BASED CO_2 MITIGATION TECHNOLOGY

Algal-CO_2-mitigation-technology is an effective and sustainable process to fix the CO_2 from industrial flue gases into organic biomass which can be transformed later into valuable products. Products such as bioethanol, biofuel, biohydrogen, amino acids, proteins, fatty acids, vitamins, minerals, pigments, dietary supplements for human, animals and aquaculture and other bio-compounds can be produced from algal biomass.

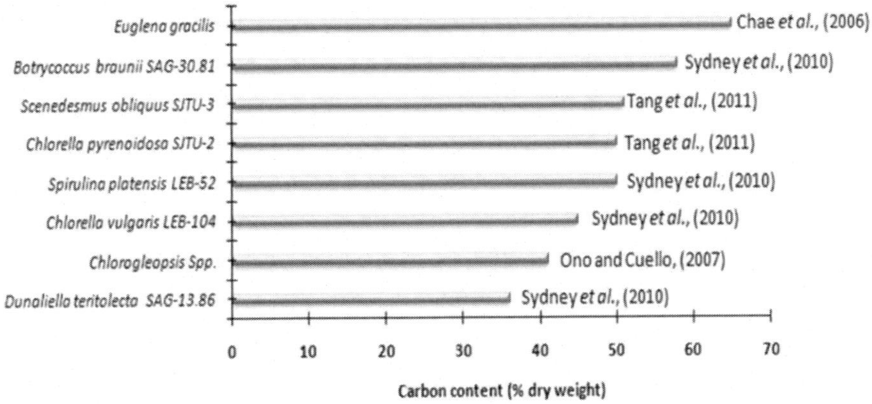

Figure 4. Total carbon content of some algal species.

(Pulz and Gross, 2004, Skjånes et al., 2007). As shown in Figure 4, carbon is a main biochemical component of algal biomass (36 to 65% of dry mass). If we assume carbon mass fraction of the algae is half of its dry biomass, then to produce 1 kg, dry algal biomass needs at least 1.83 kg CO_2 (Slade and Bauen, 2013). Thus, carbon of flue gases (CO_2 gas emissions from industries) could provide a key resource for successful algal biomass production system (Li et al., 2016; Sydney et al., 2010). CO_2 uptake mechanisms and membrane-bound transporters are vital in the translocation of HCO_3^- and CO_2 across inner membranes in cyanobacteria and eukaryotic algae. Cyanobacteria have highly efficient CCM system, which is induced under low-CO_2 conditions (Price, 2011).

In fact multi-layered CCM system of cyanobacteria have allowed retention of the ancient form of RuBisCO with a lower affinity for both CO_2 and O_2 but with a much higher catalytic turnover rates per unit protein in comparison to RuBisCOs of eukaryotic algae (Badger et al., 1998). High CO_2 concentration promotes photosynthetic efficiency of algae to reproduce within a shorter time and thus more algal biomass could be attained. However, under high-CO_2 conditions, carbonic acidification is a major factor that inhibits algal growth in high CO_2 environment.

CCMs may play an important role with activation of carbonic anhydrases to prevent inhibition of RuBisCO enzyme due to carbonic acidification (Solovchenko and Khozin-Goldberg, 2013). Care needs to be taken as some algal strains are less tolerant towards high concentration of CO_2 and high CO_2 levels may possibly inhibit their growth. Flue gases have not only CO_2, but also toxic sulfur and nitrogen oxides (SO_x and NO_x). Some researchers reported that within a narrower range, SO_x and NO_x of flue gases may serve as beneficial nutritional components for algae. Lara-Gil et al., (2014) performed toxicity assays on *Desmodesmus abundans* UTEX-2976 and *Scenedesmus* sp. UTEX-1589, under 2% (*v*/*v*) CO_2 and using nitrite, sulfite, or bisulfite salts.

Table 2. Productivity and CO_2 consumption rate of algal species at different CO_2 level

Algal species	CO2 Concentration (%)	Productivity (g.l-1.day-1)	CO_2 Consumption rate (g. L- 1 d- 1)	Reference
Aphanothece microscopica Nageli	15	0.77	1.44	Jacob-Lopes et al., (2009a)
		1.25	5.44	Jacob-Lopes et al., (2009b)
Anabaena sp.	Air	N.D	1.45	Lopez et al., (2009)
Botryococcus braunii	10	0.027	N.D	Yoo et al., (2010)
	Flue gas	0.08	N.D	Yoo et al., (2010)
	N.A	0.9	1	Murakami and Ikenouchi, (1997)
Chlorococcum littorale	20	0.53	0.9	Kurano et al., (1995)
Chlorella sp.	Air	0.68	N.D	Chiu et al., (2008)
	2	1.45		
	5	0.90		
	5	0.34	0.70	Ryu et al., (2009)
	10	0.11	N.D	Chiu et al., (2008)
		0.94	1.77	Sung et al., (1999)
		0.38	0.72	Chiu et al., (2009)
		0.61	1.15	
	15	0.09	N.D	Chiu et al., (2008)
	20	0.7	1.32	Sakai et al., (1995)
Chlorella emersonii	Air	0.04	0.08	Scragg et al., (2002)
Chlorella kessleri	6	0.09	0.16	de Morais and Costa, (2007a)
	6	0.07	0.12	de Morais and Costa, (2007b)
	18	0.09	N.D	de Morais and Costa, (2007c)
Chlorella vulgaris	10	0.11	N.D	Yoo et al., (2010)
	Air	0.04	N.D	Scragg et al., (2002)
	Air	0.02	N.D	Scragg et al., (2002)
	Air	0.04	0.08	Scragg et al. (2002)
	10	0.27	0.61	Jin et al., (2006)
	0.8-1.0	N.D	6.24	Cheng et al., (2006)
	0.09	0.15	3.45	Fan et al., (2008)
Dunaliella tertio lecta	ND	ND	0.27	Eduardo B Sydney et al., (2010)
Euglena gracilis	10	0.15	0.38	Chae et al., (2006)
Haematococcus pluvialis	16–34	0.08	N.D	Huntley and Redalje (2007)
Microcystis aeruginosa	10	0.22	0.52	Jin et al., (2006)
Microcystis ichthyoblabe	10	0.23	0.49	Jin et al., (2006)
Nannochloris sp.	15	0.32	0.60	Negoro et. al., (1991)
Nannochloropsis sp.	15	0.27	0.51	Negoro et al., (1991)
Phaeodactylum tricornutum	15	0.15	0.28	Negoro et al., (1991)
Scenedesmus sp.	10	0.19	0.46	Jin et al., (2006)
		0.22	0.41	Yoo et al., (2010)
	Flue gas	0.20	N.D	Yoo et al., (2010)
Scenedesmus obliquus	12	0.14	0.26	de Morais and Costa (2007a)
	18	0.14	N.D	de Morais and Costa (2007a)
	6	0.11	0.20	de Morais and Costa (2007b)
		0.09	0.16	de Morais and Costa (2007c)
	10	0.29	0.55	Ho et al., (2010)
Spirulina sp.	6	0.2	0.38	de Morais and Costa (2007a)
		0.21	0.39	de Morais and Costa (2007b)
	12	0.22	N.D	de Morais and Costa (2007a)

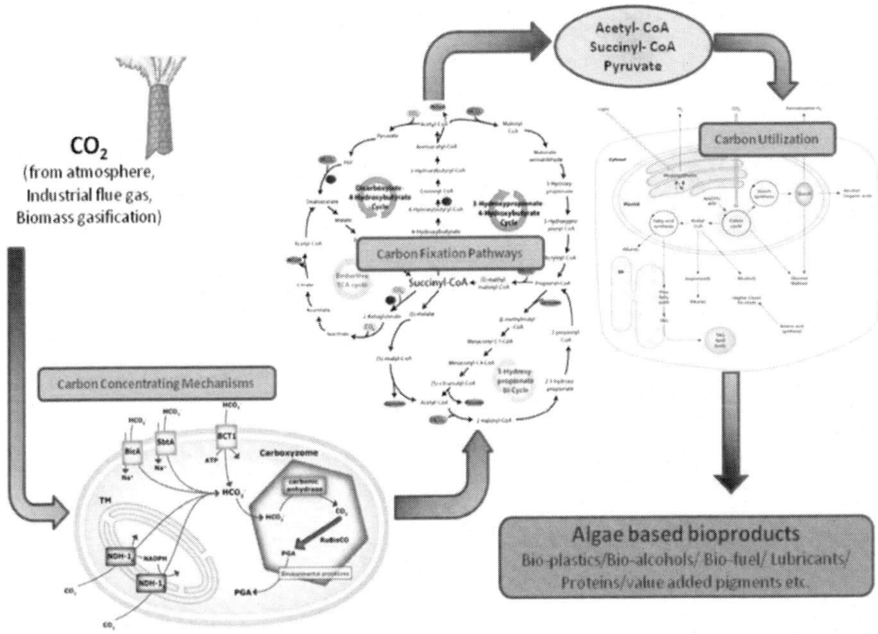

Figure 5. Carbon concentrating mechanism, carbon fixation pathways and biomass utilization strategies in algal-CO_2-mitigation-technology.

They observed that both nitrite (0–1,067 mg.L^{-1}, NO_2^- (*w*/*v*)) and sulphite (0–254 mg.L^{-1} SO_3^{2-} (*w*/*v*)), do not affect the growth of tested range, except bisulfite, which was highly toxic just above the concentration of 39 mg.L^{-1}. Thus, straight consumption of flue gas might also be probable with high flue gas tolerant algal strains. Table 2 shows the productivity and CO_2 fixation rate of some algal species with in suitable CO_2 mitigation range.

ALGAE AS SUSTAINABLE TOOL FOR GREENHOUSE GAS MITIGATION

The conceptual design of photosynthetic conversion to sequester CO_2 using algae has been shown in Figure 5. In addition to mitigating CO_2, the algal biomass can be used to produce biofuels (e.g., biodiesel, bioethanol, biohydrogen) and commercially and scientifically value-added bio-products. Brennan and Owende (2010) reported that 1 kg of dry algal biomass can utilize around 1.83 kg of CO_2. Thus, annually 54.9 – 67.7 tonnes of CO_2 can be sequestered from raceway algal pond systems, corresponding to annual dry algal biomass production rate of 30–37 tonnes per hectare. Major factors affecting the biological fixation of CO_2 can be culture strain, culture density, light, high temperatures of flue gas, pH, critical CO_2 concentration, CO_2 mass transfer, O_2 accumulation. Presence of SO_x and NO_x as well as other impurities of the fossil fuel affects carbon fixation.

Crucial parameters that influence the CO_2 fixation rate of algal culture could be better control by closed system rather than open system.

ADVANCES IN ALGAL BASED CO_2 SEQUESTRATION

To achieve full processing capabilities and develop an effective and sustainable algal technology, recent research efforts have concentrated on evaluating and optimizing the integrated system. Integrated systems seem to be quite promising for valuable biomass production with biological cleaning of flue gas. The exhaust flue gases from power plants contain 12 -13% CO_2 (Reddy and Argyle, 2010). This is responsible for more than 7% of total world CO_2 emissions from energy use (USDOE, 2000).Most studies indicate that CO_2 addition to algal cultures stimulates growth. Zeiler et al., (1995) demonstrated that green alga *Monoruphidium minutum* can efficiently utilize simulated flue gas containing high levels of carbon dioxide, as well as sulfur and nitrogen oxides, as a feedstock to produce substantial biomass. For that reason, it is beneficial if microalgae tolerant to high carbon dioxide levels be used for its fixation from flue gases. *Chlorococcum littorale*, a marine alga, has also shown exceptional tolerance to high CO_2 concentration of up to 40% (Berberoglu et al., 2009). *Chlorella* strains from hot springs, also showed to be tolerant to high temperatures up to 42°C for CO_2 fixation from industrial flue gases containing up to 40% CO_2 (Sakai et al., 1995). Over expression of selected target genes are also proven to be a powerful approach for improving at least one or more traits related to carbon fixation (Stephenson et al., 2011). Numerous efforts have been made to improve the rate of CO_2 fixation synthetic pathway via the Calvin cycle, PEP carboxylase and through introduction of metabolic engineering with varying degrees of success. These approaches can be grouped into following categories: (1) engineering of ribulose-1, 5-bisphosphate carboxylase/oxygenase (RuBisCO) for improvement in the of catalysis of carboxylation and reduction of the oxygenation reaction, (2) enhancement of activation state of RuBisCO enzymes (3) improvement in the of regeneration phase of the Calvin cycle, and (4) enrichment of CO_2 concentration around RuBisCO for suppression of the oxygenase reaction. However, till date, successful full-scale commercial technology for algal based CO_2 sequestration could not be achieved. At present most involved companies are at infancy phase start-ups and a lot investment is in research.

RECENT TRENDS IN ALGAL CO_2-MITIGATION-TECHNOLOGY

The current research efforts on algal-CO_2-mitigation-technology have been concentrated on evaluating and optimizing the integrated multidisciplinary system to achieve full processing capabilities to develop an effective algal-CO_2-mitigation-system

(Figure 5). Besides integration of different technologies, substantial progress has also been made in algal transgenics (Scaife and Smith, 2016). A wild algal cell is investing fixed carbon and its inherent free energy into growth, survival and the generation of new biomass. However metabolic engineering of algal cells could alter the photosynthetic metabolism to synthesize high value compounds from abundant and inexpensive raw material (CO_2, sunlight, and water). The rapidly growing demand for energy and the environmental concerns about CO_2 emissions make this approach more attractive (Angermayr et al., 2015). In this context genetic tools have been developed for model organisms such as the green algae *Chlamydomonas reinhardtii*, *Nannochloropsis oceanica*, *Volvox carteri* and the diatom *Phaeodactylum tricornutum*. Nuclear transformation has been successfully achieved in many industrially-relevant species such as the micro-algae *Botryococcus braunii* (Matsushima et al., 2012) *Dunaliella salina* (Ramos et al., 2009) and *Haematococcus pluvalis* (Eom et al., 2006). Different approaches have been applied for modification in green algae also. Mussgnug et al., (2007) used RNA silencing to down-regulate the entire light-harvesting antenna complexes gene family of *C. reinhardtii* to reduce energy losses by fluorescence and heat. The mutant algae exhibited an increased efficiency of cell cultivation under elevated light conditions. Random mutagenesis via 500 Gy of ^{60}Co γ irradiation was applied to improve the productivity of green alga *Chlorella* species in the presence of 15% CO_2 (Cheng et al., 2013). This study showed that biomass yield was increased to 2.41 $g.L^{-1}$ and CO_2 fixation efficiency reached 32.7%, under optimized conditions, corresponding to a CO_2 fixation rate of 1.54 g. CO_2 L^{-1}. day^{-1}. Like green algal strains, many cyanobacterial strains have also been extensively modified for sustainable production of isobutanol, ethanol, and isoprene, etc. using CO_2 and light. Gao et al., (2012) developed a bioprocessing strategy to integrate photosynthetic biomass production and ethanol production in *Synechocystis* sp. PCC 6803, using CO_2 in one single biological system. They cloned nine alcohol dehydrogenases from different cyanobacterial strains and expressed in *E. coli*. The mutant strain was able to produce significantly higher ethanol 5.50 g L^{-1} (212 $mg.L^{-1}$. day^{-1}) as compared to previous yield of 550 $mg.L^{-1}$ in the same strain (Dexter and Fu 2009). Lan and Liao, (2011) modified the CoA-dependent 1-butanol production pathway in cyanobacterium *Synechococcus elongates* for the autotrophic production of 1-butanol from CO_2. Li et al., (2013) also demonstrated autotrophic production of 1, 2-propanediol using *S. elongates*, with a reported litre of 150 $mg.L^{-1}$. Hirokawa et al., (2016) showed that modification in the synthetic metabolic pathway of *S. elongates* PCC 7942 could produce 288 $mg.L^{-1}$ of 1, 3-propanediol (1, 3-PDO) directly from CO_2 after 14 days of growth. Oliver et al., (2013) showed that genetically engineered *S. elongates* could produce high yield of 2,3-butanediol (2.38 $g.L^{-1}$).Cyanobacteria may also better host organisms for the heterologous production of terpenoids such as isoprene (Bentley et al., 2014), monoterpene β-phellandrene (Bentley et al., 2013) and sesquiterpene β-caryophyllene (Reinsvold et al., 2011), etc. Kudoh et al.,

(2014) developed a 1-deoxy-d-xylulose 5-phosphate synthase (DXS) enzyme over expressed *Synechocystis* strain and showed that the carotenoid levels increased 1.5 times. Bentley et al., (2014) advanced this approach further and heterologously expressed the entire mevalonic acid pathway genes from *Enterococcus faecalis* and *Streptococcus pneumonia*, under control of the *psbA2* promoter, led to 2.5-fold increase in isoprene production. Many researchers have showed that the heterologous expression in cyanobacteria as better biosystem to produce value added compounds. However, appropriate tools, promoters, vectors, and regulatory systems, to be use in genetic engineering of cyanobacteria and metabolic modification in pathways, are still needs to be refined.

CONCLUSION

Energy provides a vital ingredient for almost all human activities and crucial input for socio-economic development. However, overwhelming reliance on fossil fuels for energy generates a threat to alter the Earth's atmosphere and could have grave consequences for the integrity of both ecological systems and human society. Among all CO_2 mitigation routes, the photosynthetic route seems to be far superior and sustainable solution under both environmental and economic considerations when scaled up at industrial level. However various factors contribute for successful algae-based-CO_2-mitigation system such as detailed knowledge of flue gas and their effect on algal cells. Yield of value added compounds from produced algal biomass is also an important factor for commercial success. Subsequent optimization of metabolic pathways through metabolic engineering could benefit the photoautotrophic production of value added compounds along with CO_2 mitigation, commercially. However, detailed knowledge about carbon concentrating mechanism and regulation in the metabolic network of production pathways is still in its infancy. Improving these current knowledge on CCM and metabolic pathways may improve the performance of algal CO_2 mitigation technology.

ACKNOWLEDGMENTS

Authors are thankful to the Science and Engineering Research Board (SERB) and Department of Science and Technology (DST), India for their continued financial support and encouragement through the DST-Young Scientist grant for our research on mitigation using cyanobacteria and microalgae.

REFERENCES

Angermayr, S. A., Rovira, A. G. & Hellingwerf, K. J. (2015). Metabolic engineering of cyanobacteria for the synthesis of commodity products. *Trends. Biotechnol*, *33*, 352-361.

Bachram, H. (2004). Climate fraud and carbon colonialism: the new trade in greenhouse gases. *Capitalism. Nature. Socialism*, *15*, 5-20

Badger, M. R., Andrews, T. J., Whitney, S. M., Ludwig, M., Yellowlees, D. C., Leggat, W. & Price, G. D. (1998). The diversity and coevolution of RuBisCo, plastids, pyrenoids, and chloroplast-based CO_2-concentrating mechanisms in algae. *Can. J. Bot.*, *76*, 1052-1071.

Beardall, J. & Raven, J.A. (2016). Carbon acquisition by microalgae. In: Borowitzka, Michael, A., Beardall, John, Raven, John, A., (eds.). *The Physiology of Microalgae*. Springer, pp. 89-99.

Bentley, F.K., García-Cerdán, J. G., Chen, H. C. & Melis, A. (2013). Paradigm of monoterpene (β-phellandrene) hydrocarbons production via photosynthesis in cyanobacteria. *Bio. Energy. Res.*, *6*, 917-929.

Bentley, F.K., Zurbriggen, A. & Melis, A. (2014). Heterologous expression of the mevalonic acid pathway in cyanobacteria enhances endogenous carbon partitioning to isoprene. *Molecular. Plant.*, *7*, 71-86.

Berberoglu, H., Gomez, P. S. & Pilon, L. (2009). Radiation characteristics of *Botryococcus braunii*, *Chlorococcum littorale*, and *Chlorella* sp. used for CO_2 fixation and biofuel production. *J. Radioanal. Nucl. Chem.*, *110*, 1879-1893.

Brennan, L. & Owende, P. (2010). Biofuels from microalgae: A review of technologies for production, processing, and extractions of biofuels and co-products. *Renew. Sust. Energ. Rev*, *14*, 557-577.

Chen, W., Zhang, S., Rong, J., Li, X., Chen, H., He, C. & Wang, Q. (2016). Effective Biological DeNOx of industrial flue gas by the mixotrophic cultivation of an oil-producing green alga *Chlorella* sp. C2. *Environ. Sci. Technol.*, *50*, 1620-1627.

Cheng, J., Huang, Y., Feng, J., Sun, J., Zhou, J. & Cen, K. (2013). Mutate *Chlorella* sp. by nuclear irradiation to fix high concentrations of CO_2. *Bioresour. Technol*, *136*, 496-501.

Cheng, L., Zhang, L., Chen, H. & Gao, C. (2006). Carbon dioxide removal from air by microalgae cultured in a membrane-photobioreactor. *Sep. Purif. Technol.*, *50*, 324-329.

Chiu, S. Y., Kao, C Y., Chen, C. H., Kuan, T. C., Ong, S. C.& Lin, C. S. (2008). Reduction of CO_2 by a high-density culture of *Chlorella* sp. in a semi continuous photobioreactor. *Bioresour. Technol.*, *99*, 3389-3396.

Chiu, S. Y., Kao, C. Y., Chen, C. H., Kuan, T. C., Ong, S. C. & Lin, C. S. (2008). Reduction of CO_2 by a high-density culture of *Chlorella* sp. in a semi continuous photobioreactor. *Bioresour. Technol., 99*, 3389-3396.

Chiu, S. Y., Kao, C. Y., Tsai, M. T., Ong, S. C., Chen, C. H. & Lin, C. S. (2009). Lipid accumulation and CO_2 utilization of *Nannochloropsis oculata* in response to CO_2 aeration. *Bioresour. Technol, 100*, 833-838.

Davis, R., Markham, J., Kinchin, C., Grundl, N., Tan, E. C. & Humbird, D. (2016). *Process design and economics for the production of algal biomass: algal biomass production in open pond systems and processing through dewatering for downstream conversion*. Golden: National Renewable Energy Laboratory

De Morais, M. G. & Costa, J. A. V. (2007). Biofixation of carbon dioxide by *Spirulina* sp. and *Scenedesmus obliquus* cultivated in a three-stage serial tubular photobioreactor. *Biotechnol. J., 129*, 439-445.

De Morais, M. G. & Costa, J. A. V. (2007). Isolation and selection of microalgae from coal fired thermoelectric power plant for bio fixation of carbon dioxide. *Energy. Convers. Manag, 48*, 2169-2173.

De Morais, M. G. & Costa, J. A. V. (2007). Carbon dioxide fixation by *Chlorella kessleri, C. vulgaris, Scenedesmus obliquus* and *Spirulina* sp. cultivated in flasks and vertical tubular photobioreactors. *Biotechnol. letters., 29*, 1349-1352

Dexter, J. & Fu, P. (2009). Metabolic engineering of cyanobacteria for ethanol production. *Energ. Environ. Sc., 2*, 857-864.

Eom, H., Lee, C. G. & Jin, E. (2006). Gene expression profile analysis in astaxanthin-induced *Haematococcus pluvialis* using a c-DNA micro-array. *Planta., 223*, 1231-1242.

Fan, L. H., Zhang, Y. T., Zhang, L. & Chen, H. L. (2008). Evaluation of a membrane-sparged helical tubular photobioreactor for carbon dioxide bio-fixation by *Chlorella vulgaris. J. Membr. Sci*, 336-345.

Giordano, M., Beardall, J. & Raven, J.A. (2005). CO_2 concentrating mechanisms in algae: mechanisms, environmental modulation, and evolution. *Annu. Rev. Plant Biol., 56*, 99-131.

Hallmann, A. (2007). Algal transgenics and biotechnology. *Transgenic. Plant. J., 1*, 81-98.

Hannon, M., Gimpel, J., Tran, M., Rasala, B. & Mayfield, S. (2010). Biofuels from algae: challenges and potential. *Biofuels, 1*, 763-784.

Hirokawa, Y., Maki, Y., Tatsuke, T. & Hanai, T. (2016). Cyanobacterial production of 1, 3-propanediol directly from carbon dioxide using a synthetic metabolic pathway. *Metab. Eng., 34*, 97-103.

Ho, S. H., Chen, W. M. & Chang, J. S. (2010). *Scenedesmus obliquus* CNW-N as a potential candidate for CO_2 mitigation and biodiesel production. *Bioresour. Technol., 101*, 8725-8730.

Huntley, M. E. & Redalje, D. G. (2007). CO_2 mitigation and renewable oil from photosynthetic microbes: a new appraisal. *Mitigation and adaptation strategies for global change.*, *12*, 573-608.

Jacob-Lopes, E., Lacerda, L. M. C. F. & Franco, T. T. (2008). Biomass production and carbon dioxide fixation by *Aphanothece microscopica Nageli* in a bubble column photobioreactor. *Biochem. Eng. J.*, *40*, 27-34.

Jacob-Lopes, E., Revah, S., Hernandez, S., Shirai, K. & Franco, T. T. (2009). Development of operational strategies to remove carbon dioxide in photobioreactors. *Chem. Eng. J.*, *153*, 120-126.

Jacob-Lopes, E., Scoparo, C. H. G., Lacerda, L. M. C. F. & Franco, T. T. (2009). Effect of light cycles (night/day) on CO_2 fixation and biomass production by microalgae in photobioreactors. Chemical engineering and Processing: *Process. Intensification.*, *48*, 306-310.

Jajesniak, P., Ali, H. E. M. O. & Wong, T. S. (2014). Carbon dioxide capture and utilization using biological systems: opportunities and challenges. *J. Bioprocess. Biotech.*, *4*, 1.

Jin, H. F., Lim, B. R. & Lee, K. (2006). Influence of nitrate feeding on carbon dioxide fixation by microalgae. *J. Environ. Sci. Health. A*, *41*, 2813-2824.

Kudoh, K., Kawano, Y., Hotta, S., Sekine, M., Watanabe, T. & Ihara, M. (2014). Prerequisite for highly efficient isoprenoid production by cyanobacteria discovered through the over-expression of 1-deoxy-d-xylulose 5-phosphate synthase and carbon allocation analysis. *J. Biosci. Bioeng.*, *118*, 20-28.

Kuramochi, T., Ramirez, A., Turkenburg, W. & Faaij, A. (2013). Techno-economic prospects for CO_2 capture from distributed energy systems. *Renew. Sust. Energ. Rev*, *19*, 328-347.

Kurano, N., Ikemoto, H., Miyashita, H., Hasegawa, T., Hata, H. & Miyachi, S. (1995). Fixation and utilization of carbon dioxide by microalgal photosynthesis. *Ener. Con Man.*, *36*, 689-692.

Lan, E. I. & Liao, J. C. (2011). Metabolic engineering of cyanobacteria for 1-butanol production from carbon dioxide. *Metabol. Engin.*, *13*, 353-363.

Lara-Gil, J. A., Alvarez, M. M. & Pacheco, A. (2014). Toxicity of flue gas components from cement plants in microalgae CO_2 mitigation systems. *J. Appl. Phycol.*, *26*, 357-368.

Leung, D. Y. C., Caramanna, G. & Maroto-Valer, M. M. (2014). An overview of current status of carbon dioxide capture and storage technologies. *Renew. Sust. Energ. Rev.*, *39*, 426-443.

Li, P., Pan, S. Y., Pei, S., Lin, Y. J. & Chiang, P. C. (2016). Challenges and perspectives on carbon fixation and utilization technologies: An overview. *Aerosol. Air. Qual. Res.*, *16*, 1327-1344.

Lopez, C. V. G., Fernández, F. G. A., Sevilla, J. M. F., Fernández, J. F. S., Garcia, M. C. C. & Grima, E. M. (2009). Utilization of the cyanobacteria *Anabaena* sp. ATCC 33047 in CO_2 removal processes. *Bioresour. Technol.*, *100*, 5904-5910.

MacCarthy, J., Murrells, T., Pang, Y., Passant, N., Ramirez Garcia, J., Thistlethwaite, G., Webb, N., Choudrie, S., Forden, S. & Gilhespy, S. (2016). *UK Greenhouse Gas Inventory*, 1990 to 2014.

Markou, G. & Nerantzis, E. (2013). Microalgae for high-value compounds and biofuels production: a review with focus on cultivation under stress conditions. *Biotechnology Advances.*, *31*, 1532-1542.

Mathieu, P. & Bolland, O. (2013). Comparison of costs for natural gas power generation with CO_2 capture. *Energy. Procedia.*, *37*, 2406-2419.

Matsushima, D., Jenke-Kodama, H., Sato, Y., Fukunaga, Y., Sumimoto, K., Kuzuyama, T., Matsunaga, S. & Okada, S. (2012). The single cellular green microalga *Botryococcus braunii*, race B possesses three distinct 1-deoxy-D-xylulose 5-phosphate synthases. *Plant. Science.*, *185*, 309-320.

Meinshausen, M., Meinshausen, N., Hare, W., Raper, S. C. B., Frieler, K., Knutti, R., Frame, D. J. & Allen, M. R. (2009). Greenhouse-Gas emission targets for limiting global warming to 2°C. *Nature.*, *458*, 1158-1162.

Morris, I. (1980). Paths of carbon assimilation in marine phytoplankton. In primary productivity in the sea. Springer, pp. 139-159,

Muradov, N. (2014). *Liberating energy from carbon: introduction to decarbonization.* Springer

Murakami, M. & Ikenouchi, M. (1997). The biological CO_2 fixation and utilization project by rite (2)- Screening and breeding of microalgae with high capability in fixing CO_2. *Energy. Convers. Manag.*, *38*, S 493-S497.

Mussgnug, J. H., Thomas-Hall, S., Rupprecht, J., Foo, A., Klassen, V., McDowall, A., Schenk, P. M., Kruse, O. & Hankamer, B. (2007). Engineering photosynthetic light capture: impacts on improved solar energy to biomass conversion. *Plant. Biotech. J.*, *5*, 802-814.

Naam, M. (2009). Minilateralism. *Foreign Policy*, 136.

Negoro, M., Shioji, N., Miyamoto, K. & Micira, Y. (1991). Growth of microalgae in high CO_2 gas and effects of SOx and NOx. *Appl. Biochem. Biotechnol.*, *28*, 877.

Oliver, J. W. K., Machado, I. M. P., Yoneda, H. & Atsumi, S. (2013). Cyanobacterial conversion of carbon dioxide to 2, 3-butanediol. *Proceedings of the National Academy of Sciences.*, *110*, 1249-1254.

Parry, M., Arnell, N., Hulme, M., Nicholls, R. & Livermore, M. (1998). Adapting to the inevitable. *Nature.*, *395*, 741-741.

Price, G. D. (2011). Inorganic carbon transporters of the cyanobacterial CO_2 concentrating mechanism. *Photosyn. Res.*, *109*, 47-57.

Price, G. D., Badger, M. R., Woodger, F. J. & Long, B. M. (2008). Advances in understanding the cyanobacterial CO_2-concentrating-mechanism (CCM): functional components, Ci transporters, diversity, genetic regulation and prospects for engineering into plants. *J. Exp. Bot.*, *59*, 1441-1461.

Pulz, O. & Gross, W. (2004). Valuable products from biotechnology of microalgae. *Appl. Microbiol. Biotechnol.*, *65*, 635-648.

Rahman, M.A., Sinha, S., Sachan, S., Kumar, G., Singh, S. K. & Sundaram, S. (2014). Analysis of proteins involved in the production of MAAx³ s in two Cyanobacteria Synechocystis PCC 6803 and *Anabaena cylindrica*. *Bioinformation.*, *10*, 449.

Ramos, A. A., Marques, A. R., Rodrigues, M., Henriques, N., Baumgartner, A., Castilho, R., Brenig, B. & Varela, J. C. (2009). Molecular and functional characterization of a cDNA encoding 4-hydroxy-3-methylbut-2-enyl diphosphate reductase from *Dunaliella salina*. *J. Plant. Physiol.*, *166*, 968-977.

Reddy, K. J. & Argyle, M. D. (2010). *Apparatus and method to sequester contaminants.* (eds). Google Patents, USA.

Ryu, H. J., Oh, K. K. & Kim, Y. S. (2009). Optimization of the influential factors for the improvement of CO_2 utilization efficiency and CO_2 mass transfer rate. *Ind. Eng. Chem. Res.*, *15*, 471-475.

Sakai, N., Sakamoto, Y., Kishimoto, N., Chihara, M. & Karube, I. (1995). *Chlorella* strains from hot springs tolerant to high temperature and high CO_2. *Energy. Conversion. Manag*, *36*, 693-696.

Sakai, N., Sakamoto, Y., Kishimoto, N., Chihara, M. & Karube, I. (1995). *Chlorella* strains from hot springs tolerant to high temperature and high CO_2. *Energy. Convers. Manag.*, *36*, 693-696

Scaife, M. A. & Smith, A. G. (2016). Towards developing algal synthetic biology. *Biochem. Soc. Trans*, *44*, 716-722.

Scragg, A. H., Illman, A. M., Carden, A. & Shales, S. W. (2002). Growth of microalgae with increased calorific values in a tubular bioreactor. *Biomass. Bioener.*, *23*, 67-73.

Singh, S. K., Bansal, A., Jha, M. K. & Dey, A. (2012). An integrated approach to remove Cr (VI) using immobilized *Chlorella minutissima* grown in nutrient rich sewage wastewater. *Bioresour. Technol*, *104*, 257-265.

Singh, S.K., Bansal, A., Jha, M. K. & Jain, R. (2013). Production of biodiesel from wastewater grown *Chlorella minutissima*. *Indian. J. Chem. Technol*, *20*, 341-345.

Singh, S. K., Dixit, K. & Sundaram, S. (2014b). Effect of acidic and basic pretreatment of wild algal biomass on Cr (VI) biosorption. *J. Environ. Sci, Toxicol. Food. Technol*, *8*, 38-41.

Singh, S. K., Rahman, A., Dixit, K., Nath, A. & Sundaram, S. (2016). Evaluation of promising algal strains for sustainable exploitation coupled with CO_2 fixation. *Environ. Technol.*, *37*, 613-622.

Singh, S. K., Sundaram, S. & Kishor, K. (2014a). *Photosynthetic micro-organisms: Mechanism for carbon concentration*. Springer.

Singh, S. K., Sundaram, S., Sinha, S., Rahman, M. A. & Kapur, S. (2016) Recent advances in CO_2 uptake and fixation mechanism of cyanobacteria and microalgae. *Crit. Rev. Environ. Sci. Technol.*, *46*, 1297-1323.

Skjanes, K., Lindblad, P. & Muller, J. (2007) Bio CO_2: A multi-disciplinary, biological approach using solar energy to capture CO_2 while producing H_2 and high value products. *Biomolecular. Engineering.*, *24*, 405-413.

Slade, R. & Bauen, A. (2013). Micro-algae cultivation for biofuels: cost, energy balance, environmental impacts and future prospects. *Biomass. Bioenergy.*, *53*, 29-38.

Solovchenko, A. & Khozin-Goldberg, I. (2013). High-CO_2 tolerance in microalgae: possible mechanisms and implications for biotechnology and bioremediation. *Biotechnol. Lett.*, *35*, 1745-1752.

Stephenson, P. G., Moore, C. M., Terry, M. J., Zubkov, M. V. & Bibby, T. S. (2011). Improving photosynthesis for algal biofuels: toward a green revolution. *Trends. Biotechno.*, *29*, 615-623.

Sung, K. D., Lee, J. S., Shin, C. S., Park, S. C. & Choi, M. J. (1999). CO_2 fixation by *Chlorella* sp. KR-1 and its cultural characteristics. *Bioresour. Technol*, *68*, 269-273.

Sydney, E. B., Sturm, W., de Carvalho, J. C., Thomaz-Soccol, V., Larroche, C., Pandey, A. & Soccol, C. R. (2010). Potential carbon dioxide fixation by industrially important microalgae. *Bioresour. Technol*, *101*, 5892-5896.

USDOE. (July 2000). Carbon dioxide emissions from the generation of electric power in the United States. Washington, DC: U.S. Department of Energy (USDOE) and U.S. Environmental Protection Agency.

Yoo, C., Jun, S. Y., Lee, J. Y., Ahn, C. Y. & Oh, H. M. (2010). Selection of microalgae for lipid production under high levels carbon dioxide. *Bioresour. Technol*, *101*, S71-S74.

Zeiler, K. G., Heacox, D. A., Toon, S. T., Kadam, K. L. & Brown, L. M. (1995). The use of microalgae for assimilation and utilization of carbon dioxide from fossil fuel-fired power plant flue gas. *Energy. Convers. Manag.*, *36*, 707-712.

In: The Role of Photosynthetic Microbes…
Editors: K. Tripathi, N. Kumar et al.
ISBN: 978-1-53614-032-3

Chapter 6

CYANOBACTERIA: A NEW TERMINUS FOR ANTI-INFECTIOUS AGENTS

Sachin Tyagi, Preeti Singh and Rahul Kunwar Singh*
Department of Botany and Microbiology, HNB Garhwal University, Srinagar Garhwal, Uttarakhand, India

ABSTRACT

Cyanobacteria, the oxygenic photoautotrophic gram-negative bacteria, are one of the most primitive organisms on the Earth. These organisms were earlier known for their ecological and agricultural significance only. For last three decades, the scientists are being attracted to screen these organisms for different bioactivities due to their successful adaptation in various extreme conditions. Consequently, the metabolites of these organisms have not disappointed the world and exhibited diverse bioactivities. This chapter focuses on antimicrobial compounds isolated from cyanobacteria, and the bioinformatics tools being used for prediction of the novel compound structure using available genome sequences.

Keywords: bioinformatics tools, cyanobacteria, antibacterial, antiviral, antifungal, anti-infectious agent

* Corresponding Author Email: rksingh.hnb@gmail.com.

INTRODUCTION

Advancement in medical sciences along with leading integrated system biology research during the twentieth century has remarkably increased the average longevity of human being. However, the emergence of new infectious diseases and rapid development of drug resistance in pathogens of already existing diseases due to changes in lifestyle and other environmental factors are continuously threatening the human population worldwide. Recently, World Health Organization has made global priority pathogens list (GPPL) and felt an urgent need for new strategies on discovery and development of new antibiotics specifically against MDR (multi-drug resistant) and EDR (extensive drug-resistant) gram-negative bacteria.

On the other hand, the number of new therapeutic entities coming into the market is very low and the expenditure incurred on bringing a new drug to market is around 2.6 billion USD even after the technological advances as well as genomics and bioinformatics predictions (US FDA, 2015). Thus, we are left with fewer treatment options against infectious diseases as we are fast approaching the postantibiotic era (Kharstrom, 2013). Hence, the search for antimicrobial compounds from cost-effective sources is the urgent need for treating the yet incurable infectious diseases.

Natural sources provide a significant source to find new compounds with therapeutic value since ancient times. The products from these sources and their derivatives constitute 42% part of new drugs, discovered during 1981 to 2014, as they have negligible issues regarding permeability and toxicity as compared to complex synthetic compounds. Recently, the Nobel Prize for Physiology or Medicine in 2015 was awarded for the discovery and development of two natural products, avermectin/ ivermectin, and artemisinin. Among natural products, a significant number of drugs/leads are actually produced by microbes and/or microbial interactions with the "host from whom it was isolated" (Newmann and Cragg, 2015). During the golden era of microbiology, Fungi and Streptomyces were in focus of the pharmaceutical research as major sources of new antibiotics. However, the pharmaceutical industries had declined their interest in these sources due to approximately 95% rediscovery rate of already known compounds in last two decades of the twentieth century. Therefore, scientists have turned towards the new and unexplored microbial sources like cyanobacteria for the discovery of new antibiotics.

CYANOBACTERIA

Cyanobacteria, a unique group of photoautotrophic prokaryotes, are widely distributed in diverse habitats. In last three decades, these organisms have emerged as a potential source in providing new pharmaceutical agents. Earlier, these organisms were known only for their ecological and agricultural significance. Cyanobacteria have evolved about 3.5 billion years ago and are among one of the oldest known organism on earth. The idea that these organisms must have adapted to different stress conditions through the synthesis of various bioactive compounds during its evolutionary history has attracted the world to screen these organisms and their metabolites for different bioactivities. The field was initiated by Prof. WH Gerwick and Prof. RE Moore in 1980s and resulted in about 2000 new chemical entities with a broad spectrum of activities such as antibacterial, antiprotozoal, antifungal, antiviral and antitumor within the period of three decades. The high rate of novel natural compound isolation from cyanobacteria, its economic cultivation, and the presence of NRPS and PKS gene clusters in large number in its genome are the major factors responsible for making these organisms a potential source of pharmaceuticals (Singh et al., 2011).

NRPS/PKS Gene Clusters in Cyanobacteria

Natural products are actually the low molecular weight organic compounds synthesized as secondary metabolites in response to stress. A majority of cyanobacterial natural products are nonribosomal peptides or polyketides or the hybrid of two. Nonribosomal peptides (NRPs) and polyketides (PKs) are special classes of secondary metabolites with great structural diversity. Microbial syntheses of these metabolites are catalyzed by enzymes, nonribosomal peptide synthetases (NRPSs) and polyketide synthases (PKSs), respectively (Marahiel et al., 1997). These multifunctional enzymes are characterized by a modular structure, containing highly conserved regions, specifically involved in the activation and condensation of amino acids. Sometimes, the backbones of some NRPs contain acetate or propionate units, which are integrated by PKSs. Insertion of peptide and acetate or propionate units in a single molecule further enhances the diversity of these metabolites (Schwarzer and Marahiel, 2001; Du and Shen 2001). The modular structure of NRPS enzymes allows the artificial alterations in early stages of NRP biosynthetic pathways leading to a new area of research (Amoutzias et al., 2016, Singh et al., 2012). The biosynthetic gene clusters of various nonribosomal peptides and polyketides, produced by cyanobacteria, have been identified using the cloning and sequencing approaches (Chang et al., 2002; Christiansen et al., 2003; Hoffman et al., 2003, Becker et al., 2004). The PCR based screening for the presence of NRPS and PKS gene clusters in cyanobacteria has also been performed by several workers (Christiansen et al., 2001; Ehrenreich et al., 2005; Barrios-Llerna et al., 2007; Singh et al., 2015). These studies reveal the wide distribution of NRPS and PKS gene clusters throughout all the cyanobacterial orders in 63-75% of tested strains.

Table 1. List of few antibacterial compounds isolated from cyanobacteria in 21st century

S.N	Compound	Origin Source	Chemical nature	Active against	MIC/EC	Reference
1.	Aeruginazole A	*Microcystis* sp.	Cyclic peptide	*B. subtilis*	2.2 μM	Raveh et al., (2010)
2.	Ambigol A–C	*Fischerella ambigua*	Polychlorinatd polyaromatic phenols	*B. megaterium*	8mm@ 100 nmol	Wright et al., (2005)
3	Ambiguine isonitriles	*Fischerella* sp.	Alkaloids	*E. coli* *S. aureus* *B. subtilis* *C. albicans*	6 μM 0.2 μM 0.8 μM 1.0 μM	Mo et al., (2009)
4.	Bromoana-indolone	*Anabaena constricta*	Alkaloid	*B. cereus*	530 μM	Volk et al., (2009)
5.	Carbamidocyclop hanes	*Nostoc sp*	Paracyclophan-es	*M.tuberculosi* *S. aureus* *E. faecalis* *S. pneumonia*	0.8–5.4 μM 0.1-100 μM 0.2–1.1 μM 0.2–2 μM	Preisitsch et al., (2014)
6.	Crossbyanol A–C	*Leptolyn-gbya crossbyna*	Polybrominatd polyaromatic phenols	*S. aureus*	3 μM	Choi et al., (2010)
7.	Eucapsitrione	*Eucapsis sp*	Anthraquinone	*M. tuberculosis*	3–6 μM	Sturdy et al., (2010)
8.	Hapalindoles	*Hapalo-siphon fontinalis; Fischerella* sp., *Westelli-opsis* sp.	Indole alkaloids	*S. aureus* *B. subtilis* *E. coli*	36mm 30mm 31mm	Asthana et al., (2006)
9.	Lyngby-azothris	*Lyngbya* sp.	cyclic peptides	*B. subtilis* *E. coli*	18mm@16μ mol 18mm@ 65 μmol	Zainuddin et al., (2009)
10.	Norabietanes	*Microco-leous lacustris*	Diterpenes	*S. aureus* *S. epidermidis* *S. typhi* *V. cholera*	45 μM 55 μM 150 μM 850 μM	Pérez et al., (2008)
11.	Pitipeptolides A–F	*Lyngbyama-juscula/ Moorea producens*	cyclic depsipeptides	*M. tuberculosis*	0–30mm@ 60 μmol	Montaser et al., (2011)
12.	Scytoscalarol	*Scytonema* sp	Sesterterpene	*B. anthracis* *S. aureus* *E. coli* *M. tuberculosis*	6 Mm 2 μM 30 μM 110 μM	Mo et al., (2009)

(MIC=Minimum Inhibitory Concentration; EC= Effective Concentration).

ANTIBACTERIAL COMPOUNDS

The rapid and continuous development of drug resistance in bacterial pathogens is the major hurdle in the eradication of several infectious diseases like tuberculosis. Of late, the ESKAPE pathogens, (*Enterococcus faecium, Staphylococcus aureus, Klebsiella pneumoniae, Acinetobacter baumannii, Pseudomonas aeruginosa*, and *Enterobacter* spp.), capable of escaping the toxic actions of antibiotics are alarming the world towards the end of the antibiotic era and posing therapeutic challenges (Pendleton et al., 2013).

Hence scientists are screening the new sources including cyanobacteria for the discovery of new antibacterial compounds.

Several antibacterial compounds have been isolated from cyanobacteria to date. Some of these compounds have their minimum inhibitory concentration (MIC) values comparable to those of commercially available standard antibiotics. Some of these metabolites like hapalindoles, noscomin, scytoscalarol and eucapsitrione exhibit direct antibacterial effects whereas others like tumonoic acids and malyngamide C exhibit quorum sensing inhibition activities and hence may be helpful in reducing the bacterial infections (Abed et al., 2013). However, none of these compounds could reach the clinical trial stage. Table 1 summarizes the list of antibacterial compounds isolated from cyanobacteria during the 21st century.

ANTIFUNGAL COMPOUNDS

The use of antifungal drugs is leading the development of drug resistance in almost all fungal pathogens including *Candida* and *Aspergillus* species. In the recent years, multidrug resistance has emerged in fungal pathogens including different *Candida* species. On the other hand, there are only five major classes of antifungal drugs available to treat fungal infections and the treatment options are limited. This situation is alluring for the search of new antifungal compounds (Sanglard, 2016). Several cyanobacterial secondary metabolites with antifungal activity have been reported in last two decades. These compounds are active against various fungal pathogens like *Candida albicans, Aspergillus oryzae* etc, with the MIC values ranging from 1.5 to 10.5 μM. Chemical nature of these compounds are low molecular weight nonribosomal peptide, polyketides, or alkaloids. The list of few antifungal compounds isolated in recent past has been given in Table 2. Laxaphycins are well known antifungal compounds isolated from *Anabaena laxa*, effective against *Aspergillus oryzae* (Frankmölle et al., 1992). Hassallidins are a group of antifungal compounds, originally isolated from *Tolypothix,* later found to be widespread among various filamentous cyanobacteria, and are effective against *C. albicans* (Vestola et al., 2014).

ANTIVIRAL METABOLITES

The viral diseases like Dengue and AIDS are having dreadful consequences. In addition, the new emerging viruses like Ebola and Zika are making the situation more pathetic throughout the world. Hence, novel chemicals having potential antiviral

Table 2. List of few antifungal compounds reported from cyanobacteria in the 21st century

S.N	Compound	Origin sources	Chemical nature	Active against	MIC/EC	References
1.	Ambiguine isonitriles	*Fischerella* sp.	Alkaloids	*C. albicans*	1.0 µM	Mo et al., (2009)
2.	Balticidins	*Anabaena cylindrica*	Glycosylat ed. Lipo- peptide	*C. maltosa*	6 nmol	Bui et al., (2014)
3.	Hapalindoles	*Hapalosiphon fontinalis*	Indole alkaloids	*C. albicans*	0.7 µM	Kim et al., (2012)
4.	Hassallidins A and B	*Hassallia* sp.; widely spread among filamentous cyanobacteria	Glycosylated lipopeptide	*A. fumigatus* *C. albicans*	3.5 µM	Neuhof et al., (2006)
5.	Laxaphycins	*Anabaena laxa, Moorea producens, Anabaena Torulosa*	cyclic peptides	*A. oryzae*	20 µM	Bonnard et al., (2007)
6.	Lobocycl amide A–D	*Lyngbya confervoides*	Cyclic lipo-peptides	*C.* albicans	10 µM	MacMillan et al., (2002)
7.	Scytoscalarol	*Scytonema* sp.	sesterterpene	*C.* albicans	4 µM	Mo et al., (2009)

(MIC = Minimum Inhibitory Concentration; EC = Effective Concentration)

activities are urgently needed to tackle the situation. During last three decades, several compounds with antiviral activities have been isolated from cyanobacteria. These antiviral compounds, on the basis of their chemical structures, may be classified into three groups namely, sulfoglycolipids, lectins, and polysaccharides.

Sulfoglycolipids with antiviral activities have been purified from genera *Lyngbya, Phormidium, Scytonema, Anabaena, Calothrix*, and *Oscillatoria*. The long chain fatty acids of these sulfoglycolipids are essential for its activity. These compounds inhibit the HIV infection via the restriction over DNA polymerase activity of the virus. Lectins are another antiviral class of compounds produced by cyanobacteria and are found to be active against a variety of organisms. Cyanovirin N, a 101 amino-acid residue peptide, is the first natural anti-HIV compound isolated from *Nostoc ellipsosporum*. The compound is being developed as a vaginal gel for inhibiting the sexual transmission of HIV by Cellegy Pharmaceuticals, San Francisco. This compound is found to be active against influenza virus, respiratory syncytial virus, enteric virus and coronaviruses (Boyd et al., 1997). Besides, microvirin and scytovirin, isolated from *Microcystis aeruginosa* and *Scytonema varium,* respectively, are the other lectins having anti-HIV activity. Nostoflan is a complex acidic polysaccharide isolated from *Nostoc flagelliforme* exhibiting broad-spectrum antiviral activity against HSV-1 & 2, human cytomegalovirus, and influenza A virus (Kanekiyo et al., 2005). Recently, Gupta et al., (2014) have isolated three new compounds from *Trichodesmium erythraeum.* These compounds, named as aplysiatoxins, are found to be active against chikungunya virus and do not fall into any of the three categories mentioned above.

Genome Driven Computational Approach

Nowadays, advancement in molecular biology and genomics research has led to computation of whole genome sequence of several organisms. The analyses of these whole genome sequences using bioinformatics have enabled us to characterize several unknown natural products, which can be produced by particular organisms. This technique is known as computational genome mining. The said technology is a computer-based assistance to make strategies in natural product research, which require minimum laboratory resources and reduces the number of hit and trial experiment. However, this requires extensive expertise in the field. Although this approach is more theoretical than the actual findings, this is able to fill the gap between experimental investigations and theoretical resources. In this approach, generally new biosynthetic gene clusters encoding natural products are identified and the characteristics of the encoded products are predicted. Genome-based methods use several databases, tools, and software to identify and characterize the secondary metabolites and their biosynthetic gene clusters.

Shih et al., (2013) have compared the whole genome sequences of 54 cyanobacterial strains and identified 21,000 new proteins which indicate the diversity of secondary metabolite biosynthesis gene clusters. The data of this finding was referred to as "CyanoGEBA" dataset. Any wet lab experiment to determine the bioactivities of metabolites is time-consuming. The list of few tools and software used for prediction of metabolites is given in Table-3. Although the biosynthesis of a metabolite is influenced not only by the gene sequence but also by several other factors, such as activators and repressors of particular gene/gene-clusters, the optimum growth conditions, and post-transcriptional modification. The genome mining technology is very helpful for selection and screening of potential strains to avoid excessive work by the researcher. Another outcome of this screening is the possibility of revealing previously silent and unidentified gene clusters.

Virtually in most cases, the prediction of natural compounds with bioinformatics tools find several natural compounds which are previously unidentified, thus they have to be isolated. Several biosynthetic pathways have been identified in cyanobacterial genomes using genome mining tools. These pathways are able to produce various novel metabolites theoretically, which are still to be isolated in the real-time situation. Isolation and characterization of these predicted metabolites is a big challenge. Recent researches in drug discovery may use combined or integrated system biology efforts which leads to great success in compression of individual approach. Wase and Wright (2008) systematically explained the integrated use of various approaches for natural product drug discovery from cyanobacteria.

Table 3. Databases and tools used for metabolites search and prediction

S.N.	Name of databases	Website and link for access
Databases of gene clusters:		
1.	ClusterMine360	http://www.clustermine360.ca/
2.	ClustScan Database	http://csdb.bioserv.pbf.hr/csdb/ClustScanWeb.html
3.	DoBISCUIT	http://www.bio.nite.go.jp/pks/
4.	MIBiG	http://mibig.secondarymetabolites.org
5.	Recombinant ClustScan Database	http://csdb.bioserv.pbf.hr/csdb/RCSDB.html
Software for predicting substrate specificities:		
1.	NRPSpredictor/NRPS predictor2	http://nrps.informatik.uni-tuebingen.de
2.	SEQL-NRPS	https://services.birc.au.dk/seql-nrps/
3.	LSI-based A-domain function predictor	http://bioserv7.bioinfo.pbf.hr/LSIpredictor/AdomainPrediction.jsp
4.	NRPS/PKS substrate predictor	http://www.cmbi.ru.nl/NRPS-PKS-substrate-predictor/
5.	PKS/NRPSWeb Server/Predictive Blast Server	http://nrps.igs.umaryland.edu/nrps/
Tools for mining of secondary metabolite gene clusters:		
1.	2metDB	http://secmetdb.sourceforge.net/
2.	antiSMASH	http://antismash.secondarymetabolites.org
3.	CLUSEAN	https://bitbucket.org/tilmweber/clusean
4.	ClusterFinder	https://github.com/petercim/ClusterFinder
5.	NaPDoS	http://napdos.ucsd.edu/
6.	SMURF	http://jcvi.org/smurf/index.php
7.	GNP/Genome Search	http://magarveylab.ca/gnp/#!/genome
8.	GNP/PRISM	http://magarveylab.ca/prism
Software for analysis of type I PKS and NRPS pathways:		
1.	ClustScan Professional	http://bioserv.pbf.hr/cms/index.php?page=clustscan
2.	NP.searcher	http://dna.sherman.lsi.umich.edu/
3.	NRPS-PKS/SBSPKS	http://www.nii.ac.in/~pksdb/sbspks/master.html
4.	SEARCHPKS	http://linux1.nii.res.in/~pksdb/DBASE/pagesearchpks.html
Databases of bioactive compounds:		
1.	ChEBI	https://www.ebi.ac.uk/chebi/
2.	ChEMBL	https://www.ebi.ac.uk/chembl/
3.	Antibioticome	http://magarveylab.ca/antibioticome
4.	ChemSpider	http://www.chemspider.com/
5.	Novel Antibiotics Database	http://www.antibiotics.or.jp/journal/database/database-top.htm
6.	PubChem	http://pubchem.ncbi.nlm.nih.gov/
7.	NORINE	http://bioinfo.lifl.fr/norine
High-throughput metabolic modeling tools:		
1.	antiSMASH	http://www.secondarymetabolites.org
2.	FAME	http://f-a-m-e.fame-vu.vm.surfsara.nl/ajax/page1.php
3.	Merlin	http://www.merlin-sysbio.org/
4.	CoReCo	https://github.com/esaskar/CoReCo
5.	MetaFlux in Pathway Tools	http://bioinformatics.ai.sri.com/ptools/

(Sources: Weber et al., 2016, Ziemert et al., 2016, Trosset et al., 2015)

CONCLUSION

Cyanobacteria, one of the oldest known organisms on the Earth, are serving as a source of novel bioactive compounds for last three decades. Till date, about 2000 new compounds with different pharmaceutical activities have been isolated from these organisms. A majority of these bioactive metabolites are endo-metabolites, i.e., they have been extracted from the biomass. The investigations on exo-metabolites may further increase the number of bioactive metabolites from these organisms. Another reason behind the use of cyanobacteria as a source of pharmaceutical agents is their cultivation in simple inorganic media. With the advancement of genomics and bioinformatics tools, the NRPS and PKS gene clusters have been identified in cyanobacterial genomes. Of late, they have been predicted to have more potential of producing potent pharmaceutical agents biosynthesized by these gene clusters, which are still to be isolated and characterized. Since these organisms are known to produce the pharmaceutical compounds in response to different stress factors, cyanobacterial diversity from different untouched extreme habitats must also be explored.

Besides the great potential and advantages as a drug source, these organisms have few hurdles in their exploration. It is very tough to maintain these organisms in axenic culture due to contamination of heterotrophic bacteria. Further, due to the phototrophic mode of nutrition, the cultivation cost of these organisms becomes higher. Harvesting of cyanobacterial culture is also difficult due to its small size. Moreover, the slow growth rate and longer generation time make the cultivation process of cyanobacteria time-consuming. Due to these few hurdles in cultivation, very few cyanobacterial compounds have entered clinical trials and no one has been approved by the FDA. In our opinion, the research in the field of cyanobacterial pharmaceutics is still in its infancy and demands more scientific consideration with interdisciplinary approaches.

ACKNOWLEDGMENTS

ST and PS are thankful for providing research fellowships provided by UGC, New Delhi. RKS is also thankful to UGC, New Delhi for financial assistance in form of Start-up grant.

CONFLICT OF INTEREST

No conflict of interest among the authors.

REFERENCES

Abed, R. M. M. (2013). Quorum-sensing inhibitory compounds from extremophilic microorganisms isolated from a hypersaline cyanobacterial mat. *J. Indus. Microbiol. Biotechnol*, 40(7): 759-772.

Amoutzias, G., Chaliotis, A., and Mossialos, D. (2016). Discovery strategies of bioactive compounds synthesized by nonribosomal peptide synthetases and type-I polyketide synthases derived from marine microbiomes. *Marine Drugs*, 14(4): 80-99.

Asthana, R. K., Srivastava, A., Singh, A. P., Singh, S. P., Nath, G., Srivastava, R., and Srivastava, B. S. (2006). Identification of an antimicrobial entity from the cyanobacterium *Fischerella* sp. isolated from bark of *Azadirachta indica* (Neem) tree. *J. Appl. Phycol*, 18: 33-39.

Barrios-Llerena, M. E., Burja, A. M., and Wright, P. C. (2007). Genetic analysis of polyketide synthase and peptide synthetase genes in cyanobacteria as a mining tool for secondary metabolites. *J. Indus. Microbiol. Biotechnol,* 34(6): 443-456.

Becker, J. E., Moore, R. E., Moore, and B. S. (2004). Cloning, sequencing, and biochemical characterization of the nostocyclopeptide biosynthetic gene cluster: molecular basis for imine macrocyclization. *Gene,* 325: 35-42.

Bonnard, I., Rolland, M., Salmon, J. M., Debiton, E., Barthomeuf C, and Banaigs B. (2007). Total structure and inhibition of tumour cell proliferation of laxaphycins. *J. Med. Chem*, 50: 1266-1279.

Boyd, M. R., Gustafson, K. R., McMahon, J. B., Shoemaker, R. H., et al. (1997). Discovery of cyanovirin-N, a novel human immunodeficiency virus-inactivating protein that binds viral surface envelope glyco-protein gp120: potential applications to microbicide development. *Antimicrob. Agents. Chemother*, 41: 1521-1530.

Bui, T. H., Wray, V., Nimtz, M., Fossen, T., Preisitsch, M., Schröder, G., Wende, K., and Heiden, S. E., Mundt, S. (2014). Balticidins A–D, antifungal hassallidin-like lipopeptides from the Baltic Sea cyanobacterium *Anabaena cylindrical* Bio33. *J. Nat. Prod,* 77: 1287-1296.

Chang, Z., Flatt, P., Gerwick, W. H., Nguyen, V., Willis, C. L., and Sherman, D. H. (2002). The barbamide biosynthetic gene cluster: a novel marine cyanobacterial system of mixed polyketide synthase (PKS)-non-ribosomal peptide synthetase (NRPS) origin involving an unusual trichloroleucyl starter unit. *Gene*, 296: 235–247.

Choi, H., Engene, N., Smith, J. E., Preskitt, L. B., and Gerwick, W. H. (2010). Crossbyanols A–D, toxic brominated polyphenyl ethers from the Hawaiian bloom-forming Cyanobacterium *Leptolyngbya crossbyana*. *J. Nat. Prod,* 73: 517-522.

Christiansen, G., Dittmann, E., Ordorika, L., Rippka, R., Herdman, M., and Borner, T. (2001). Nonribosomal peptide synthetase genes occur in most cyanobacterial genera as evidenced by their distribution in axenic strains of the PCC. *Arch. Microbiol,* 176: 452-458.

Christiansen, G., Fastner, J., Erhard, M., Borner, T., and Dittmann, E. (2003). Microcystin biosynthesis in *Planktothrix*: genes, evolution, and manipulation. *J. Bacteriol*, 185: 564-572.

Du, L., and Shen, B. (2001). Biosynthesis of hybrid peptide-polyketide natural products. *Curr Opin Drug Dis Dev,* 4: 215-228.

Ehernich, I. M., Waterbury, J. B, and Webb, E. A. (2005). Distribution and diversity of natural product genes in marine and freshwater cyanobacterial cultures and genomes. *App. Env. Microbiol,* 71(11): 7401-7413.

Frankmölle, W. P., Knübel, G., Moore, R. E., and Patterson, G. M. L. (1992). Antifungal cyclic peptides from the terrestrial blue-green alga *Anabaena laxa*. II. Structures of laxaphycins A, B, C, D and E. *J. Antibiot,* 45:1458–1466.

Gupta, D. et al. (2014). Anti-Chikungunya Viral Activities of Aplysiatoxin-Related Compounds from the Marine Cyanobacterium *Trichodesmium erythraeum. Marine Drugs*, 12(1): 115-127.

Hoffmann, D., Hevel, J. M., Moore, R. E., and Moore, B.S. (2003). Sequence analysis and biochemical characterization of the nosto-peptolide A biosynthetic gene cluster from *Nostoc* sp. GSV224. *Gene* 311: 171-180.

Kahrstrom, C. T. (2013). Entering a post-antibiotic era? *Nature Reviews Microbiology* 11: 146-146.

Kanekiyo, K., Lee, J.B., Hayashi, K., Takenaka, H., Hayakawa, Y., Endo, S., and Hayashi, T. (2005). Isolation of an antiviral polysaccharide, nostoflan, from a terrestrial cyanobacterium, *Nostoc flagelliforme. J. Nat. Prod*, 68:1037-1041.

Kim, H., Lantvit, D., Hwang, C. H., Kroll, D. J., Swanson, S. M., Franzblau, S. G., and Orjala, J. (2012). Indole alkaloids from two cultured cyanobacteria, *Westiellopsis* sp. and *Fischerella muscicola. Bioorg. Med. Chem*, 20: 5290-5295.

MacMillan, J. B., Ernst-Russell, M. A., deRopp, J. S., and Molinski, T. F. (2002). Lobocyclamides A–C, lipopeptides from a cryptic cyano-bacterial mat containing *Lyngbya confervoides. J. Org. Chem*, 67: 8210- 8215.

Marahiel, M. A., Stachelhaus, T., and Mootz, H. D. (1997). Modular peptide synthetases involved in non-ribosomal peptide synthesis. *Chem. Rev,* 97: 2651-2673.

Mo, S., Krunic, A., Chlipala, G., Orjala, J. (2009). Antimicrobial ambiguine isonitriles from the cyanobacterium *Fischerella ambigua. J. Nat. Prod*, 72: 894-899.

Mo, S., Krunic, A., Pegan, S.D., Franzblau, S. G., Orjala, J., (2009). An antimicrobial guanidine-bearing sesterterpene from the cultured cyanobacterium *Scytonema* sp. *J Nat Prod*, 72: 2043-2045.

Montaser, R., Paul, V. J., Luesch, H. (2011). Pitipeptolides C–F, anti-mycobacterial cyclodepsipeptides from the marine cyanobacterium *Lyngbya majuscule* from Guam. *Phytochem,* 72: 2068-2074.

Neuhof, T., Schmieder, P., Preussel, K., Dieckmann, R., Pham, H., Bartl, F., and von Döhren, H. (2005). Hassallidin A, a glycosylated lipopeptide with antifungal activity from the cyanobacterium *Hassallia sp. J. Nat. Prod,* 68: 695-700.

Newman, D. J., and Cragg, G. M. (2015). Natural products as sources of new drugs from 1981 to 2014. *J. Nat. Prod*, 79(3): 629-661.

Niedermeyer, T. (2015). Anti-infective Natural Products from Cyano-bacteria. *Planta Med*, 81(15): 1309-1325.

Pendleton, J. N., Gorman, S. P., and Gilmore, B. F. (2013). Clinical relevance of the ESKAPE pathogens. *Exp. Rev. Anti. Infect. Ther,* 11(3): 297-308.

Pérez Gutiérrez, R. M., Martinéz, Flores. A., Vargas Solís, R., Carmona, Jiminez. J. (2008). Two new antibacterial norabietane diterpenoids from cyanobacteria, *Microcoleous lacustris. J. Nat. Med*, 62: 328-331.

Preisitsch, M., Harmrolfs, K., Pham, H. T., Heiden, S. E., Füssel, A. et al. (2014). Anti-MRSA-acting carbamidocyclophanes H–L from the Vietnamese cyanobacterium *Nostoc* sp. CAVN2. *J. Antibiot*, 68: 165-177.

Raveh, A., Carmeli, S. (2010). Aeruginazole A, a novel thiazole-containing cyclopeptide from the cyanobacterium *Microcystis* sp. *Org. Lett*, 12: 3536-3539.

Sanglard, D. (2016). Emerging threats in antifungal resistant fungal pathogens. *Front. Med,* 3(11), doi: 10.3389/fmed.2016.00011.

Schwarzer, D., and Marahiel, M. A. (2001) Multimodular biocatalysts for natural product assembly. *Naturwissenschaften,* 88: 93-101.

Singh, R. K., Singh, P., Sharma, R., Mohapatra, T. M. (2012). Nonribosomal peptide synthesis in microbes. In: *Recent Advances in Microbiology* Volume 1. Eds. Singh R. K., Sharma R, Tiwari S. P. Nova Science Publishers, Inc. New York, USA.

Singh, R. K., Tiwari, S. P., and Mohapatra, T. M. (2015). Characterization of cyanobacteria with nonribosomal peptide synthetase gene clusters from tropical environment of Varanasi city, India. *Trends Biosci*, 8(18): 4841-4847.

Singh, R. K., Tiwari S. P., Rai, A. K., and Mohapatra, T. M. (2011). Cyanobacteria: an emerging source for drug discovery. *J. Antibiot*, 64(6): 401–412.

Sturdy, M., Krunic, A., Cho, S., Franzblau, S., and Orjala, J. (2010). Eucapsitrione, an anti-Mycobacterium tuberculosis anthraquinone derivative from the cultured freshwater cyanobacterium *Eucapsis* sp. *J. Nat. Prod*, 73:1441-1443.

Trosset, J. Y., and Carbonell, P. (2015). Synthetic biology for pharmaceutical drug discovery. *Drug. Des. Dev. Ther*, 6285-6302.

Vestola, J., Shishido, T. K., Jokela, J., Fewer, D. P., Aitio, O., and Permi, P. et al. (2014). Hassallidins, antifungal glycolipopeptides, are wide-spread among cyanobacteria and are the end-product of a nonribosomal pathway. *Proc. Natl. Acad. Sci. USA*, 111: E1909-E1917.

Volk, R. B., Girreser, U., Al-Refai, M., and Laatsch, H. (2009). Bromo-anaindolone, a novel antimicrobial exo-metabolite from the cyanobacterium *Anabaena constricta*. *Nat. Prod. Res,* 23: 607-612.

Wase, N. V., and Wright, P. C. (2008). Systems biology of cyanobacterial secondary metabolite production and its role in drug discovery. *Exp. Opin. Drug. Dis*, 3(8): 903-929.

Weber, T., and Kim, H. U. (2016). The secondary metabolite bioinformatics portal: Computational tools to facilitate synthetic biology of secondary metabolite production. *Synth. Syst. Biotechnol*, 1(2): 69-79.

WHO. (2017) *List of bacteria for which new antibiotics are urgently needed.* World Health Organization. Available at: http://www.who.int/ mediacentre/news/releases/2017/bacteria-antibiotics-needed/en/.

Wright, A. D., Papendorf, O, and König G. M. (2005). Ambigol C and 2, 4-dichlorobenzoic acid, natural products produced by the terrestrial cyanobacterium *Fischerella ambigua*. *J. Nat. Prod*, 68: 459-461.

Zainuddin, E. N., Jansen, R., Nimtz, M., Wray, V., Preisitsch, M., Lalk, M., and Mundt S. (2009). Lyngbyazothrins A–D, antimicrobial cyclic undecapeptides from the cultured cyanobacterium *Lyngbya* sp. *J. Nat. Prod*, 72: 1373-1378.

Ziemert, N., Alanjary, M. and Weber, T. (2016). The evolution of genome mining in microbes - a review. *Nat. Prod. Rep*, 33(8): 988-1005.

In: The Role of Photosynthetic Microbes…
Editors: K. Tripathi, N. Kumar et al.
ISBN: 978-1-53614-032-3

Chapter 7

UTILIZATION OF MICROALGAL BIOMASS AS A SOURCE OF BIOENERGY

Pankaj Kumar Rai[1,*], Abudukeremu Kadier[2, 3], Manish Kumar[4] and Sureshwar Prashad Singh[5]

[1]Department of Biotechnology and Bioinformatics Centre, Barkatullah University, Bhopal, Madhya Pradesh, India

[2]Department of Chemical and Process Engineering, Faculty of Chemical Engineering and Built Environments, National University of Malaysia (UKM), UKM, Bangai, Selangor, Malaysia

[3]Research Centre for Sustainable Process Technology (CESPRO), Faculty of Engineering and Built Environment, National University of Malaysia (UKM), UKM Bangi, Selangor, Malaysia

[4]Amity Institute of Biotechnology, Amity University, Gwalior, Madhya Pradesh, India

[5]Centre of Advanced Study in Botany, Banaras Hindu University, Varanasi, Uttar Pradesh, India

ABSTRACT

The consumption of fossil fuels and accumulation of greenhouse gases in the environment attract researchers worldwide to focus more on developing an alternative renewable and potentially carbon neutral biofuel. Microalgae have the potential to solve

[*] Corresponding Author Email: pankajraibhu@gmail.com.

both the issues simultaneously by CO_2 sequestration from the environment to increase its biomass and use of this biomass for biofuels production. Knowing the potential of photosynthetic microalgae, research studies focusing their utilization as feedstock for biofuel production has been carried out recently by many groups. Microalgae have added advantage that it can grow in waste water, are photosynthetic and unlike first and second generation biofuels it does not put strain on food market or changes in land use. This chapter reviews microalgae use for biofuels generation along with their large scale biomass production, harvesting techniques, pretreatment/processing and conversion of microalgae to biofuel using biochemical process.

Keywords: biomass, bioenergy, bioethanol, biofuel, biogas, greenhouse gas, microalgae

INTRODUCTION

Microalgae are photosynthetic, prokaryotic or eukaryotic micro-organism exhibiting a wide range of habitat and are capable to grow and thrive in harsh conditions ranging from fresh to marine water bodies (Markou and Georgakakis, 2011; Rai, 2016). Microalgae can be cultivated on large scale to produce biomass for a wide range of applications including animal and human nutrition, health sector, cosmetics and in agriculture. An important application discovered during the last two decades is the utilization of algal biomass for energy generation. The micro algal biomass generated can be converted into energy through a number of processes. Microalgae have higher photosynthetic efficiency than plants or trees, approximately 183 Giga tons of CO_2 is fixed to produce 100 Giga tons of microalgal biomass (Najafi et al., 2011). Microalgae cultivation utilizing atmospheric CO_2 will help in reducing CO_2 concentration in the environment at faster rate and biomass thus generated could be used for generation of biofuel. Biofuels are liquid or gaseous fuels that are produced from readily available waste materials or biomass. They are renewable, sustainable, biodegradable, carbon neutral and are hence environment friendly. Biofuels like biohydrogen, biodiesel, bioethanol and biogas appears to be an attractive option to the fossil fuels. Biomass energy is the largest renewable energy feedstock representing 77.4% of global renewable energy supply (Carlos and Khang, 2008). Global biofuel production has increased from 4.8 billion gallons in 2000 to about 16 billion in 2007, but still far behind the global transportation fuel supply (Coyle, 2008). Algae are emerging as an economical and environmentally sustainable, renewable source of biomass for the production of biofuels (John et al., 2011). Algae contain lipids/oils which could be used as raw material for biodiesel production (Petrou and Pappis, 2009). They also have carbohydrates, which can be converted into bioethanol, biohydrogen and biogas. The rapid growth rate and ability to grow on wide range of waste water using atmospheric CO_2 as the carbon source make algae most suitable candidate for biofuel production (Suganya et al., 2016). Biofuel production processes from microalgae may be divided into four main stages; cultivation, harvesting, processing/pretreatment, oil extraction for biodiesel production or fermentation for biohydrogen and bioethanol production.

MICROALGAE CULTIVATION

Although microalgae such as blue green algae have been used as food source by humans but actual algae cultivation is not old (Spolaore et al. 2006). The first successful unialgal culture was *Chlorella vulgaris* reported by Beijerinck in 1980 (Milano et al., 2016). Commercial microalgae cultivation was started in Japan using *Chlorella* sp. in 1960. There are several studies on microalgae cultivation on large scale but they are not always resulted in commercial production (Spolaore et al., 2006). The most common systems used for microalgae cultivation are open and closed systems. In open systems, cultivation is carried out in open environments such as raceway ponds, lagoons, deep channels, shallow circulating units. An open system has relatively low cost of construction, installation and maintenance and is easy to operate (Kroger and Miller, 2012; Richardson et al. 2012). The drawbacks of the open system are the contamination of undesired algal species, evaporation and lack of control on the physicochemical parameters (Koller et al., 2012). In closed systems, microalgae cultivation is carried out in photobioreactors (PBRs) made up from transparent glass and are fitted with artificial light source or exposed under sunlight. Frequently used PBRs are tubular (Ras et al., 2013) flat tank (Samson and LeDuy, 1983) and bubble column (Zamalloa et al., 2012). The advantages of closed systems are the ability to control growth parameters and avoidance of culture contamination. Higher operating and maintenance costs are the major drawback of the closed cultivation system as compared to open system. Microalgae required light and carbon source for their growth. Light is needed for CO_2 fixation and low intensity light slow down the CO_2 fixation process. The carbon source is primary requirement for the microalgae growth (Suali and Sarbatly, 2012; Ramraj et al., 2015; Singh et al., 2015). The high concentration of CO_2 results in faster growth rate and biomass yield but it also results in the reduction of pH of the growth medium and ultimately may inhibit the growth (Pires et al., 2012). Like other microorganisms, growth of microalgae is also affected by carbon and nitrogen ratio (C:N) (Gigova and Ivanova, 2015). Microalgae species having fast growth rate, requires ammonium as a primary N-source (Green and Durnford, 1996). However, when the N is limiting in the growth medium, the intermittent feeding of nitrate can enhance the growth rate of microalgae (Jin et al., 2006). Thus, the concentration and duration of C and N source should be applied to microalgal culture in such a way that it increases the biomass productivity. The next important nutrient for microalgal growth is phosphorous. It should be used in significant amount in the form of phosphate. For the effective cultivation of microalgae, trace elements like Mg, Ca, Mn, Zn, Cu and Mo and vitamins need to be added to the growth medium (Becker, 1994).

HARVESTING OF MICROALGAE

After cultivation, the microalgal biomass produced need to be separated from the cultivation medium. The small size of microalgae makes the harvesting process difficult but for the further utilization of biomass for biofuel production, water content should be removed. Harvesting of microalgae is usually costly and accounts about 20-30% of total biofuel production cost (Takeda, 1996; Jin et al., 2006). Therefore, it is important to choose appropriate harvesting technique to minimize the production cost of the biofuel. The harvesting techniques applied are highly dependent on the type of microalgae and cell density (Ozkurt, 2009). Microalgal harvesting can be divided into two-steps. The first step is separation of algal biomass from the growth medium by using techniques like flocculation, flotation or sedimentation (Jankowska et al., 2016). The second step is cell biomass concentration using the techniques like centrifugation and filtration.

Flocculation

Flocculation is a process in which solute in suspension come together and form aggregates called flocs and settle down. Microalgal cells carry negative charge and this similar charge does not let them come closer in the cultivation medium (Brennan and Owende, 2010). Microalgae can be harvested by flocculation only by disrupting this stable system (Chen et al., 2011). Chemicals called flocculants when introduced into the system disturbs the stability of the microalgal cells causing them to aggregate (Salim et al., 2011). Flocculation combined with other techniques like sedimentation or filtration increases harvesting efficiency (Rawat et al., 2011). Flocculants that are commonly used for harvesting microalgal cells are ferric chloride, ferric sulphate, aluminum sulphate and other organic salts (Brennan and Owende, 2010; Pragya et al., 2013). A technique of autoflocculation is also used for microalgal cell harvesting. Limited carbon supply or certain abiotic factors can induce autoflocculation (Rawat et al., 2011). Cultivation of microalgae in sunlight with limited CO_2 supply causes autofloccultion. Sometimes NaOH is added to initiate autoflocculation as it helps to maintain the desired pH for flocculation (Chen et al., 2011). High concentration of algal biomass results into frequent cell to cell encounter thus help flocculation. Slow mixing also results in the same, by bringing algal cells together. But, if the shearing forces are high, it can also disrupt the flocs (Pragya et al., 2013).

Filtration

In this technique, after the desired microalgal biomass is achieved, the whole cultivation medium with the algal cells is passed through a membrane filter. The membrane filter traps microalgal cells and allows only water to pass through. There are various types of filtration such as microfiltration, dead end filtration, vacuum filtration, pressure filtration, ultrafiltration and tangential flow filtration (TFF) used for recovery of algal cells from the cultivation medium (Pragya et al. 2013). Using TFF about 70-89% algae were recovered while maintaining the structure, properties and motility of the filtered algae (Chen et al., 2011). Although, the technique appears to be cheapest, membrane replacement and blockage and backwashing are the major drawbacks.

Centrifugation

Centrifugation involves the centrifugal force for the separation of two substances having different density. The algae more dense than cultivation medium migrate away from the axis while the latter move towards the axis. Thus, algae can be separated by removing the supernatant. Centrifugation is very rapid and efficient showing 80-90% recovery within 2-5 min (Harun et al., 2010). Rapid and high efficiency of biomass harvesting make centrifugation one of the most preferred technique among others. However, the requirements of electricity and maintenance cost make it economically not feasible (Rawat et al., 2011). Moreover, centrifuging large volume of biomass consumes a lot of time (Pragya et al., 2013).

Sedimentation

In the sedimentation technique, microalgal cells settle down at the bottom of the cultivation system under gravity and form a concentrated layer leaving clear liquid above. As there is no need of external energy source, this is a highly energy efficient process (Rawat et al., 2011).High density and large size microalgal species, like *Spirulina* settle down using this technique. However, it is also a time consuming process (Uduman et al., 2010).

MICROALGAEBIOMASS CONVERSION TO BIOFUELS

Biomass produced from cultivation of microalgae after harvesting can be converted into biofuels through various processes classified as biochemical and thermochemical conversion process (Figure 1). The biochemical conversion processes are transesterification and fermentation, which produce biodiesel, ethanol and hydrogen, respectively. The thermo chemical processes include pyrolysis, liquefaction, gasification and hydrogenation. Pyrolysis and liquefaction produces bio-oil, whereas gasification produces 'syngas', a mixture of gases comprising of carbon monoxide, carbon dioxide, and hydrogen. Hydrogenation is a process to improve the biofuel or feedstock properties.

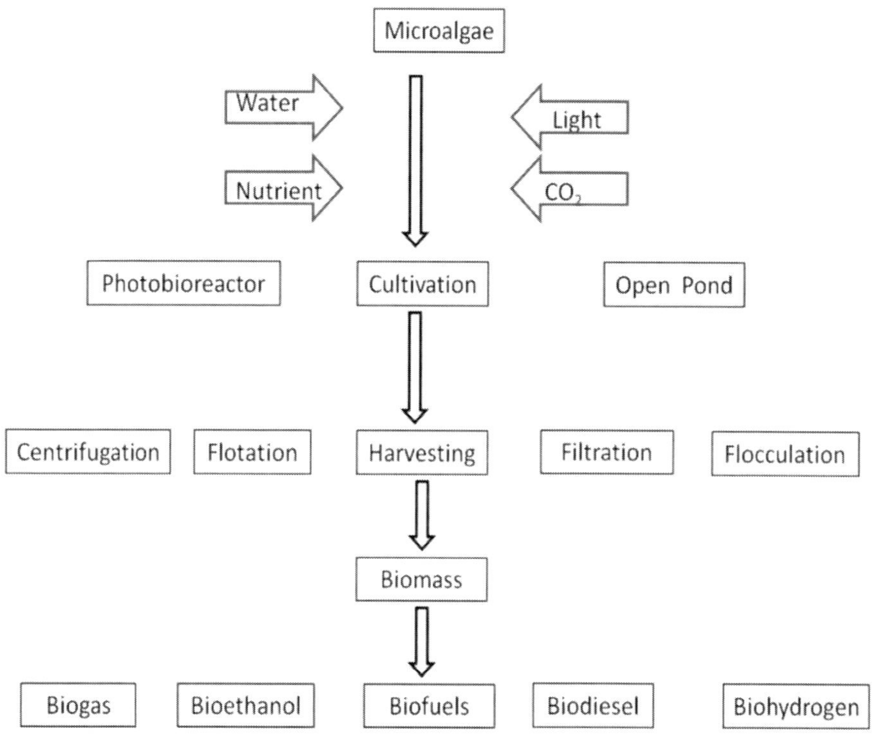

Figure 1. Schematic representation of steps involved in biofuel production from microalgae biomass.

Biodiesel Production

Biodiesel is one of the most common biofuel produced from algal biomass (Mata et al., 2010). Biodiesel is produced from algal biomass through transesterification (Milano et al. 2016). Transesterification is a process of lipid and mono alcohol reaction in the presence of a catalyst to produce mixture of fatty acids methyl ester (FAME) and glycerol as byproduct (Razzak et al., 2013). The transesterification process is carried out in a reactor, where blended methanol and catalyst react with the triglyceride present in the algal oil (Suali and Sarbatly, 2012; Chisti, 2008). There are several reports on biodiesel production from microalgae using transesterification. Many microalgal species are forced to grow in such a condition that leads to the accumulation of more lipids. The average lipid content range from 1-70% but under certain optimum conditions it may reach up to 90% of dry cell weight (Li et al. 2008; Chisti, 2007; Spolaore et al., 2006). The lipid content in the microalgae is specific to strain and growth condition specific. The most common algae used for biodiesel production are *Botryococcus, Chlorella, Dunaliella, Nanochloropsis, Scendesmus* and *Spirulina.* One of the most effective methods to improve lipid content in microalgae is nitrogen limitation, which not only increase lipid content but also change composition from free fatty acids to triacylgycerol (TAG), more useful for biodiesel production (Widjaja et al., 2009).

Bioethanol Production

Ethanol fermentation is the breakdown of sugars into ethanol. To decrease the production cost of ethanol various types of substrates are being used. Algal biomass contains carbohydrates which can be used as substrate for ethanol production. Prior to its use as substrate, biomass is hydrolyzed, grounded to release the sugars in the medium and hydrolysate is transferred to fermentors where yeast cells convert the sugar into ethanol (McKendry, 2002). The absence of lignin in algal biomass makes the hydrolysis easier. High starch containing organism like *Chlorella vulgaris* are considered as good substrate for ethanol production due to its starch (37% dry weight) content. Up to 65% ethanol conversion efficiency was established for *C. vulgaris* (Hirano et al., 1997). Pretreatment of algal biomass enhances the fermentation efficiency by more than 33% (Chen and Oswald, 1998).To release the sugar from biomass, enzymatic and acid hydrolysis methods were used to hydrolyze *Chlorella vulgaris* (Ho et al., 2013). They obtained high glucose yield (90.4%) indicating the superior potential of *C. vulgaris* as substrate for ethanol production. Lipid extracted algal biomass of *Chlorococcum* sp. was used for bioethanol production and 60% higher ethanol yield was reported (Harun et al., 2010). John et al., (2011) reported that the strains such as *Saccharina latissima*, *Laminaria hyperborean*, *Chlamydomonas reinhardtii* and *Spirulina fusiformis* after oil extraction can be used as substrate for bioethanol production. To use the algal biomass more efficiently and to make bioenergy generation cost effective, lipid from the biomass should be extracted for biodiesel production first and the remaining biomass should be utilized for bioethanol production. Thus, a combination of biodiesel and bioethanol can improve the overall performance of the system.

Biohydrogen Production

Owing to the non-polluting nature of hydrogen, its production through biological routes has gained considerable attention over the last few decades (Rai et al., 2012, 2014a, 2016). Biohydrogen production through fermentation using algal biomass as substrate is currently a major area of research due to process simplicity and possibility to cut down the price of feedstock and to make the overall process cost- effective (Rai et al. 2016, 2017). There are many reports on biohydrogen production using algal biomass as substrate (Kawaguchi et al., 2001; Nguyen et al., 2010; Nayak et al., 2014). Microalgal biomass is mainly composed of 8- 30% carbohydrates, 40- 60% proteins and 5- 60% lipids (Uggetti et al., 2014). The composition of organic matter varies from species to species under cultivation conditions. Carbohydrate content of *Arthrospira maxima* can be increased from approximately 20% to 50% of dry weight under sodium stresses (Carrieri et al., 2010). Whereas, under nitrogen starvation conditions carbohydrate content of *Arthrospira platensis* could be increased upto 50% of dry weight (Aoyama et al., 1997). Algal biomass requires pre- treatment to convert polymeric carbohydrates into simple sugars. Pre- treatment methods commonly used to hydrolyze the algal biomass are physical (milling, sonication, and microwave), chemical (acid and alkali treatment) and biological (enzymatic) or combination of these two processes. Pre-treatment with acid at high temperature is reported to be more effective hydrolysis approach (Rai et al., 2014b). However, the formation of inhibitory compounds like furan derivatives (furfurals and 5-hydroxyl methyl furfural) inhibits the cellular machinery of the fermenting microorganisms and ultimately results in the decline in H_2 yield (Sambusiti et al., 2015). Pre- treatment increases the H_2 production cost thus affect the overall economics of biohydrogen production from algal biomass (Rai et al., 2016b). Kawaguchi et al., (2001) used combination of two algae *Dunaliella tertiolecta* and *Chlamydomonas reinhardtii* biomass as substrate for biohydrogen production. Nguyen et al., (2010) used *C. reinhardtii* pre-treated biomass (1.5% HCl, 121°C, 20 min and enzyme treatment) as substrate for biohydrogen production using hyper thermophilic strain *Thermotoga neapolitana*. When algal biomass was used directly the H_2 yield ranged from 1.8 -2.2 mol H_2/mol glucose. The pre- treated biomass (HCl + heat) and enzymatic treatment resulted in increased H_2 yield of 58% (v/v) and 64% (v/v), respectively. Microalgal biomass of *C. vulgaris* ESP6 was used by Liu et al., (2012) as substrate for hydrogen production. The algal biomass was pretreated with acid or alkaline/enzymatic treatment prior to its use. Using pre-treated algal biomass *Clostridium butyricum* CGS5 under optimal condition produced cumulative H_2 (1476 ml/L), production rate (246 ml/L/h) and yield 1.15 mol/mol reducing sugar. The results indicated the suitability of algal biomass as a potential substrate for biohydrogen production. Kumar et al., (2013) utilized biomass of *Chlorella sorokiniana* as substrate for biohydrogen production by *Enterobacter cloacae* IIT BT-08. The pre- treated biomass (2% HCl, v/v and heat) resulted H_2 yield of 9±2 mol

H_2/Kg COD reduced and was found fitting with Gompertz equation. Biomass of *Anabaena* PCC7120 was utilized by thermophilic mixed microflora (Nayak et al., 2014). Algal biomass was subjected to various pre- treatment methods prior to its utilization as substrate. Maximum H_2 production (1600 ml/L) was reported upon pre-treatment with amylase before fermentation. De- oiled algal cake (after lipid extraction) was used as feed stock for biohydrogen production by Subhash and Venkatamohan (2014) using selectively enriched acidogenic consortia. The results of this study tested the feasibility of microalgae as a potential feedstock for simultaneous production of two energy forms viz., biodiesel and biohydrogen.

Biogas Production

Biogas is produced from the algal biomass through anaerobic digestion (Jankowska et al., 2016). Anaerobic digestion is a biochemical process in which organic material is converted to methane and CO_2 and traces of other gases (Brennan and Owende, 2010). Anaerobic digestion is most favourable for the organic matter having high moisture content (80-90% moisture) (McKendry, 2002). Therefore, it can be useful for CH_4 production from wet algal biomass. The first report on the anaerobic digestion of microalgae biomass was Golueke et al., (1957). They utilized the algal biomass *C. vulgaris* and *Scenedesmus* sp. grown for wastewater treatment for anaerobic digestion. The *Microcystis* sp. found mostly as algal biomass in eutrophicated lakes (Singh et al., 2015). Algal biomass was utilized as substrate for biogas production and ~70 mL/g VS (volatile solids) gas was produced from untreated *Microcystis* sp. (Zeng et al., 2010). *Dunaliella tertiolecta* a saline microalga was also investigated for its potential for anaerobic digestion by Lakaniemi et al., (2011). They reported low production rate of 24 mL/g VS mainly attributed to the effect of salinity. A mixed unidentified microalgal consortia was also used for methane/biogas production by De Schamphelaire and Verstraete (2009). They found the mixed consortia a promising substrate for methane production as the maximum gas production was 600 mL/g VS. The authors also suggested that a prior concentration of biomass is required for the optimal performance of anaerobic digestion. Some microalgae have high protein content that results in low C:N ratio which can adversely affects the performance of the anaerobic digestion (Suganya et al., 2016). To overcome this problem Yen and Brune, (2007) added waste paper to the algal biomass of *Scenedesmus* and *Chlorella* sp. and obtained very high methane production rate (1.17 mL/L/day) using 1:1 waste paper/algal biomass mixture compared to anaerobic digestion of pure algal biomass. High protein content in algal biomass resulted in increased ammonium production ultimately inhibiting anaerobic microorganisms (Brennan and Owende, 2010). At a high C:N, the amount of total ammonia- nitrogen can be too low required for the anaerobic microorganisms. Some reports suggested that a minimum of 50 to 200 mg/L of nitrogen as ammonia is essential for the requirements of the microorganisms associated with anaerobic digestion (Parkin and Owen, 1986).

CONCLUSION

Biofuel production from algal biomass is still at an infancy stage. Current efforts in biofuel production from microalgal biomass are not enough. Several researchers/companies are claiming to have developed the technology for efficient biofuels production from algal biomass. However, these claims need to be verified scientifically. Till today, cost- effective and efficient cultivation of microalgae is a challenging and the system still needs improvement to increase the yield of biomass per unit area. By integrating the algae cultivation with CO_2 sequestration from flue gas or waste water treatment and/or with extraction of high value compounds, the cultivation cost of algal biomass could be brought down. Harvesting of algal biomass require huge amount of energy input leading to increase in the production cost. Cost effective harvesting methods are required to make the biofuel production feasible and economical. There is no single best method for microalgal biomass harvesting. Optimum conditions for most of the microalgal species are also not known for their large scale production which requires immediate attention.

ACKNOWLEDGMENTS

PKR thankful to head of the Department of Biotechnology and Bioinformatics Centre, Barkatullah University, for providing necessary facilities.

CONFLICT OF INTEREST

No conflict of interest among the authors.

REFERENCES

Aoyama, K., Uemura, I., Miyake, J., and Asada, Y. (1997).Fermentative metabolism to produce hydrogen gas and organic compounds in a cyanobacterium *Spirulina platensis*. *J. Ferment. Bioeng,* 83:17-20.

Becker, E, W. (1994). *Microalgae biotechnology and microbiology.* Cambridge: Cambridge University Press.

Brennan, L., and Owende, P. (2010).Biofuels from microalgae-A review of technologies for production, processing, and extractions of biofuels and co-products. *Renew. Sustain. Energy. Rev,* 14: 557-577.

Carlos, R. M., and Khang, D. B. (2008). Characterization of biomass energy projects in Southeast Asia. *Biomass. Bioenergy,* 32:525-532.

Carrieri, D., Momot, D., Brasg, I. A., Ananyev, G., Lenz, O., Bryant, D. A., and Dismukes, G. C. (2010). Boosting auto fermentation rates and product yields with sodium stress cycling: application to production of renewable fuels by cyanobacteria. *Appl Environ Microbiol* 76:6455-6462.

Chen, C. Y., Yeh, K. L., Aisyah, R., Lee., D. J., and Chang, J. S. (2011). Cultivation, photobioreactor design and harvesting of microalgae for biodiesel production: a critical review. *Bioresour. Technol* 102:71-81.

Chen, P. H., and Oswald, W. J. (1998).Thermochemical treatment for algal fermentation. *Environ. Int,* 24:889-897.

Chen, Y., Cheng, J. J., and Creamer, K. S. (2008). Inhibition of anaerobic digestion process: a review. *Bioresour. Technol* 99:4044-4064.

Chisti, Y. (2007). Biodiesel from microalgae. *Biotechnol. Advan,* 25: 294-306.

Chisti, Y. (2008). Biodiesel from microalgae beats ethanol. *Trends Biotechnol,* 26:126-130.

Coyle, W. (2008). The future of biofuels: a global perspective. *Oregon. Wheat,* 26-30.

De, Schamphelaire, L., and Verstraete, W. (2009).Revival of the biological sunlight-to-biogas energy conversion system. *Biotechnol. Bioeng,* 103:296-304.

Gigova, L. G., and Ivanova, N. J. (2015). Microalgae respond differently to nitrogen availability during culturing. *J. Biosci,* 40:365-374.

Golueke, C. G., Oswald, W. J., Gotaas, H. B. (1957). Anaerobic digestion of algae. *Appl. Microbiol.* 5:47-55.

Green B. R., Durnford D. G. (1996). The chlorophyll-carotenoid proteins of oxygenic photosynthesis. *Rev. Plant. Phys,* 47:685-714.

Harun, R., Singh, M., Forde, G. M., and Danquah, M. K. (2010). Bioprocess engineering of microalgae to produce a variety of consumer products. *Ren. Sustain. Energy. Rev,* 14:1037-1047.

Hirano, A., Ueda, R., Hirayama, S., and Ogushi, Y. (1997).CO_2 fixation and ethanol production with microalgal photosynthesis and intra-cellular anaerobic fermentation. *Energy,* 22:137-142.

Ho, S. H., Huang, S. W., Chen, C. Y., Hasunuma, T., Kondo, A., and Chang, J. S. (2013). Characterization and optimization of carbohydrate production from an indigenous microalga *Chlorella vulgaris* FSPE. *Bioresour. Technol,* 135: 157-165.

Jankowska, E., Sahu, A. K., and Oleskowicz-Popiel, P. (2017). Biogas from microalgae: Review on microalgae's cultivation, harvesting and pre-treatment for anaerobic digestion. *Renew. Sustain. Energy Rev,* 75:692-709.

Jin, H. F., Lim, B. R., and Lee, K. (2006). Influence of nitrate feeding on carbon dioxide fixation by microalgae. *J. Environ. Sci. Heal,* 41:2813-2824.

John, R. P., Anisha, G. S., Nampoothiri, K. M., and Pandey, A. (2011). Micro and macro algal biomass: a renewable source for bioethanol. *Bioresour. Technol,* 102:186-93.

Kawaguchi, H., Hashimoto, K., Hirata, K., and Miyamoto, K. (2001). H_2 production from algal biomass by a mixed culture of *Rhodobium marinum* A-501 and *Lactobacillus amylovorus*. *J. Biosci. Bioeng,* 91:277-282.

Koller, M., Salerno, A., Tuffner P, Koinigg M., Bochzelt H., and Schober, S., et al. (2012). Characteristics and potential of microalgal cultivation strategies: a review. *J. Clean. Prod,* 37:377-88.

Kroger, M., Muller-Langer, F. (2012). Review on possible algal-biofuel production processes. *Biofuels* 3:333-349.

Kumar, K., Roy, S., and Das, D. (2013). Continuous mode of carbon dioxide sequestration by *C. sorokiniana* and subsequent use of its biomass for hydrogen production by *E. cloacae* IIT-BT 08. *Bioresour. Technol,* 145:116-122.

Lakaniemi, A. M., Hulatt, C. J., Thomas, D. N., Tuovinen, O. H, and Puhakka, J. A. (2011). Biogenic hydrogen and methane production from *Chlorella vulgaris* and *Dunaliella tertiolecta* biomass. *Biotechnol. Biofuels,* 4:34.

Li, Y., Wang, B., Wu, N., and Lan, C. Q. (2008). Effects of nitrogen sources on cell growth and lipid production of *Neochloris oleoabundans. Appl. Microbiol. Biotechnol,* 81:629-636.

Liu, C. H., Chang, C. Y., Cheng, C. L., Lee, D. J., and Chang, J. S. (2012). Fermentative hydrogen production by *Clostridium butyricum* CGS5 using carbohydrate-rich microalgal biomass as feedstock. *Int. J. Hydrogen. Energy,* 37: 15458-15464.

Markou, G., and Georgakakis, D. (2011). Cultivation of filamentous cyanobacteria (blue-green algae) in agro- industrial wastes and wastewaters: A review. *Appl. Energy,* 88 3389-3401.

Mata, T. M., Martins, A. A., and Caetano, N. S. (2010). Microalgae for biodiesel production and other applications: A review. *Renew. Sustain. Energy. Rev,* 14: 217-232.

McKendry, P. (2002). Energy production from biomass (part 2): conversion technologies. *Bioresour. Technol,* 83:47-54.

Milano, J., Ong, H. C., Masjuki, H. H., Chong, W. T., Lam, M. K., Loh, P. K., and Vellayan, V. (2016).Microalgae biofuels as an alternative to fossil fuel for power generation. *Renew. Sustain. Energy. Rev,* 58:180-197.

Najafi, G., Ghobadian, B., and Yusaf, T. F. (2011). Algae as a sustainable energy source for biofuel production in Iran: A case study. *Renew. Sustain. Energy. Rev,* 15 (8):3870-3876.

Nayak, B. K., Roy, S., and Das, D. (2014). Biohydrogen production from algal biomass (*Anabaena* sp. PCC 7120) cultivated in airlift photobioreactor. *Int. J. Hydrogen. Energy.* 39:7553-7575.

Nguyen, T. A. D., Kim, K. R., Nguyen, M. T., Kim, M. S., Kim, D., and Sim, S. J. (2010). Enhancement of fermentative hydrogen production from green algal biomass of *Thermotoga neapolitana* by various pre-treatment methods. *Int. J. Hydrogen. Energy.* 35:13035-13040.

Ozkurt, I. (2009). Qualifying of safflower and algae for energy. *Energy. Educ. Sci. Technol. Part A,* 23:145-151.

Parkin, G. F. and Owen, W. F. (1986). Fundamentals of anaerobic digestion of wastewater sludges. *J. Environ. Eng*, 24 (8): 867-920.

Petrou, E. C., and Pappis, C. P. (2009). Biofuels: a survey on pros and cons. *Energy. Fuels,* 23:1055-1066.

Pires, J. C. M., Alvim-Ferraz, M. C. M., Martins, F. G., and Simões, M. (2012). Carbon dioxide capture from flue gases using microalgae: engineering aspects and biorefinery concept. *Renew. Sustain. Energy. Rev,* 16:3043-3053.

Pragya, N., Pandey, K. K., and Sahoo, P. K. (2013).A review on harvesting, oil extraction and biofuels production technologies from microalgae. *Renew. Sustain. Energy. Rev,* 24:159-171.

Rai, P. K, and Singh, S. P. (2013). Biological production of clean energy: Hydrogen. In *Recent Advances in Microbiology.* S. P. Tiwari, Rajesh Sharma and Rajeev Gaur, (eds), 55-84. Nova Science Publishers Inc. New York, USA.

Rai, P. K. (2013). *Hydrogen production from dairy and agro wastes by integrating dark- and photo- fermentation.* PhD thesis, Banaras Hindu University, Varanasi-22005, India.

Rai, P. K. (2016). Recent advances in substrate utilization for fermentative hydrogen production. *J. Appl. Biol. Biotechnol,* 4, 59-67.

Rai, P. K., Asthana, R. K., Singh, S. P. (2014a). Optimization of photo- hydrogen production based on cheese whey spent medium. *Int. J. Hydrogen. Energy,* 39: 7597-7603.

Rai, P. K., Singh, S. P. (2016). Integrated dark- and photo- fermentation: Recent advances and provisions for improvement. *Int. J. Hydrogen. Energy,* 41, 19957-19971.

Rai, P. K., Singh, S. P., and Asthana, R. K. (2014b). Biohydrogen production from sugarcane bagasse by integrating dark- and photo-fermentation. *Bioresour. Technol,* 152:140-146.

Rai, P. K., Singh, S. P., and Asthana, R. K. (2011). Dairy waste based fermentative H_2 production. *J. Microbial. World,* 13: 207-213.

Rai, P. K., Singh, S. P., and Asthana, R. K. (2011). Prospects of utilizing dairy waste for bio-hydrogen production. *Int. J. Biotechnol. Biosci,* 1: 263-270.

Rai, P. K., Singh, S. P., and Asthana, R. K. (2012). Biohydrogen Production from cheese whey wastewater in a two-step anaerobic process. *Appl. Biochem. Biotechnol,* 167: 1540-1549.

Ramraj, R., Tsai, D. D. W., and Chen, P. H. (2015). Carbon dioxide fixation of fresh water micro-algae growth on natural water medium. *Ecol. Eng,* 75:86-92.

Ras, M., Lardon, L., Sialve, B., Bernet, N., and Steyer, J. P. (2011). Experimental study on a coupled process of production and anaerobic digestion of *Chlorella vulgaris. Bioresour. Technol,* 102:200-206.

Rawat, I., Kumar, R. R., Mutanda, T., and Bux, F. (2011). Dual role of microalgae: phycoremediation of domestic wastewater and biomass production for sustainable biofuels production. *Applied Energy,* 88:3411-3424.

Razzak, S. A, Hossain, M. M., Lucky, R. A., Bassi, A. S., and de Lasa, H. (2013). Integrated CO_2 capture, waste water treatment and biofuels production by microalgae culturing- a review. *Renew. Sustain. Energy. Rev,* 27:622-53.

Richardson, J. W., Johnson, M. D., and Outlaw, J. L. (2012). Economic comparison of open pond raceways to photo bio-reactors for profitable production of algae for transportation fuels in the Southwest. *Algal. Res* 1:93-100.

Salim, S., Bosma, R., Vermue, M. H., Wijffels, R. H. (2011). Harvesting of microalgae by bio- flocculation. *J. Applied. Phycol,* 23:849-55.

Sambusiti, C., Bellucci, M., Zabaniotou, A., Beneduce, L., and Monlau, F. (2015). Algae as promising feedstocks for fermentative biohydrogen production according to a biorefinery approach: A comprehensive review. *Renew. Sustain. Energy Rev,* 44: 20-36.

Samson, R., and LeDuy, A. (1983). Influence of mechanical and thermochemical pre-treatment's on anaerobic digestion of *Spirulina maxima* algal biomass. *Biotechnol Lett,* 5:671-676.

Singh, D., Yadav, K., Deepshildia, Singh, R. S. (2015). Biofixation of carbon dioxide using mixed culture of microalgae. *Indian. J. Biotechnol,* 14:228-232.

Singh, S., Rai, P. K., Chau, R., Ravi, A. K., Neilan, B. A, and Asthana R K. (2015). Temporal variations in microcystin- producing cells and microcystin concentrations in two fresh water ponds. *Water Research* 69, 131-142.

Spolaore, P., Joannis-Cassan, C., Duran, E., and Isambert, A. (2006). Commercial applications of microalgae. *J. Biosci. Bioeng,* 101:87-96.

Suali, E., and Sarbatly, R. (2012). Conversion of microalgae to biofuels. *Renew. Sustain. Energy Rev,* 16:4316-442.

Subhash, G., and Venkata Mohan, S. V. (2014). Deoiled algal cake as feedstock for dark fermentative biohydrogen production: An integrated biorefinery approach. *Int. J. Hydrogen. Energy.* 39: 9573-9579.

Suganya, T., Varman, M., Masjuki., H. H., and Renganathan, S. (2016). Macroalgae and microalgae as a potential source for commercial applications along with biofuels production: A biorefinery approach. *Renew. Sustain. Energy. Rev,* 55:909-941.

Takeda, H. (1996). Cell wall sugars of some *Scenedesmus* species. *Photochemistry,* 42:673-675.

Uduman, N. Qi Y., Danquah, M. K., and Hoadley, A. F. (2010). Marine microalgae flocculation and focused beam reflectance measurement. *Chem. Eng. J,* 162:935-940.

Uggetti, E., Sialve. B., Trably. E., and Steyer, J. P. (2014). Integrating microalgae production with anaerobic digestion: a biorefinery approach. *Biofuels. Bioprod. Biorefin.* 8:516-529.

Widjaja, A., Chien, C-C., and Ju, Y-H. (2009). Study of increasing lipid production from fresh water microalgae *Chlorella vulgaris. J. Taiwan. Inst. Chem. Eng,* 40:13-20.

Yen, H. W., and Brune, D. E. (2007). Anaerobic co-digestion of algal sludge and waste paper to produce methane. *Bioresour. Technol,* 98:130-134.

Zamalloa, C., Boon, N., and Verstraete, W. (2012). Anaerobic digestibility of *Scenedesmus obliquus* and *Phaeodactylum tricornutum* under mesophilic and thermophilic conditions. *Appl. Energy,* 92:733-738.

Zeng, S. J., Xianzheng, Y., Xianzheng, S., and Yanling, Q. (2010). Effect of inoculum/substrate ratio on methane yield and ortho phosphate release from anaerobic digestion of *Microcystis* spp. *J. Hazard. Matt,* 178:89-93.

In: The Role of Photosynthetic Microbes...
Editors: K. Tripathi, N. Kumar et al.
ISBN: 978-1-53614-032-3

Chapter 8

A NEW INSIGHT INTO THE COMMERCIAL APPLICATIONS OF HALOTOLERANT GREEN ALGAE *DUNALIELLA*

Pradeep Kumar Rai, Anuradha Rai and Surendra Singh*

Centre of Advanced Study in Botany,
Banaras Hindu University, Varanasi, India

ABSTRACT

Recently, the cultivation of microalgae as a renewable source of fuel and energy has been receiving much attention globally. *Dunaliella* is probably the most halotolerant and dominant photosynthetic eukaryote that thrives in most natural and artificial hypersaline niches occupied almost exclusively by halophilic archaebacteria. This alga is cultivated commercially as a source of many valuable macromolecules such as proteins, lipids and pigments. It is one of the best sources for the commercial production of carotene, glycerol and protein. Different species of *Dunaliella* have the ability to produce a wide variety of high-value bioactive compounds, fine biochemicals, and bulk compounds such as lipids, pigments, natural dyes and biomass. The ability to grow at high salt concentrations coupled with the production of high amount of β-carotene, glycerol and protein have made this microalga an attractive candidate for commercial application in medicine, the food industry and agriculture.

Keywords: agriculture, algal-meal, biofuels, carotenoids, *Dunaliella*, food Industry, halotolerant, medicine

---

* Corresponding Author Email: surendrasingh.bhu@gmail.com.

INTRODUCTION

Microalgal cultivation as a commercial source of renewable fuel and energy has been receiving much attention globally (Borowitzka, 1992; Bosma and Wijffels, 2003; Benemann, 2008). In spite of higher productivity, their cultivation does not require arable land or freshwater (Ben-Amtoz et al., 2009, 2003). This combination of high biomass productivity with no demands of arable land and freshwater supports the idea for large-scale production of microalgal biomass without affecting the agriculture (Ben-Amtoz et al., 2009). At present, microalgae are exploited for the production of various molecules such as pigments, lipids and proteins that can be incorporated into important industries (Spolaore et al., 2006). Microalgal biomass is used as nutraceutical or food supplement for human consumption and animal feed (Aksu et al., 1997; Ben-Amtoz, 2003; Stottrup and McEvory 2003). Microalgae can also be used as natural food colorants (Borowitzka, 1999; Soletto et al., 2005).

Microalgae are chlorophyll bearing fastest growing autotrophs on the earth and can utilize commonly available nutrients for their growth. India, being a tropical country provides most favourable conditions for algal growth. *Dunaliella* is a green, unicellular and biflagellated motile alga belonging to the family polyblepharidaceae and class chlorophyceae (Avron and Ben-Amotz, 1992). Teodoresco (1905) was the first to describe the genus *Dunaliella.* At present, 29 species and many varieties of *Dunaliella* are recognized (Massyuk 1973). *D. salina*, *D. primolecta*, *D. viridis*, *D. bioculata*, *D. tertiolecta*, *D. acidophyla*, *D. parva* and *D. media* are the best-known species of *Dunaliella. Dunaliella* is morphologically similar to *Chlamydomonas* and is characterized by an ovoid cell volume, single large cup shaped chloroplast with a single-centered starch-surrounded pyrenoid, a few vacuoles, a nucleus and a nucleolus (Oren, 2005). The cell shape of this alga is spherical, cylindrical, pear or spindle-shaped with bilateral or dorsoventral symmetry. Under stress conditions, their cells often become spherical. Cell size also varies with growth conditions and light intensity (Massyuk, 1973). The cells divide lengthwise in the motile state. Under certain conditions, they may also develop into a palmella stage and become embedded in a thick layer of mucilage or form aplanospores with a thick rough wall. They are the richest natural source of β–carotene. In addition to chlorophyll *a* and *b*, this green alga also possesses lutein and xanthin like carotenoids. Like other members of green alga, *Dunaliella* cells have membrane-bound nucleus, an eyespot, Golgi apparatus, vacuoles, and mitochondria (Ben-Amotz and Avron 1989, 1999). *Dunaliella* lacks cell wall, and the cells are enclosed with plama membrane having mucous coating (Borowitzka and Borowitzka 1988). Lack of a rigid cell wall in *Dunaliella* permits rapid flexibility to cell volume changes and cell membrane elasticity in response to extra cellular changes in osmotic pressure. Sexual reproduction is frequent in field condition but rare in cultures. *Dunaliella* reproduce either by fusion of two motile cells to form a zygote or by longitudinal division of the motile cells. The zygote is green or red in color, and is surrounded by a thick smooth wall of sporopollenin (Borowitzka et al., 1982). Sexual zygote formation has been reported in five *Dunaliella* spp., namely *D. salina, D. parva, D. peircei, D. euchlora* and *D. minuta* (Lerche, 1937).

Dunaliella is probably the most halotolerant and dominant photosynthetic eukaryote, which thrives in most natural and artificial hypersaline niches occupied almost exclusively by halophilic archaebacteria. They show different degree of adaptation to a variety of salt concentrations (100 mM to 5.5 M NaCl). Under hypersaline environments, *Dunaliella* removes Na^+ ions by a novel redox-driven sodium pump (Avron and Ben-Amotz 1992; Katz and Pick 2001). *Dunaliella* species such as *D. acidophila* can grow at pH ranging from 0–1, while *D. antarctica* grows at subzero temperatures.

A number of *Dunaliella* species possesses enormous prospect and potential for valuable application in medicine, pharmacology, the food industry and agriculture. They produce a wide range of valuable compounds such as lipids, natural dyes, oils, pigments, sugars, antioxidants, high-value bioactive compounds, and other fine biochemicals and biomass (Oren, 2005; Li et al., 2008; Raja et al., 2008; Tafreshi and Shariati, 2009). The ability of *Dunaliella* to grow at a very high salt concentration and to synthesize high amount of β-carotene, glycerol and protein has made this organism an attractive candidate for commercial exploitation (Apte and Behrens, 1999; Çelekli and Donmez, 2006; Takagi et al., 2006). Commercial valuable products of *Dunaliella* are discussed below.

CAROTENOIDS

Carotenoids are lipid-soluble colored compounds. The different colours of plant leaves, flowers and fruits, as well as the color of feathers, crustacean shells, fish flesh, skin etc. are due to the carotenoids (Johnson and Schroeder, 1995). Most of the carotenoids are derived from a 40-carbon polyene chain. The polyene system provides distinct molecular structure, chemical properties and light-absorbing characteristics of carotenoides. In algae and higher plants, carotenoids play multiple and essential roles in photosynthesis and are involved in light harvesting, function of photosynthetic complexes, scavenging reactive oxygen species (ROS) and in the dissipation of excess energy (Demming-Adams and Adams, 2002).

Dunaliella is the richest natural source of carotenoids and contains a varied range of carotenoids with wider applications. The carotenoids of *Dunaliella* include β-carotene, phytoene, phytofluene, lutein and zeaxanthin (Tafreshi and Shariati, 2009). β-Carotene is an important carotenoid found in green leafy plants, fruits and vegetables.

Table 1. Important applications of β-carotene isomers extracted from *Dunaliella*

Isomers	Applications	Reference
9-*cis*	Food colouring agent	Berset, (1990)
9-*cis*	Prevents cancer of various parts of digestive, respiratory and	Raja et al., (2004)
	reproductive system	
9-*cis*	Involved in neoplastic transformation and	Bertram and Bortkiewicz
	control of growth	(1995)
all-*trans*	Source of vitamin A in animal feed	Gomez and Gonzalez, (2004)
		Omenn et al., (1994)
all-*trans*	Reduces the incidence of cancer and cardiovascular diseases	
	Enhances yolk colour in vitamin A deficient	Terao, (1989)
9-*cis* and all-*trans*	Chicken	
9-*cis* and all-*trans*	Possesses antioxidant activity by quenching singlet state	Poppel and Goldbohm,
	oxygen, scavenges peroxyradicals and inhibits lipid	(1995)
	peroxidation	
9-*cis* and all-*trans*		
9-*cis* and all-*trans*	Influences intracellular communication	Sies and Stahl, (1997)
	Control immunity by increasing the	Williams et al., (2000)
9-*cis* and all-*trans*	percentage of monocytes and enhances natural	
	killer cell activity in elderly men	Williams et al., (2000)
9-*cis* and all-*trans*		
9-*cis* and all-*trans*	The metabolites of carotenoids influence the metabolic	Avron et al., (1987)
	pathway	
9-*cis* and all-*trans*		Mathews-Roth
	Hepatoprotective	(1987)
9-*cis* and all-*trans*		
	Beneficial against scleroderma, a connective	Peto et al.,(1981)
	tissue disorder	
	Decreases sensitivity to sun and protects against	Ben-Amotz et al., (1989)
	erythropoietic protoporphyria	
	Provides retinal in rat and chicks	

Vitamin A produced by the oxidation of β-carotene by the liver enzymes is essential for the proper functioning of vision and epithelial tissues. Thus, β-carotene of *Dunaliella* has a great potential, and can be used as animal and human dietary supplement (Borowitzka and Borowitzka 1988; Pulz and Gross 2004). It is noted that β-carotene rich diet of *Dunaliella* minimizes the risk of degenerative diseases and reduces the tumour occurrence in animals and humans (Comstock et al., 1991).

Dunaniella salina has been considered as an important biological source of β-carotene (Ben-Amotz and Avron, 1990). The β-carotene producing strains of *Dunaliella* predominate over all other organisms in hypersaline lakes, which are generally low in available nitrogen and exposed to high solar radiation. The β-carotene in *Dunaliella* accumulates within distinct oily globules in the inter-thylakoid spaces of the periphery chloroplast. β-carotene of *Dunaliella* is composed mainly of two stereoisomers: all-*trans* and 9-*cis*, with the rest a few other mono-*cis* and di-*cis* stereoisomers of the β-carotene.

In *Dunaliella* biomass, the quantity of β-carotene and its isomers (9-*cis*-to-all *trans* ratio) is governed by the growth conditions, especially cells division time and light intensity (Ben-Amotz and Avron, 1990; Garcia-Gonzalez et al., 2005). Some important applications of β-carotene isomers extracted from *Dunaliella* are given in Table 1. *D. bardawil* has the ability to accumulate 47 g of phytoene per litre of the culture (Leon et al., 2005). Phytoene plays an important role in cancer-prevention and antioxidant activities in mammalian cells. According to Nishino (1998), lutein is also an important carotenoid, and is a pigmentation factor in poultry and fish. Mares-Perlman et al., (2002) reported that lutein acts against cataract and macular degeneration. Zeaxanthin like lutein is a carotenoid found in highest concentration in the macular region of the eyes, and zeaxanthin of *Dunaniella* acts as nutraceuticals against macular degradation (Jin et al., 2003).

GLYCEROL

Glycerol (1,2,3-propanetriol) is a simple alcohol and has a key role in petroleum, food, pharmaceutical and textile industries. Glycerol is naturally produced by microbial fermentation or chemically from petrochemical or soap industries. Glycerol can be produced autotrophically using *D. tertiolecta* and *D. bardawil* at the expense of CO_2 as a carbon and light as an energy source. At higher NaCl concentration, these microalgae can accumulate glycerol intracellularly up to 50% of its dry weight (Ben-Amotz and Avron 1980, 1983, 1990). Grizeau and Navarro, (1986) obtained glycerol (5g/L) extracellularly when calcium alginate immobilized *D. tertiolecta* cells were cultured in a bioreactor containing hypersaline (4M NaCl) medium.

Glycerol may be produced biotechnologically using a number of microorganisms including bacteria, yeasts, moulds and microalgae. The major disadvantages of fermentative processes using bacteria, yeasts, and molds have been the low yield of glycerol from carbohydrates, and difficulties in the recovery of glycerol from the fermentation broth (Taherzadeh et al., 2002). *Dunaliella* possesses the unique ability to accumulate large amounts of glycerol as compatible solute in response to extracellular osmotic stress. Oren (2005) has suggested glycerol as a possible secondary product of *Dunaliella* cultivation. Under nutrient deprived conditions, various *Dunaliella* species have the capacity to store large amounts of triacylglycerides (TAGs), which make *Dunaliella* as a valuable source of biodiesel production (Davidi et al., 2012). No significant change in intracellular Na^+, K^+ and Cl^- contents in *Dunaliella* at varying salt concentrations indicates that glycerol is the only osmotic element in this alga (Pick et al., 2002). Glycerol has an important function in the survival of alga, as it enables them to survive at subzero temperatures in hypersaline lakes by acting as an effective anti-freeze agents (Franzmann, 1991).

PROTEIN

The proteins from *Dunaliella* biomass can be utilized for making bread and other products (Finney et al., 1984). Whole cells can also be utilized as animal, fish and poultry feed safely (Mokady et al., 1989). *D. salina* has become popular as a food-grade green microalga particularly due to its lipid and protein contents, glycerol concentration and β-carotene content (up to 4% of its dry weight), and their exceptional ability to grow under brackish conditions (Barrow and Shahidi, 2007). Due to the presence of high levels of protein and β-carotene (vitamin A), it is recognized as safe and excellent poultry and aquaculture feed or food (Tafreshi and Shariati, 2009). *D. tertiolecta* is suggested to be an ideal single cell protein (Fabregas and Herrero 1985). Recombinant proteins from *D. salina* and many expression systems are currently available. It has been reported that the production of *Dunaliella* can be introduced in aquaculture industry in either liquid paste or dry powder form (Ben-Amotz et al., 2009 Feng et al., 2014). Stottrup and McEvoy, (2003) reported that *Dunaliella* in aquaculture acts as a sole source of food for filter feeders, a food additive for many fish and crustacean species as well as enrichment for artemia and rotifers. *Dunaliella* cells contain about 0·2% xanthophyll (mostly lutein), and meal from this alga can be used for increasing the colour of egg yolk (Ben-Amotz et al., 1986; Moulton and Burford, 1990). Better health and fertility of cattle fed with *Dunaliella* rich meals have been attributed to extra β-carotene (Borowitzka and Borowitzka, 1989).

DRIED ALGAL MEAL

Ben-Amotz and Avron (1990) reported that *Dunaliella* cells contain about 40% proteins. After removal of β- carotene and glycerol, the *Dunaliella* biomass is known as dried algal meal. This dried algal meal contained less concentration of nucleic acids (RNA and DNA) in comparison to yeast and bacteria used as SCP source (Mokady, 1992). The presence of relatively higher content of glycerol, β- carotene, lipid and protein in *Dunaliella* makes it popular as food- grade green microalga. Proteins from *Dunaliella* can also be used as a supplement in bread and other products (Finney et al., 1984). Recently, *Dunaliella* has been introduced to the aquaculture industry in two forms, liquid paste and dry powder due to its high levels of protein, carotenoids, minerals, fatty acids and vitamins (Ben-Amtoz et al., 2009). The alga is also used as a source of natural pigments for the culture of prawns, salmonid fish and ornamental fish (Stottrup and McEvoy, 2003).

Biofuel

These microalgae may be considered as an alternative source for the production of biofuels in place of fossil diesel (Chisti, 2007). Microalgae offer following potential advantages:

1. Rapid growth, which provides high biomass and requires less time.
2. Contain high content of commercially valuable chemicals such as β-carotene, starch, glycerol, and lipids (Avron and Ben-Amotz, 1992; Oren, 2005).
3. High accumulation of oils.
4. After oil extraction, the resulting algal biomass can either be processed into livestock feed or burnt for energy co-generation, ethanol and methanol production (Wang et al., 2008).
5. Have high CO_2 fixation and O_2 evolution rates, thus reduce the emission of greenhouse gases.
6. These microalgae require less water in comparison to traditional crops and can be harvested daily instead of seasonally (Campbell, 1997; Chisti, 2008). Some important crops (Jatropha, Soybean and Castor), fish oil and animal fat have not been used to produce biodiesel in large quantities sustainably. However, these microalgae can be used sustainably as an alternative to produce various biofuels such as biodiesel derived from lipid (Schenk et al., 2008; Scott et al., 2010), bioethanol (Dismukes, et al., 2008), biomethane (Sialve et al., 2009; Alzate et al., 2012) and biohydrogen (Hemschemeier et al., 2009). Algal biodiesel is non-toxic, biodegradable, contains no sulfur, and has reduced emissions of particulate matter, CO, hydrocarbons and Sox. Some microalgal species can accumulate high lipid contents, which vary between 1 to 70%, while in certain species it may reach up to 90% (w/w) under certain conditions (Li et al., 2008). *D. salina* with the biodiesel yield of 66% is a suitable candidate for biodiesel production.

There are several alternative ways to use *D. salina* as fuels and these include:

a. Drying the biomass followed by its direct combustion for power generation or other thermochemical conversions (pyrolysis) to generate gas or oils.
b. Chemical or physical separation of lipids from the biomass for production of biodiesel.

c. Fermentation of biomass for the production of ethanol, methane or other fuels by microbial action (Ben-Amotz et al., 2009).

BIOACTIVE COMPOUNDS

Microalgae are rich source of novel and biologically active primary and secondary metabolites. These metabolites may be potential source of bioactive compounds such as β-carotene, glycerol, protein, carbohydrate and omega 3-fatty acids, which are used in the pharmaceutical, cosmetic and food industries. They are also implicated as renewable sources of bioactive lipids having high amount of polyunsaturated fatty acids (PUFA) such as eicosoapentaenoic acid, a-linolenic acid and docosapentaenoic acid etc. These bioactive lipids function as antioxidants in therapeutic and dietetics uses, which decrease the risk of cancer, aging, inflammation, stroke and neurodegenerative diseases (Ben-Amotz et al., 1982; Abd El-Baky et al., 2004; Murthy et al., 2005). Kumar et al., (2013) have studied the bioactive metabolites of *D. salina* against various human pathogens such as *Vibrio cholerae, Klebsiella pneumoniae, Escherichia coli, Pseudomonas aeruginosa* etc. and found it inhibitive to *V. cholerae*. The cells of *D. parva* immobilized in Ca-alginate are used in the production of 1,2-propanediol, which is an expensive glycol from hydroxyacetone (Hatanaka et al., 1999).

The carbohydrate content is considered as an important parameter in production of ethanol and other solvents from microalgae. The basic idea of ethanol generation from *Dunaliella* is fermentation of starch or glycerol that is accumulated at high salinities. Shirai et al., (1998) have cultivated *Dunaliella* on a highly saline (3% NaCl) soy sauce waste, disrupted the cells and saccharified the intracellular starch using glucoamylase (from *Rhizopus* species). The saccharified starch was then fermented by *Saccharomyces cerevisiae* yielding ethanol (11 mg per gram of dry weight). Nakas et al., (1983) found that glycerol produced by *Dunaliella* can be fermented to solvents including ethanol. However, the yield of ethanol depended on the fermentation conditions.

In Japan and other Far East countries, dry *Dunaliella* is distributed popularly to consumers who are more familiar with and are accustomed to *Chlorella* and *Spirulina.* Their cultivation is based on autotrophic growth in media containing CO_2 as the carbon source and inorganic nutrients (Ben-Amotz and Shaish, 1992). Figure 1 shows the cultivation of *Dunaliella* for commercial application. *Dunaliella* production is estimated to be about 1200 t $year^{-1}$ (Pulz and Gross, 2004) on a global scale (Borowitzka and Borowitzka, 1988). Several companies are actively engaged in the production of *Dunaliella*. The following are the commercial producers of *Dunaliella.*

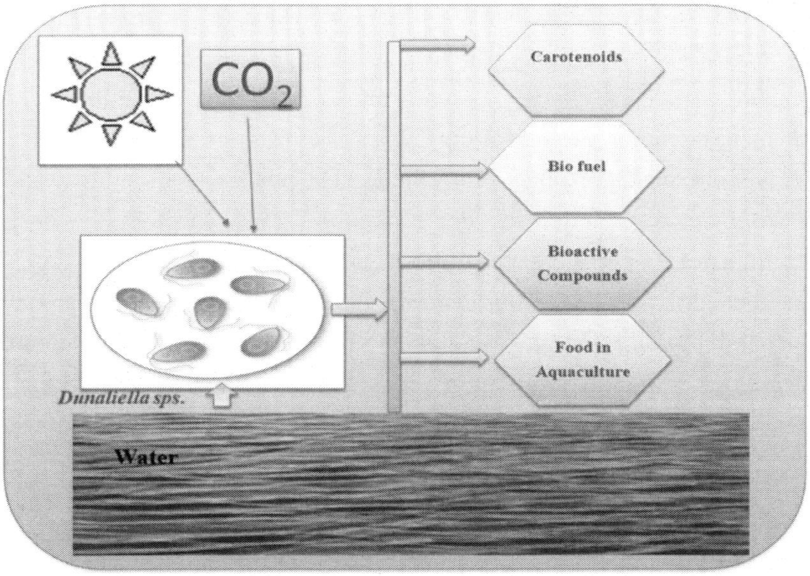

Figure 1. Schematic representation of the cultivation of *Dunaliella* and its commercial application.

Table 2. Companies engaged in commercial production of *Dunalialla*

	Name of Company	Country
1.	Nature Beta Technologies (NBT) Ltd., Eilat 88106, Israel, a subsidiary of Nikken Sohonsha Co. Gifu	Japan
2.	Cyanotech Corp, Kailua-Kona, HI 96740	USA
3.	Nutrilite, Calipatria, CA 92233, USA, a division of Amway Corporation	USA
4.	Western Biotechnology Ltd. Bayswater, W. A. 6053, Australia, a subsidiary of Coogee Chemicals Pty. Ltd	Australia
5.	Photo Bioreactors, PBL Ltd., Murcia 30003	Spain
6.	Tianjin Lantai Biotechnology, Inc. Nankai, Tianjin, in collaboration with the Salt Scientific Research Institute of Light Industry Ministry, P. R	China

Conclusion

India has a large coastal area which can provide the major selective environment for the growth and development of *Dunaliella* on a commercial scale. In this review, we have emphasized on the possible applications of *Dunaliella*, which has the ability to grow under extreme environment with considerable growth and higher biomass production. Owing to their ability to grow in extreme environment, *Dunaliella* has wide applications in pharmaceutical, food and petrochemicals industries. It is considered as an excellent source of high-value products including nutraceuticals and bioactive molecules that may lead to the discovery of new drugs. It is potentially used in the food industry as single cell protein (SCP) and a source of essential amino acids. Due to its higher content of long chain hydrocarbons, its crude lipids can be used in large-scale production of biofuel and as an alternative source of energy. Looking at its enormous potentialities, traditional commercial manufacturers are culturing *Dunaliella* on a commercial level for the production of β-carotene and other biotechnological purposes. Its ability to survive at extreme environment is exploited for identification of genes to generate transgenic crop plants with improved salt-tolerance. Introduction of new approaches are however, required for the cultivation of *Dunaliella* in diverse culture conditions (autotrophic, heterotrophic and mixotrophic) as well as in different phase systems to make them economically more competitive in future. It is desirable that scientists must seriously focus on the development of reliable approaches to genetically transform and metabolically engineer *Dunaliella*, so that it can be efficiently used as a source of high-value proteins such as enzymes, antibiotics, and vaccines.

ACKNOWLEDGMENTS

One of the authors (PKR) thanks the University Grant Commission (UGC) for providing financial support in the form of UGC Research Fellowship. Authors are highly thankful to Prof. Ashwani K Rai for improving the manuscript, Head, Department of Botany, Banaras Hindu University for providing necessary facilities.

CONFLICT OF INTEREST

No conflict of interest among the authors.

REFERENCES

Aksu, Z., Acikel, U. and Kutsal, T. (1997). Application of multicomponent adsorption isotherms to simultaneous biosorption of iron (III) and chromium (VI) on *Chlorella vulgaris*. *J. Chem. Tech. Biotechnol,* 70:368-378.

Alzate, M. E., Munoz, R., Rogalla, F., Fdz-Polanco, F. and Perez-Elvira, S. I. (2012). Biochemical methane potential of microalgae: influence of substrate to inoculums ratio, biomass concentration and pretreatment. *Bioresour. Technol,* 123:488–494.

Apte, K. and Behrens, P. (1999). Commercial developments in microalgal biotechnology. *J. Phycol,* 35(2): 215-226.

Avron, M. and Ben-Amotz, A. (1992). *Dunaliella: physiology, biochemistry, and biotechnology,* CRC Press, Boca Raton, Florida.

Avron, M., Edelstein, S. and Ben-Amotz, A. (1987). *Feed supplement.* UK Patent application number 2189675A.

Barrow, C. and Shahidi, F. (2007). *Marine nutraceuticals and functional foods, lipids.* CRC Press, Boca Raton, Florida.

Ben-Amotz, A. (1999). *Dunaliella* β-carotene: From science to commerce. In: Seckbach J (eds) *Enigmatic Microorganisms and Life in Extreme Environments.* Kluwer Publication, Netherlands, pp 401–410.

Ben-Amotz, A. 2003. Industrial production of microalgal cell-mass and secondary products-major industrial species. In: Richmond A (eds) *Handbook of Microalgal Cultures, Biotechnology and Applied Phycology.* Blackwell Publication, UK, pp. 273-280.

Ben-Amotz, A. and Avron, M. (1983). On the factors which determine massive beta-carotene accumulation in the halotolerant alga *Dunaliella bardawil. Plant. Physiol,* 72: 593–597.

Ben-Amotz, A. and Avron, M. (1989). The biotechnology of mass culturing of *Dunaliella* for products of commercial interest. In: Cresswell R. C., Ress T. A. V., Shah, N. (eds) *Algal and Cyanobacterial Biotechnology.* Longman Scientific and Technical Press, London, pp. 90–114.

Ben-Amotz, A. and Shaish, A. (1992). B-carotene biosynthesis. In: Avron M., Ben-Amotz A. (eds) *Dunaliella*: *Physiology, Biochemistry, and Biotechnology.* CRC press, Boca Raton, Florida, pp. 205–216.

Ben-Amotz, A. and Avron, M. (1980). Glycerol, beta-carotene and dry algal meal production by commercial cultivation of *Dunaliella.* In: Shelef G., Soeder C. J. (eds) *Algae Biomass, Production and Use.* Elsevier, North Holland Biomedical Press, Amsterdam, pp. 603–661.

Ben-Amotz, A. and Avron, M. (1990). The biotechnology of cultivating the halotolerant alga *Dunaliella. Trends. Biotechnol,* 8: 121-126.

Ben-Amotz, A., Edelstein, S., Avron, M. (1986). Use of the beta-carotene rich alga *Dunaliella bardawil* as a source of retinol. *Br. Poult. Sci,* 27(4): 613-619.

Ben-Amotz, A., Katz, A. and Avron, M. (1982). Accumulation of β-carotene in halotolerant algae: Purification and characterization of β-carotene globules from *Dunaliella bardawil. J. Phycol,* 18(4): 529-537.

Ben-Amotz, A., Polle, J. E. and Rao, D. V. (2009). *The alga Dunaliella. biodiversity, physiology, genomics and biotechnology.* Science Publishers, Enfield, NH.

Ben-Amotz, A., Shaish, A. and Avron, M. (1989). Mode of action of the massively accumulated β-carotene of *Dunaliella bardawil* in protecting the alga against damage by excess irradiation. *Plant. Physiol,* 91:1040–1043.

Benemann, J. R. (2008). *Opportunities and challenges in algae biofuels production: a position paper.* http://www.fao.org/uploads/media/algae _positionpaper.pdf

Berset, C. (1990). The research situation with regard to natural food colourants. *Ind Aliment. Agric,* 107:1163–1166.

Bertram, J. S. and Bortkiewicz, H. (1995). Dietary carotenoids inhibit neoplastic transformation and modulate gene expression in mouse and human cells. *Am. J. Clin. Nutr,* 62 (6l):1327S-1336S.

Borowitzka, L. J. and Borowitzka, M. A. (1989) β- Carotene (Provitamin A) production with algae. In: Vandamme E (eds) *Biotechnology of Vitamins, Pigments and Growth Factors*. Springer, Netherlands, pp. 15-26.

Borowitzka, M. and Borowitzka, L. (1988). *Micro-algal biotechnology,* Cambridge University Press, Cambridge.

Borowitzka, M. A. (1992). Algal biotechnology products and processes-matching science and economics. *J. Appl. Phycol,* 4:267–279.

Borowitzka, M. A. (1999). Commercial production of microalgae: ponds, tanks, and fermenters. *Progr. Industri. Microbiol,* 35: 313-321.

Borowitzka, M. A. and Borowitzka, L. J. (1988). *Dunaliella* In: Borowitzka M. A., Borowitzka L. J. (eds) *Micro-algal biotechnology*. Cambridge University Press, Cambridge, UK, pp 27–58.

Borowitzka, M. A., Post, F., Borowitzka, L. J. (1982). The life cycle of *D. salina*. Australasian Society for Phycology and Aquatic Botany, Annual Meeting.

Bosma, R. and Wijffels, R. H. (2003). Marine biotechnology in education: A competitive approach. *Biomol. Engin,* 20: 125-131.

Campbell, C. J. (1997). *The coming oil crisis*. Multi-science Publishing Company and petro consultants S. A., Essex, England.

Çelekli, A. and Donmez, G. (2006). Effect of pH, light intensity, salt and nitrogen concentrations on growth and β-carotene accumulation by a new isolate of *Dunaliella* sp. *World. J. Microbiol. Biotechnol,* 22(2): 183-189.

Chidambara Murthy, K. N., Vanitha, A., Rajesha, J., Mahadeva Swamy, M., Sowmya, P. R. and Ravishankar, G. A. (2005). *In vivo* antioxidant activity of carotenoids from *Dunaliella salina,* a green microalga. *Life. Sci,* 76: 1381–1390.

Chisti, Y. (2007). Biodiesel from microalgae. *Biotechnol. Advan,* 25(3): 294-306.

Chisti, Y. (2008). Biodiesel from microalgae beats bioethanol. *Trends. Biotechnol* 26:126–131.

Comstock, G. W., Helzlsouer, K. I. and Bush, T. L. (1991). Prediagnostic serum levels of carotenoids and vitamin E as related to subsequent cancer in Washington County, Maryland. *Am. J. Clin. Nutr,* 53, 260S–264S.

Davidi, L., Katz, A. and Pick, U. (2012). Characterization of major lipid droplet proteins from *Dunaliella*. *Planta,* 236(1): 19-33.

Demming-Adams, B. and Adams, W. W. III, (2002). Antioxidants in photosynthesis and human nutrition. *Science,* 298:2149–2153.

Dismukes, G. C., Carrieri, D., Bennette, N., Ananyev, G. M. and Posewitz, M. C. (2008). Aquatic phototrophs: efficient alternatives to land-based crops for bio- fuels. *Curr. Opin. Biotechnol,* 19:235–240.

El-Baky, H. A., El-Baz, F. K., and El-Baroty, G. S. (2004). Production of antioxidant by the green alga *Dunaliella salina. Int. J. Agri. Biol,* 6:49-53.

Fabregas, J. and Herrero, C. (1985). Marine microalgae as a potential source of single cell protein (SCP). *Appl. Microbiol.Biotechnol,* 23(2): 110-113.

Feng, S., Li, X., Xu, Z. and Qi, J. (2014). *Dunaliella salina* as a novel host for the production of recombinant proteins. *Appl. Microbiol. Biotechnol,* 98:4293–4300.

Finney, K., Pomeranz, Y. and Bruinsma, B. (1984) Use of alga *Dunaliella* as a protein supplement in bread. *Cereal. Chemist,* 61: 402–406.

Franzmann, P. D. (1991). The microbia of saline lakes of the Vestfold hills, Antarctica. In: Rodriguez-Valera (eds) *General and Applied Aspects of Halophilic Microorganisms*. Plenum Press, New York, pp. 9-14.

Garcia-Gonzalez, M., Moreno, J., Manzano, J.C., Canavate, J. P., Florencio, F. J. and Guerrero, M. G. (2005). Production of *Dunaliella salina* biomass rich in 9-cis β-carotene and lutein in a closed tubular photobioreactor. *J. Biotechol,* 115:81–90.

Gomez, P. I. and Gonzalez, M. A. (2004). Genetic variation among seven strains of *Dunaliella salina* (Chlorophyta) with industrial potential, based on RAPD banding patterns and on nuclear ITS rDNA sequences. *Aquacultu,* 233:149–162.

Grizeau, D. and Navarro, J. M. (1986). Glycerol production by *Dunaliella tertiolecta* immobilized within calcium alginate beads. *Biotechnol. Lett,* 8: 261–264.

Hatanaka, Y., Kudo, T., Miyataka, M., Kobayashi, O., Higashihara, M. and Hiyama, K. (1999). Asymmetric reduction of hydroxyacetone propanediol in immobilized halotolerant microalga *Dunaliella parva. J. Biosci. Bioeng,* 88:281–286.

Hemschemeier, A., Melis, A. and Happe, T. (2009). Analytical approaches to photo-biological hydrogen production in unicellular green algae. *Photosyn. Res,* 102:523–540.

Jin, E. S., Feth, B. and Melis, A. (2003). A mutant of the green alga *Dunaliella salina* constitutively accumulates zeaxanthin under all growth conditions. *Biotechnol. Bioeng,* 81:115–124.

Johnson, E. A. and Schroeder, W. A. (1995). Microbial carotenoids. *Adv. Biochem. Eng. Biotechnol,* 53:119–178.

Katz, A. and Pick, U. (2001). Plasma membrane electron transport coupled to Na^+ extrusion in the halotolerant alga *Dunaliella. Biochim. Biophys. Acta,* 1504: 423–431.

Kumar, K., Dooslin, S., Mercy, B., Alexis, V. and Rajan, R., (2013). Evaluation of bioactive metabolites from halophilic microalga *Dunaliella salina* by GC-MS analysis. *Inte. J. Pharma. Pharmaceu. Sci,* 5(4):296-303.

Leon, R., Vila, M., Hernanz, D. and Vilchez, C. (2005). Production of phytoene by herbicide-treated microalga *Dunaliella bardawil* in two-phase systems. *Biotechnol. Bioeng,* 92:695–701.

Lerche, W. (1937). Untersuchungen über entwicklung und fortpflanzung in der Gattung *Dunaliella. Arch. Protistenkd,* 88:236-268. [Studies on development and reproduction in the genus *Dunaliella.*]

Li, Y., Wang, B., Wu, N. and Lan, C. Q. (2008). Effects of nitrogen sources on cell growth and lipid production of *Neochloris oleoabundans. Appl. Microbiol. Biotechnol,* 81(4): 629-636.

Mares-Perlman, J. A., Millen, A. E., Ficek, T. L. and Hankinson, S. E. (2002). The body of evidence to support a protective role for lutein and zeaxanthin in delaying chronic disease. Overview. *J. Nutr,* 132:5185–5245.

Massyuk, N. P. (1973). *Morphology, taxonomy, ecology and geographic distribution of the genus Dunaliella Teod. and prospects for its potential utilization.* Naukova Dumka, Kiev, pp. 242.

Mathews-Roth, M. M. (1987). Photoprotection by carotenoids. *Fed. Proc,* 46:1890–1893.

Mokady, S. (1992). Nutritional, toxicological, and therapeutic aspects. In: Avron M., Ben-Amotz A. (eds) *Dunaliella: Physiology, Biochemistry and Biotechnology*, CRC Press, Boca Raton, Florida, pp. 217–229.

Mokady, S., Abramovici, A. and Cogan, U. (1989). The safety evaluation of *Dunaliella bardawil* as a potential food supplement. *Food. Chemi. Toxicol,* 27: 221-226.

Moulton, T. P. and Burford, M. A. (1990). The mass culture of *Dunaliella viridis* (*Volvocales, Chlorophyta*) for oxygenated carotenoids: laboratory and pilot plant studies. *Hydrobiol,* 204-205(1): 401-408.

Nakas, J. P., Schaedle, M., Parkinson, C. M., Coonley, C. E. and Tanenbaum, S. W. (1983). System development for linked-ferment-ation production of solvents from algal biomass. *Appl. Environ. Microbiol,* 46 (5):1017-1023.

Nishino, H. (1998). Cancer prevention by carotenoids. *Mutat. Res,* 402: 159–163.

Omen, G. S., Goodman, G. and Thornquist, M. (1994). The β-carotene and retinol efficacy trial (CARET) for chemoprevention of lung cancer in high risk populations: smokers and asbestos exposed workers. *Cancer Res,* 54:2038–2043.

Oren, A. (2005). A hundred years of *Dunaliella* research: 1905-2005. *Saline Systems,* 1(1): 2. 1-14.

Peto, R., Doll, R., Buckley, J. D. and Sporn, M. B. (1981). Can dietary β-carotene materially reduce human cancer rates? *Nature* 290:201–208.

Pick, U. (2002). Adaptation of the halotolerant alga *Dunaliella* to high salinity. In: Lauchli A., Luettge U. (eds) *Salinity: environment-plants-molecules*, Kluwer Acad. Pub, Dordrecht, Netherlands, pp. 97-112.

Poppel, G. V. and Goldbohm, R. A. (1995). Epidemiological evidences for β- carotene and cancer prevention. *Am. J. Clin. Nutr* 62:1393S–1402S.

Pulz, O. and Gross, W. (2004). Valuable products from biotechnology of microalgae. *Appl. Microbiol. Biotechnol,* 65:635–648.

Raja, R., Anbazhagan, C., Lakshmi, D. and Rengasamy, R. (2004). Nutritional studies on *Dunaliella salina* (*Volvocales, Chlorophyta*) under laboratory conditions. *Seaweed Res. Util.* 26:127–146.

Raja, R., Hemaiswarya, S. N., Kumar, A., Sridhar, S. and Rengasamy, R. (2008) A perspective on the biotechnological potential of microalgae. *Crit. Rev. Microbiol.* 34(2): 77-88.

Schenk, P. M., Thomas-Hall, S. R., Stephens, E., Marx, U. C., Mussgnug, J. H., Posten, C., Kruse, O. and Hankamer, B. (2008). Second generation biofuels: high-efficiency microalgae for biodiesel production. *Bioenergy Res,* 1(1): 20-43.

Scott, S. A., Davey, M. P., Dennis, J. S., Horst, I., Howe, C. J., Lea-Smith, D. J. and Smith, A. G. (2010) Biodiesel from algae: challenges and prospects. *Curr. Opin. Biotechnol,* 21:277–286.

Shirai, F., Kunii, K., Sata, C., Teramoto, Y., Mizuki, E., Murao, S. and Nakayama, S. (1998). Cultivation of microalgae in the solution from the desalting process of soy sauce waste treatment and utilization of algal biomass for ethanol fermentation. *World J. Microbiol. Biotechnol,* 14:839-843.

Sialve, B., Bernet, N. and Bernard, O. (2009). Anaerobic digestion of microalgae as a necessary step to make microalgal biodiesel sustainable. *Biotechnol. Adv,* 27:409–416.

Sies, H. and Stahl, W. (1997) Carotenoid and intracellular communications via gap junction. *Int. J. Vitam. Nutr. Res,* 67:364–367.

Soletto, D., Binaghi, L., Lodi, A., Carvalho, J. C. M. and Converti, A. (2005). Batch and fed-batch cultivations of *Spirulina platensis* using ammonium sulphate and urea as nitrogen sources. *Aquacultur,* 243: 217–224.

Spolaore, P., Joannis-Cassan, C., Duran, E. and Arsene, I. (2006). Commercial applications of microalgae. *J. Biosci. Bioeng,* 101 (2): 87–96.

Stottrup, J. G. and McEvoy, L.A. (2003). *Live Feeds in Marine Aquaculture*: Blackwell Science, Oxford, UK, pp. 318.

Tafreshi, A. and Shariati, M. (2009). *Dunaliella* biotechnology: methods and applications. *J. Appl. Microbiol,* 107(1): 14-35.

Taherzadeh, M. J., Adler, L. and Liden, G. (2002). Strategies for enhancing fermentative production of glycerol--a review. *Enzyme. Microbial. Technol,* 31(1-2): 53-66.

Takagi, M., Karseno and Yoshida T. (2006). Effect of salt concentration on intracellular accumulation of lipids and triacylglyceride in marine microalgae *Dunaliella* cells. *J. Biosci. Bioengin,* 101(3): 223-226.

Teodoresco, E. (1906). Organisation et developpement du *Dunaliella*, nouveau genre de Volvocacée-Polyblepharidée. *Rev. Gén. Bot*, 18:353-371. [Organization and development of *Dunaliella*, a new genus of Volvocaceae-Polyblepharidae.]

Terao, J. (1989). Antioxidant activity of β-carotene-related carotenoid in solution. *Lipids,* 24:659–661.

Wang, B., Li, Y., Wu, N. and Lan, C. (2008). CO_2 bio-mitigation using microalgae. *Appl. Microbiol. Biotechnol,* 79(5): 707-718.

Williams, A., Pryor, B. S., Wilhelm, S., Cheryl, L. and Rock, R. D. (2000). B-carotene from biochemistry to clinical trials. *Nutr. Rev,* 1:39–53.

In: The Role of Photosynthetic Microbes…
Editors: K. Tripathi, N. Kumar et al.
ISBN: 978-1-53614-032-3

Chapter 9

BIOREMEDIATION BY MICROALGAE

Behl Kannikka, Charan Pasupuleti Sesha and Nigam Subhasha*

Algal Biotechnology Lab,Amity Institute of Biotechnology,
Amity University, Noida, Uttar Pradesh, India

ABSTRACT

A substantial upsurge in technology has been observed for efficient removal of hazardous contaminants and effective waste management techniques. Although several effective physicochemical methods have been developed, they require improvement. The focus, of late, has shifted towards exploiting biological, eco-friendly and cheaper modes of technology. Bioremediation is one such technique that utilizes microorganisms to degrade and transform contaminants into lesser harmful forms. Microalgae have emerged as an exciting and prospective contender for the removal of such toxic pollutants. The chapter highlights different mechanisms such as biosorption, bioaccumulation, biotransformation used by microalgae to effectively bioremediate a wide variety of toxic substances such as pesticides, heavy metals, radioactive compounds and petroleum products among other common pollutants.

Keywords: biosorption, contamination, environment, industrialization, microalgae

*Corresponding Author Email: snigam@amity.edu.

INTRODUCTION

Environmental pollution has become one of the major threats, continuously increasing manifold in magnitude with each passing year. Over the years, polluting waste materials have been conventionally disposed in landfills, by incineration, combustions, composting or by gasification (Karigar 2011). Although these methods are moderately effective they have quite a few long-term drawbacks. The challenge is to tackle the problem of environmental degradation in a safe, environmentally sound manner, with rational cost implications. Therefore, the focus is now shifting more and more towards using biological methods to solve this problem.

The technology that is being intensively studied currently is bioremediation. This is one such technology that utilizes micro-organisms to eradicate or neutralize any hazardous wastes from a contaminated site of soil and water (Priyadarshini et al., 2011). The process of detoxification by this mechanism targets the harmful chemicals by mineralization, transformation or alteration (Shannon and Unterman, 1993). Using biological materials for treatment is definitely more effective than conventional methods because such strategies can be used directly at the site of contamination, without having to transfer the polluted wastes. Oil and other petroleum products, solvents and pesticides are the major contaminants which can be treated using the process of bioremediation.

Bioremediation mainly depends on microorganisms that enzymatically attack the pollutants and convert them into innocuous products. This process accelerates the growth of microbes that use these contaminants as a source of food and energy and then degrade them into other compounds. It is more effective under certain environmental conditions that promote fast growth of microbes and degradation. However, it functions with better efficiency when the environmental conditions are manipulated, to further enhance the microbial growth and degradation (Gaur et al., 2013).

Most bioremediation systems usually function well under aerobic conditions, although anaerobic conditions may also support microbial growth and degradation. In such cases, enzymes are utilized to treat these polluted sites to regain their healthy condition again. This process is vastly accepted and favored, as it results in better degradation due to low cost and economic inputs, as compared to the more conventional methods. Thus, it may be safely suggested that bioremediation is more cost-effective and eco-friendly comparatively, and can be used as an effective solution for the treatment of contaminated sites.

Generally, the process of bioremediation can be executed *in-situ* as well as *ex situ*. This assessment is mostly dependent upon the nature and type of contaminants, the objective of the remedial action, availability of funds/time and viewpoints of the local people and concerned authorities. When the pollution impacts the shallow soil layers, excavation and consequent treatment at other sites (*ex situ* bioremediation) is often suggested either by making Biopiles (wherein contaminant material is piled on the ground surface) or by making biocells (wherein contaminant material is deposited and treated underground at a clean site) and both of which must be lined to contain the contaminant.

In situ bioremediation includes handling the polluted material at the site while *ex situ* involves the removal of the polluted material to be treated elsewhere (Aggarwal et al., 1990). This bioremediation can also be described as the process in which organic pollutants are degraded biologically under environmental conditions to either carbon dioxide, water or a less complex transformation product. The *in situ* bioremediation is a low-cost, low maintenance, environment-friendly and sustainable approach to treat and rectify polluted sites. With the need for removal from the site of the contaminated samples, for treatment, the expenditure of *ex situ* bioremediation process becomes higher compared to *in situ* methods.

While both *in situ* and *ex situ* remediation depend principally on microbial metabolism, *in situ* bioremediation methods are preferred to those of *ex situ* for ecological restitution of contaminated soil and water environments (Jorgensen, 2007). The three different types of *in situ* bioremediation process are (i) bioattenuation which depends on the natural process of degradation, (ii) biostimulation where intentional stimulation of degradation of chemicals is achieved by addition of water, nutrient, electron donors or acceptors, and (iii) bioaugmentation where the microbial members with proven capabilities of degrading or transforming the chemical pollutants are added (Madsen, 1991).

Ex situ bioremediation is a biological process involving the physical extraction of the contaminated media to another location for treatment. If the contaminant is present in the soil, then the soil is excavated and placed in a lined above-ground treatment area and aerated followed by processing to enhance the degradation of the organic contaminants by the indigenous microbial population. If the contamination has reached the groundwater, it must be pumped out and any contaminated soil must also be removed. There are two phases in *ex situ* bioremediation: solid phase and slurry phase. The most common types of solids bioremediation are (1) land farming or land treatment, (2) composting and (3) biopiles cells or mounds.

Micro-algae are photosynthetic organisms that are found in both marine and fresh waters. Their photosynthetic mechanism is like land-based plants, but due to a simple cellular structure, and an aqueous environment where they have efficient access to water, carbon dioxide, and other nutrients, microalgae are generally more efficient in converting solar energy into biomass (Prabha et al., 2016). They are the source of oxygen and the primary step in the food chains of aquatic systems. Algae, over the course of evolution, have fostered a wide range of polymers which can easily scavenge out for materials of interest, even when present in very low concentration in the environment. This capacity of the microalgae to absorb toxic materials makes them a perfect candidate for use in the process of bioremediation. Algae can efficiently utilize excess nitrogen, phosphorous and other metals present in industrial effluents for their growth and biomass production. More importantly, since microalgae are photosynthetic in nature, they do not require an extra carbon source, unlike other counterparts like bacteria, fungi, and yeast, that do require a start-up nutrient as an energy source (Priyadarshini et al., 2011).

Different Mechanisms of Bioremediation Followed by Microalgae

Different mechanisms of bioremediation are in use, these include bio-sorption, metal-microbe interactions, bioaccumulation, biomineralization, biotransformation, and bioleaching (Dixit et al., 2015).

Biosorption

Biosorption uses biological materials to eliminate contaminants using different mechanisms like absorption, adsorption, precipitation, surface complexation and ion exchange. It also relies heavily on several factors such as environmental conditions, biosorbents, substances to be biosorbed and certain metabolic processes. Both the mechanisms of adsorption and absorption are included under biosorption. The contaminants to be removed by the process of biosorption can be organic, inorganic, soluble and insoluble compounds. Highly mobile metal ions, e.g., potassium, magnesium have been easily removed using this process. Moreover, heavy metals, phenolic compounds, pesticides and a wide variety of textile dyes are also being removed effectively through this process.

Types of Biosorbents

Biosorbents can be broadly divided into two categories: living biomass and non-living biomass. Living biomass consists of bacteria, fungi, and algae (micro-algae, macro-algae, brown algae, red seaweeds) while non-living biomass includes industrial and agricultural wastes, natural residues and another biomass. Of late, non-living biomass has acquired considerable popularity as it does not require supplementation of additional nutrients and has a minimal maintenance cost. The easy availability of non-living biomass from any industrial waste is another advantage making the techniques more economical in use. The living biomass, on the other hand, depends on the proper maintenance of microbial cultures and requires constant monitoring of environmental factors as well as nutrient supplies. The process, however, is certainly more effective in reducing the long-term impact on the environment at the site as well as also having the potential to be really cost effective in real terms.

Biosorption Mechanism

The biosorption of metals by living cells is a two-step process. First, the metal ions are adsorbed to the surface of cells by interaction between metal–functional groups displayed on the cell's surface, such as carboxyl, amino, phosphate, hydroxyl, thiol, and sulfide functional groups among a few. This first step, also known as passive biosorption, is metabolism-independent and proceeds swiftly within several minutes by any one or a combination of the following metal-binding mechanisms: coordination, complexation, ion exchange, physical adsorption (e.g., electrostatic) or inorganic microprecipitation. Passive biosorption is a dynamic equilibrium of reversible adsorption-desorption.

The mechanisms of biosorption depicted in Figure 1 can be classified based on cellular metabolism and site of biosorption (Singh et al., 2012). Cellular metabolism can be further categorized into metabolism dependent and non-metabolism dependent biosorption.

Metabolism dependent biosorption occurs only in viable cells. It also plays an important role in defending microbes when dealing with toxic metals. Metabolism-dependent biosorption may be further divided into:

(i) Transport through the cell membrane: The microorganisms share the same mechanism to transport heavy metals across the membrane, as well as, to transport the metabolic ions like sodium, magnesium etc. Such type of transport is not involved with the metabolic activity and comprises of two steps:
   (a) Metabolism-independent binding, where the metal binds to the cell wall.
   (b) Metabolism dependent cellular uptake, which incorporates the transport of the metal ions across the cell membrane.

(ii) Metabolism independent precipitation: The precipitation is generally related to the defense mechanism of the microorganism. The reaction in the presence of toxic metals produces compounds which cause precipitation.

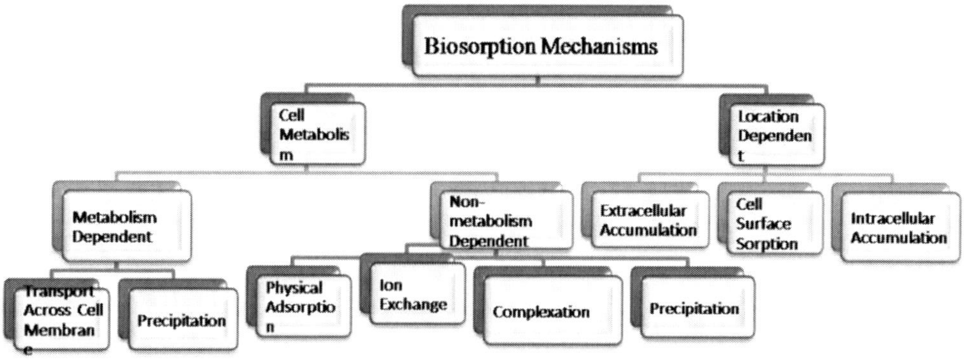

Figure 1. Mechanism of biosorption.

Non-Metabolism dependent biosorption includes the interaction of the functional groups present on the microbial cell surface and the metals (Fulekar, 2010). It can be categorized further into:

(i) Physical/Surface adsorption: A physical phenomenon that involves electrostatic forces and van der Waal's forces. Dead biomass of algae too, has shown substantial adsorption of heavy metals like cadmium, copper, cobalt, zinc, uranium, through electrostatic interactions (Figure 2).

(ii) Ion exchange: Where, the polysaccharides present in the microbial cell wall aid in the exchange of bivalent ions. These polysaccharides act as counter ions. Marine algae and alginates can absorb heavy metals by exchange of counterions.

(iii) Complexations: Complexations and adsorption of metals are due to the presence of carboxyl groups in microbial polysaccharides and polymers. When the metal ions interact with the functional groups present on the surface of the microorganism, they form cell surface complexes and are readily removed from the solution. Organic acids are known to play a vital role in removing toxic metals by solubilizing and leaching which results in the production of metallo-organic molecules.

(iv) Precipitation: Metabolism-independent precipitation is a resultant of a chemical interaction between the metal ions and the surface of the microorganism.

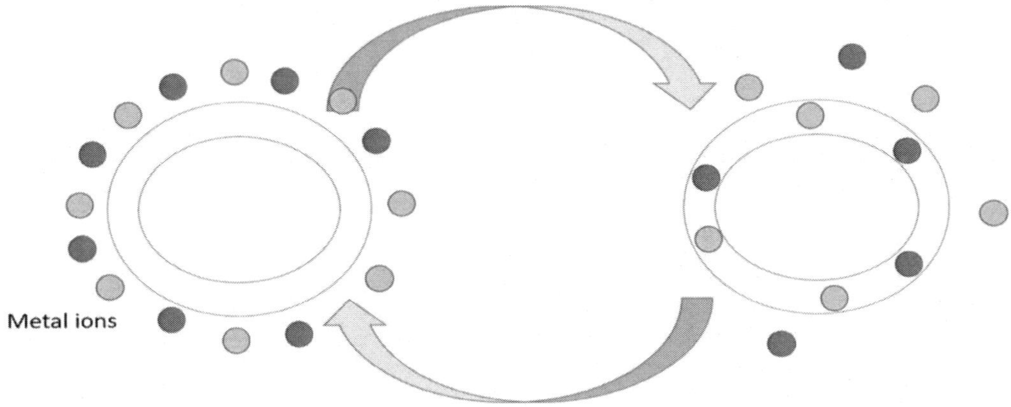

Figure 2. Biosorption of metals. (Source: Chojnacka K., 2009)

Advantages and Disadvantages of Biosorption

There are certainly, both advantages and disadvantages of biosorption by algae. Biosorption by algae is known to be an eco-friendly and a cost-effective method of removing contaminants. Algae have the potential to efficiently remove contaminants present even at low concentrations and produce no toxic sludge after bioremediation. In addition, microalgae also provide adequate opportunities for biosorption of metals so that they can be efficiently recovered. These biological biosorbents can be recycled again and have shown excellent removal efficiency. It can also be used *in situ.*

While biosorption by algae has several advantages, it also has certain limitations. Firstly, in some cases, the dead algal biomass raises concerns of its own toxicity at the site of bioremediation. Since the biosorbents are living microorganisms, there is no control over the biological characteristics of these biosorbents. Moreover, early saturation of biosorbents in some cases is also commonly observed.

BIOACCUMULATION

As already discussed, absorption of metals is a two-step process. The first- biosorption is a typical, reversible physicochemical process, involving binding of sorbate to sorbent of biological origin.

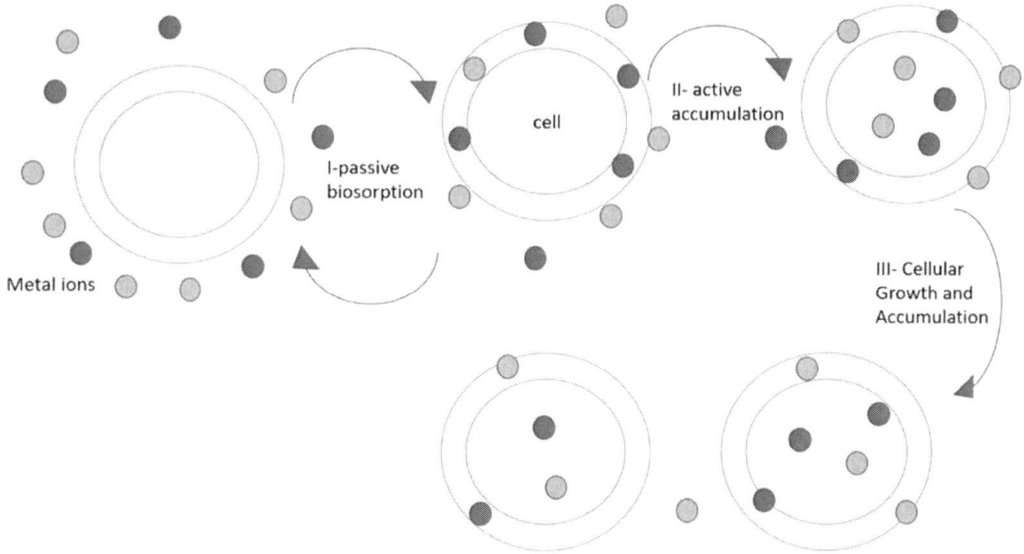

Figure 3. The stages of bioaccumulation of metal ions. Source: (Chojnacka K., 2009).

The second process, bioaccumulation, deals with the removal of metal ions, bound to the surface of the cell in the first, passive stage, which is identical with biosorption (Figure 3). These metal ions are then transported (generally by an active transport system, which requires additional cellular energy) to the cell's interior. Thus, a part of the metal cation is bound with the cell's surface and the remaining portion stays inside the cell. The process, therefore, is irreversible and makes it more difficult to recover the metal because simple elution processes would remove only the part of metal which was bound to the surface of the cell. In this case, destructive methods can be used including the combustion of the biomass and reuse of metal-laden ash.

The process of bioaccumulation itself enables removal of metal cations to very low levels (lower than biosorption), but it is more complex because it involves working with living organisms. The process installations should consist of the same equipment as for the cultivation of the biomass and it is required that the effluent simultaneously serves as a growth medium. There is also a danger that the metal would pose toxicity to the cells. For this reason, it is necessary to select bioaccumulating organisms very carefully, since they should have minimum growth requirements and should be tolerant to the toxins.

Molecular mimicry is one such mechanism, whereby, the metal ions either compete for binding to multivalent ion carriers, or, after binding to low molecular weight thiols (such as cysteine), enter the cell by active transport. In another type of mechanism, metal ions bound to chelating proteins (metallothionein) may enter the cell by endocytosis (Van Ho et al., 2002; Zalups and Ahmad, 2003). Metal ions can also enter the cells if the cell wall is disrupted by natural or artificial force (Wang and Chen, 2009). After entering, the metal ions are compartmentalized into different sub-cellular organelles. However, it seems that the intracellular accumulation is an energy-driven process dependent on active metabolism. Even though many parameters partake in accumulation, it is clear that different species of algae accumulate heavy metal ions to various degree. Based on location, the location-dependent biosorption can be classified from where the metals removed from the solution accumulate.

Table 1. Characteristic differences between biosorption and bioaccumulation

Biosorption	Bioaccumulation
Reversible, passive process	Partially reversible, active process
Metals are bound to dead biomass	Metals are bound to the cell surface and accumulated in living biomass
Nutrient independent adsorption process	Nutrient dependent absorption process
Fast single stage process	Slow double-stage process
No scope for toxic effects	Scope for toxic effects on cell growth

They include: (i) Extracellular accumulation/precipitation, (ii) Cell surface sorption/precipitation, and (iii) Intracellular accumulation. Extracellular accumulation/ precipitation can be enabled using feasible microorganisms, by cell surface sorption or by complexation, which can be ensured both by living and dead microorganisms, while intracellular accumulation entails microbial activity. Though both dead, as well as living cells, are capable of metal accumulation, the difference between the two lies in the mechanisms involved, which depends upon the extent of metabolic dependence (Wang and Tay, 2010). The fundamental differences between the characteristics of biosorption by dead biomass and bioaccumulation using living cells are given in Table 1. An array of wastewaters and other industrial effluents can easily be treated by both biosorption and bioaccumulation. Metals that can be treated include Ag, Al, Au, Cd, Co, Cr, Cu, Fe, Hg, Mn, Mo, Ni, Pb, Pt, Se, U, V, and Zn.

BIOTRANSFORMATION

The role of microorganisms in the biotransformation of heavy metals into nontoxic forms is well-studied, and understanding the molecular mechanism of metal accumulation has numerous biotechnological implications for bioremediation of metal-contaminated sites. Biotransformation can strongly modify the bioaccumulation of chemicals in an organism.

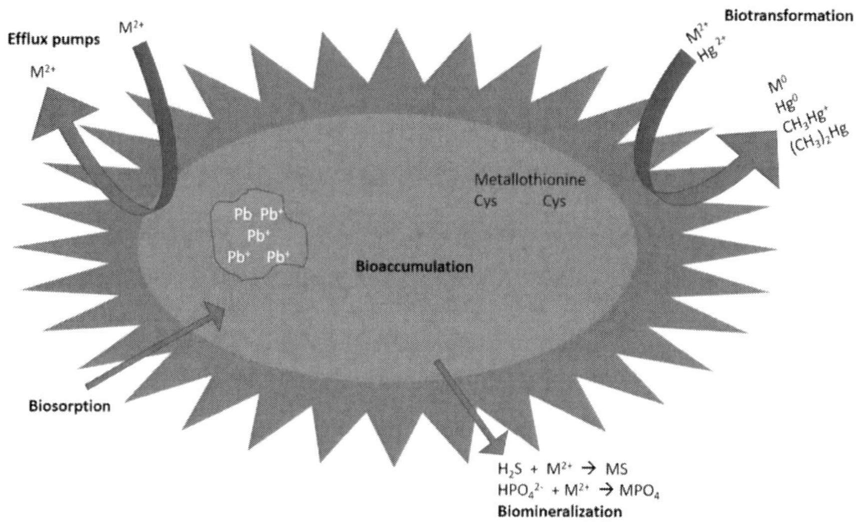

Figure 4. Different mechanisms of bioremediation by microalgae.

Biotransformation of various contaminants is a sustainable way to remediate the polluted environment. These bioremediation and biotransformation methods harness the naturally occurring, microbial catabolic diversity to degrade, transform or accumulate a colossal variety of compounds which include hydrocarbons (oils), polychlorinated biphenyls (PCBs), polyaromatic hydrocarbons (PAHs), pharmaceutical substances, radionuclides and metals (Dixit et al., 2015). Algae are important components of the freshwater aquatic environment and have the potential to remediate arsenic-contaminated waters in wetlands through adsorption and biotransformation of inorganic arsenic. Microalgae have been shown to accumulate arsenic (V), with bioconcentration factors ranging from 200–4000 (Figure 4).

BIOMINERALIZATION

Biomineralization is a process by which living organisms yield minerals. Biomineralization processes usually result in the hardening or stiffening of the mineralized materials. This includes the formation of silicates in algae and diatoms, carbonates in invertebrates, and calcium phosphates and carbonates in the hard tissues of vertebrates (Elekwachi et al., 2014).

Mineralization by algae can be classified into two categories, (i) Extracellular mineralization, and (ii) Intracellular mineralization. Extracellular mineralization comprises crystal formation on the outer side of the cell wall and in the cell wall as well. The crystals present on the outer surface are randomly oriented, signifying that they are not growing within or on the oriented molecule. If the outer cell wall is covered in a sheath, this may be due to the presence of crystals that develop within the sheath or between the wall and the sheath. The intracellular mechanism, on the other hand, is related to the spaces sheltered between the cells. It is very common in invertebrates and higher plants. The only known algal group that uses this method of calcium carbonate deposition, is *Coccolithophoridae* sp.

BIOLEACHING

Bioleaching involves utilization of microbes to dissolve metals instead of chemical solutions. Bioleaching has been used to dissolve metals such as nickel, copper, zinc, cobalt, gold, lead, arsenic and others. The process works by using specific bacteria that can essentially "eat" the metal content out of ore. It is typically used for small quantities of metal ore or when environmental treatment is a concern. The bioleaching process is also being researched for its use in the mining of small ore bodies. Bioleaching of uranium and copper has been widely commercialized. Algal species such as *Scenedesmus quadricauda* and *Nostoc linkia* have been able to successfully leach out uranyl ions from uranium ores. However, the major drawback lies in the large-scale bioleaching, which may cause environmental problems when the dump is not managed properly. This also results in seepage of leach fluids containing a large number of metals and low pH into nearby natural water supplies and groundwater.

Leaching Process

There are three types of commercial methods used in leaching: (i) Slope leaching, (ii) Heap leaching, and (iii) *In-situ* leaching.

i. Slope Leaching: Around 10,000 tonnes of ores are ground first to get a fine powder. It is then dumped into large piles down a mountainside leaching dump. Water containing microbial inoculum is continuously sprinkled over the pile. The water is then collected at the bottom and is used to extract metals.
ii. Heap Leaching: The metal ore is dumped into large heaps called the leach dump. Further steps of treatment are same as described for slope leaching.
iii. *In-situ* Leaching: In this process, the ores remain in its original position on the earth. Surface blasting of rock is done just to increase the permeability of water. Thereafter, water enriched with the microbial inoculum is pumped through the drilled passage to the ores. The acidic water seeps through the rock and collects at the bottom. Again, from the bottom, water is pumped, the mineral is extracted and water is recycled for further reuse.

BIOREMEDIATION OF PESTICIDES

A pesticide is a synthetic or naturally derived substance or a mixture of substances used to destroy, suppress or alter the life cycle of any pest. Pesticides such as chlorogenic compounds, aldrin, heptachlor, endosulfan, DDT (Figure 5) are emerging as potential environmental threats as they accumulate inside living tissues resulting in serious environmental threats. Increased use of pesticides has resulted in problems like cardiopulmonary, neurological and skin disorders, fetal deformities, miscarriages, lowering the sperm count of applicators etc. (Bag, 2000). Some microorganisms can use these pesticides as an energy source for their growth and simultaneously effectively mineralize and detoxify the compounds.

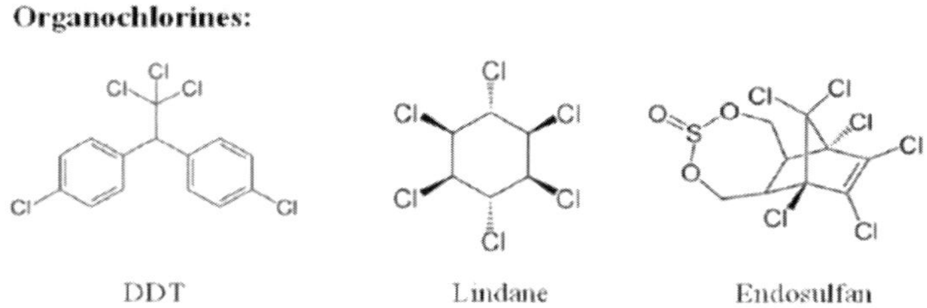

Figure 5. Some common structures of pesticides.

An alternate mode of degradation is when microorganisms use their enzymatic activities to degrade bonds in the pesticide molecules without capitalizing any benefit from it (Larsson, 1990). Different species of microalgae have different sensitivity to pesticides. The response to the pesticide is reliant upon factors such as the concentration of the pesticide, and its exposure time. The biodegradation of pesticides majorly depends on two factors: the optimum condition of the microorganisms used, and on the chemical nature of the pesticide. Studies and research on bioremediation of pesticides by algae have shown that not only microalgae are capable of bioaccumulating pesticides, but they are also capable of biotransforming some of these environmental pollutants. Different microalgae species employed in bioaccumulation and biotransformation of pesticides is depicted in Table 2.

BIOREMEDIATION OF HEAVY METALS

Heavy metals are naturally occurring elements that have a high atomic weight and a density at least 5 times greater than that of water. The presence of heavy metal ions such as lead, copper, cadmium, zinc, and nickel have further contaminated the already polluted industrial wastewater. Biosorption and bioaccumulation of such heavy metal ions in the

Table 2. Different microalgae species in bioaccumulation and biotransformation of pesticides

Microalgae	Bioaccumulation	Biotransformation
Chlamydomonas sp.	Mirex	Lindane, naphthalene, phenol
Chlorella sp.	Toxaphene, methoxychlor	Lindane, Chlordimeform
Chlorococcum sp.	Mirex	
Cylindrotheca sp.	DDT	
Dunaliella sp.	Mirex	DDT, naphthalene
Euglena gracilis	DDT, parathion	Phenol
Scenedesmus obliquus	DDT, parathion	Naphthalene sulfonic acid
Selanastrum capricornatum	Benzene, toluene, chlorobenzene, 1,2-dichlorobenzene	Benzo[a]pyrene

(Source: Priyadarshini et al., 2011)

food chain are causing major health problems (Sridhara et al., 2008). Physicochemical methods to remove these heavy metals include chemical precipitation (Charerntanyarak, 1999), ion exchange (Da Dabrowski et al., 2004), electro-kinetic (Yuan and Weng, 2006), membrane processing (Qdais and Moussa, 2004) and adsorption (Lee et al., 2012; Goharshadi and Moghaddam, 2015).

A major drawback of these methods is that while they show promising results in laboratory scale, the high costs of the chemicals and inadequate removal of the ions at the industrial scale make it difficult to operate efficiently and economically. Biosorption of heavy metal ions in wastewater by algae can offer a better alternative as it is an ecologically safer, cheaper, and a more efficient means to degrade and eradicate metal ions from wastewater. Indeed, algae can be used for sorption of toxic and radioactive metal ions (Pohl and Schimmack, 2006), and to recover precious metal ions like gold and silver (Darnall et al., 1986; Mata et al., 2009a).Ion exchange is one vital mechanism for the biosorption of heavy metal ions by algal biomass (Michalak and Chojnacka, 2010; Mehta et al., 2002a).

Heavy metal ion accumulation by microorganisms generally occurs in two phases (Monteiro et al., 2011a, 2012). The first phase occurs on the cell surface and involves fast inactive biosorption, independent of cellular metabolism. Algal cell walls are encircled by a three-dimensional network of macromolecules (polysaccharides and proteins) that carry negatively charged functional groups (carboxyl, hydroxyl, phosphate) or amine groups. Metal ions are easily absorbed by algal cell walls as they are present generally in cationic form. The biosorption depends on the pH condition and acidic pH (3-5) is favorable for biosorption. The second phase consists of active sorption of heavy metal ions into the cytoplasm of algal cells. This phase is dependent on cell metabolism and is known as intracellular ion uptake (Tabatabaei et al., 2013). Brown algae are potential candidates for the biosorption of heavy metal ions based on the comparison of different algal strains and biomass-metal ion affinity. Table 3 shows different microalgal species used for bioremediation of heavy metals.

Table 3. Different microalgae species used for bioremediation of heavy metals

Metal	Algae
Al(III)	*Laminaria japonica*
As(III)	*Ulothrix cylindricum*
Au(III)	*Fucusvesiculosus*
Cd(II)	*Ascophyllum nodosum, Asparagopsis armata, Chlorella vulgaris, Chondrus crispus, Cladophora fracta, Chlamydomonas reinhardtii, Codium vermilara, Laminaria japonica, Fucusspiralis, Spirogyra insignis, Ulva lactuca.*
Cr(II)	*Laminaria japonica*
Cr(III)	*Chlorella miniata, C. sorokiniana, Rhizoclonium heiroglyphicum, Spirogyra condensate, Spirogyra* sp.
Cr(VI)	*Chlorella vulgaris, Chlamydomonas reinhardtii, Dunaliella* sp., *Scenedesmus incrassatuluss, Spirogyra* sp., *Ulva lactuca*
Cu(II)	*Ascophyllum nodosum, Asparagopsis armata, Chlorella vulgaris, Chondrus crispus, Cladophora fascicularis, C. crispate, Cladophorasp, Codium vermilara, Fucus spiralis, F. vesiculosus, Laminaria japonica, Sphaero pleasp, Spirogyra insignis, S. neglecta, Ulothrix zonata, Ulva fasciata.*
Hg(II)	*U. lactuca, Chlamydomonas reinhardtii*
Ni(II)	*Ascophyllum nodosum, Asparagopsis armata, Chlorella miniata C. sorokiniana, C. vulgaris, Chondrus crispus, Codium vermilara, Fucus spiralis, F. vesiculosus, Sphaeroplea* sp.
Pb(II)	*Ascophyllum nodosum, Asparagopsis armata, Chondruscrispus, Cladophora fascicularis, C. fracta, Cladophora* sp, *Chlamydomonas reinhardtii, Codium vermilara, Fucus spiralis, F. vesiculosus, Laminaria japonica, Spirogyra insignis, S. Neglecta*
Se(IV)	*Cladophora hutchinsiae*
U(VI)	*Chlorella vulgaris*
Zn(II)	*Ascophyllum nodosum, Asparagopsis armata, Chondrus crispus, Cladophora crispate, Codium vermilara, Fucus spiralis, Aminaria japonica, Scenedesmus obliquus, Spirogyra insignis*

(Source: Zeraatkar et al., 2016; Romera et al., 2007)

Table 4. Degradation of various petroleum compounds by algal species

Algae	Compound
Selanastrum capricornatum	Benzene, toluene, naphthalene, phenanthrene, pyrene
Cyanobacteria	Benzene, toluene, naphthalene, phenanthrene, pyrene
Microcystis aeruginosa	Acrylonitrile

(Source: Priyadarshini et al., 2011)

BIOREMEDIATION OF PETROLEUM HYDROCARBONS

The accumulation of petroleum hydrocarbons in the environment results in serious problems, affecting the stability of aquatic ecosystems and health. Remedial action on a contaminated site can include *in-situ* or *ex-situ* action. These remediation methods include excavation and landfill disposals or incineration. However, these methods are expensive and only transfer the contamination from one place to another. Bioremediation can easily degrade a wide variety of organic pollutants including petroleum hydrocarbons. This is fast emerging as a promising and an environmentally friendly new technology. But it has certain drawbacks, such as lack of local micro-flora that can efficiently degrade a broad range of petroleum compound having the capability of microorganisms to resist the toxic metals present at the contamination site. Another major area of concern is the breakdown of hydrocarbons since they are more hydrophobic in nature and in turn less bioavailable.

Many microalgae can accumulate lipids due to excess of photosynthate enzyme and some species can accumulate the number of lipids under heterotrophy or environment stress, such as nutrient deficiency or salt stress (Takagi et al., 2006). Algal species such as *Scenedesmus quadricauda* and *Chlorella* sp. have shown a positive effect on bioremediation of industrial wastewater. They utilize this wastewater as a cheap source of nutrient for growth and subsequent oil accumulation. Microalgae species *Protothecazopfi* can efficiently bioremediate crude oil and mixed hydrocarbons and has also exhibited extensive biodegradation of n-alkanes and iso-alkanes as well as in aromatic hydrocarbon compounds. Different groups of microalgae are capable of degradation of various petroleum compounds as shown in Table 4.

Bioremediation of Polycyclic Aromatic Hydrocarbons (PAHS)

Polycyclic aromatic hydrocarbons (PAH) are a group of chemicals that have two or more fused aromatic rings in linear, angular or cluster arrangements. Some PAHs are possible human carcinogens, and hence their distribution and exposure to humans have been the focus of much attention. PAHs are relatively stable and recalcitrant in soil and less easy to degrade than many other organic compounds. They are difficult to remove from contaminated soil using other treatments. Research on the degradation of PAHs by biological organisms like bacteria, algae, fungi etc. has shown fruitful results. Table 5 shows degradation of various polycyclic aromatic hydrocarbons by a different group of microalgae. These microorganisms have catabolic abilities that can effectively remediate PAH contaminated water and soil sites. Currently, bioremediation has been shown to be effective in remediating soils polluted with low molecular weight PAHs however, the high molecular weight PAHs are generally recalcitrant to microbial attack (Park et al., 1990; Erickson et al., 1993; Cerniglia, 1992).

Bioremediation of Radioactive Compounds

Uranium has become one of the most serious heavy metal threats to the environment because of its high toxicity and radioactivity. Large quantities of uranium have found their ways into the environment through the activities associated with the nuclear industry. Conventional methods to remove heavy metals from industrial effluents are often unsuccessful and costly when applied to diluted and very diluted effluents (Aksu 1998). That marine algais capable of absorbingradionuclides, such as radium, thorium, and uranium has been known for a long time. The biosorption of uranium by *Cystoseira indica*, a brown algae biomass has been reported as far back as the dawn of the 21st century (Khani et al., 2006).

Table 5. Bioremediation of polycyclic aromatic hydrocarbons by different microalgae species

PAH Compound	Microalgae species
Fluoranthene	*Chlorella vulgaris, Scenedesmus platydiscus, Scenedesmus quadricauda and Raphidocolis captricornutum*
Pyrene	*Chlorella vulgaris, Scenedesmus platydiscus, Scenedesmus quadricauda and Raphidocolis captricornutum*
Naphthalene	*Oscillatoria* sp., *Microcoleus chthonoplastes, Nostoc* sp., *Anabaena* sp., *Agmenellum quadruplicatum, Coccochloris elabens, Aphanocapsa* sp., *Chlorella sorokiniana, Chlorella autotrophica, Dunaliella tertiolecta, Chlamydomonas angulosa, Ulva fasciata, Cylindrotheca* sp., *Amphora* sp., *Nitzschia* sp., *Navicula* sp., *Porphyridium cruentum*
Phenanthrene	*Oscillatoria* sp., *Agmenellum quadruplicatum*
Benzo[a] pyrene	*Selenastrum capricornutum*

(Source: Juhasz, L. A., et al., 2000)

Table 6. Bioremediation of radioactive compounds by different species of microalgae

Algae	Radioactive compound	References
Stigonema ocellatum, Oedogonium sp., *Egeria densa, Desmids*	Strontium	Shin-ya Fukuda et al., (2014) (Minna R. Krejci, (2011)
Batracho spermum, virgato-decaisneanum, Chloroidium saccharophilum	Cesium	Shin-ya Fukuda et al., (2014)
Nostoc commune, Scytonema javanicum, Stigonema ocellatum, Ophiocytium sp.	Iodine	Shin-ya Fukuda et al., (2014)
Cystoseira indica	Uranium	Edgington et al., (1970).

(Source: Priyadarshini et al., 2011)

Uranium bioleaching has become a popular way of disposal and extraction of uranium from contamination sites. Although less amount of uranium is obtained, bioleaching is still effective in the extraction of uranyl ions. To get one tonne of uranium, a thousand tonnes of uranium ore must be handled. *In situ* uranium leaching is gaining vast acceptance. However, uranium leaching from ore on a large scale is widely practiced in the USA, South Africa, Canada, and India. Different groups of microalgae used in bioremediation of various radioactive compounds shown in Table 6.

Bioremediation of Explosives

Trinitrotoluene (TNT) is a pervasive pollutant that has been widely used as an explosive around the world for decades. Water used for washing during explosives manufacturing and loading operations is contaminated with the formulation being handled. Because the cleanup of areas contaminated by explosives is now a public health concern, considerable effort has been invested in finding economical remediation technologies. Biological treatment processes are often considered since these are usually the least expensive means of destroying organic pollution. Marine red alga *Portieria hornemannii* has reportedly shown capabilities to biodegrade TNT from seawater (Cruz-Uribe, 2007).Green alga *Acrosiphoniacoalita* and red algae *Porphyra yezoensis* species possess a metabolic route to remove the explosive compound 2,4,6-trinitrotoluene (TNT) from seawater (Rylott, 2011). They can effectively reduce TNT to 2-amino-4,6-dinitrotoluene and 4-amino-2,6-dinitrotoluene, that are less toxic in nature. Table 7 shows biodegradation of various explosives by different species of microalgae.

Table 7. Bioremediation of explosives by different species of microalgae

Microalgae	Explosive	Reference
Acrosiphonia coalita, *Porphyra yezoensis,* *Portieria hornemannii*	TNT	Octavio Cruz-Uribe et al., (2007)

(Source: Priyadarshini et al., 2011)

Conclusion

Bioremediation is a pollution control technology that can be channelized to utilize biological organisms to degrade or transform a wide variety of toxins into less harmful forms. It is a comparatively inexpensive and competent scheme to decontaminate pollutants that are gradually increasing in popularity of use to efficiently reduce environmental pollution. Several physicochemical methods like precipitation, ion-exchange, evaporation etc. have been developed to remove contaminants. However, these methods have certain drawbacks and hence are unsuccessful in effective removal of the contaminants. Developing a sustainable and effective biological treatment system has become the need of the hour. Such systems, if planned pragmatically, are economically cheaper and positively more environmentally friendly than the current physicochemical methods. In this context,algae aregaining more importance as bioremediation agents and are already being used effectively in bioremediation processes more and more. A wide variety of microalgae, macroalgae, green, red and brown algae have shown remarkable potential for the removal of different pollutants such as industrial wastes, organic and inorganic contaminants, heavy metals, oil effluents, petroleum and aromatic compounds, explosives and radioactive compounds as well from wastewaters. Algae offer tremendous potential in the successful and efficient purification of contaminants due to its low cost and having negligible or low secondary pollution problems.

ACKNOWLEDGMENTS

The authors are thankful to Amity Institute of Biotechnology, Amity University, Noida, Uttar Pradesh, India.

CONFLICT OF INTEREST

No conflict of interest among the authors.

REFERENCES

Aggarwal, P. K., Means, J. L., Hinchee, R. E., Headinoton, G. L. Gavaskar A. R. (1990). *Methods to select chemicals for In Situ biodegradation of Fuel Hydrocarbons.* ESL-TR-90-13.

Aksu Z. Biosorption of heavy metals by microalgae in batch and continuous system edited by Y. S. Wong, N. F. Y. Tam (Eds.), *Wastewater Treatment with Algae,* Springer (1998), pp. 37–53.

Bag, D. (2000). Pesticides and health risks. *Econ. Politic. Wkly,* 6(16): 20 21.

Cerniglia, C. E., van Baalen, C., and Gibson, D. T. (1992a). Metabolism of phenanthrene by the cyanobacterium *Agmenellum quadruplicatum* PR-6. *Appl. Environ. Microbiol,* 58, 1352-1359.

Charerntanyarak, L. (1999). Heavy metals removal by chemical coagulation and precipitation. *Water. Sci. Technol.* 39 (10), 135e138.

Chojnacka, K. (2009). *Biosorption and bioaccumulation in practice*, Nova Science Publishers.

Cruz-Uribe, O., Cheney, D. P., and Rorrer, G. L. (2007). Comparison of TNT removal from seawater by three marine macroalgae. *Chemosphere,* 67(8):1469-1476.

Da Dabrowski, A., Hubicki, Z., Podkoscielny, P., and Robens, E. (2004). Selective removal of the heavy metal ions from waters and industrial wastewaters by ion-exchange method. *Chemosphere,* 56 (2): 91-106.

Darnall, D. W., Greene, B., Henzl, M. T., Hosea, J. M., McPherson, R. A., and Sneddon, J., and Alexander, M. D. (1986). Selective recovery of gold and other metal ions from an algal biomass. *Environ. Sci. Technol.* 20 (2): 206-208.

Dixit, R., Wasiullah, Malviya, D., Pandiyan, K., Singh, U. B., Sahu, A., Shukla, R., Singh B. P., Rai, J. P., Sharma, P. K., Lade, H., and Paul, D. (2015). *Bioremediation of Heavy Metals from Soil and Aquatic Environment: An Overview of Principles and Criteria of Fundamental Processes Sustainability* 7(2):189-212.

Edgington, D. N., Gordon, S. A., Thommes, H. L. M., and Almodovar, L R. (1970). The concentration of radium, thorium, and uranium by tropical marine algae. *Limnol. Oceanogr.* 15, 945-955.

Elekwachi, C. O., Andresen, J., and Hodgman, T. C. (2014). Global use of bioremediation technologies for decontamination of ecosystems. *J. Bioremed. Biodeg,* 5: 225.

Erickson, D. C., Loehr, R. C., and Neuhauser, E. F. (1993). PAH loss during bioremediation of manufactured gas plant site soil. *Water. Research,* 27, 911-919.

Fukuda, S., Iwamoto, K., Atsumi, M. Yokoyama, A., Nakayama, T., Ishida, K., et al. (2014). Global searches for microalgae and aquatic plants that can eliminate radioactive cesium, iodine,and strontium from the radio-polluted aquatic environment: a bioremediation strategy. *J. Plant. Res*, (2014) 127: 79. doi: 10.1007/s10265-013-0596-9.

Fulekar, H. (2010). *Bioremediation Technology: Recent Advances,* Springer, Netherlands

Gaur, N., Flora, G., Yadav, M., and Tiwari, A. (2013). A review with recent advancements on bioremediation based abolition of heavy metals. *Environ. Sci. Processes Impacts.*

Goharshadi, E. K., and Moghaddam, M. B. (2015). Adsorption of hexavalent chromium ions from aqueous solution by graphene nanosheets: kinetic and thermodynamic studies. *Int. J. Environ. Sci. Technol.* 12 (7): 2153-2160.

Jorgensen K. S. (2007). *In situ* bioremediation. *Adv. Appl. Microbiol,* 61: 285–305.

Karigar, C. S., and Rao, S. S. (2011). Review of microbial enzymes in the bioremediation of pollutants: A review. *Enzyme research.*

Khani, M., Keshtkar, A., Meysami, B., Zarea, M., and Jalali, R. (2006). Biosorption of uranium from aqueous solutions by nonliving biomass of marine algae. *Cystoseiraindica. Electronic Journal of Biotechnology,* 9 (2).

Krejci, M. R., Finney, L., Vogt, S., and Joester, D. (2011). Selective sequestration of strontium in desmid green algae by biogenic co-precipitation with barite. *Chem. Sus. Chem*, 4: 470–473.

Larsson, T. (1990). *Sidoeffekter av svampbekampningsmedel.* SLU Info/ Redaktionen. Fakta – Mark/Vaxter, no 6 [*Side Effects of Mushroom Pollutants*. SLU Info / Editors. Facts - Mark / Growth, no 6]

Lee, S. M., Laldawngliana, C., and Tiwari, D. (2012). Iron oxide nano-particles-immobilized-sand material in the treatment of Cu(II), Cd(II) and Pb(II) contaminated wastewaters. *Chem. Eng. J*, 195e196 (0), 103-111.

Madsen, E. L. (1991). Determining *in situ* biodegradation: facts and challenges. *Environ. Sci. Technol*, 25:1663-1673.

Michalak, I., and Chojnacka, K. (2010). Interactions of metal cations with anionic groups on the cell wall of the macro-alga. *Vaucheria* sp. *Eng. Life Sci*, 10 (3):209-217.

Monteiro, C., Castro, P. L., and Malcata, F. X. (2011a). Microalga-mediated bioremediation of heavy metal contaminated surface waters. In: Khan, M. S., Zaidi, A., Goel, R., Musarrat, J. (eds.), *Biomanagement of Metal-contaminated Soils*. Springer, Netherlands, pp. 365-385.

Monteiro, C. M., Castro, P. M. L., and Malcata, F. X. (2012). Metal uptake by microalgae: underlying mechanisms and practical applications. *Biotechnol. Prog*, 28 (2), 299-311.

Octavio, Cruz-Uribe., Donald, P. Cheney., Gregory, L., and Rorrer. (2007). Comparison of TNT removal from seawater by three marine macroalgae. *Chemosphere*, 67 (8): 1469-1476.

Park, K. S., Sims, R. C., and Dupont, R. (1990). Transformations of PAHs in soil systems. *J. Environ. Eng*, (ASCE) 116, 632-640.

Pohl, P., and Schimmack, W. (2006). Adsorption of radionuclides (134Cs, 85Sr, 226Ra, 241Am) by extracted biomasses of cyanobacteria (*Nostoc Carneum*, *N. Insular*, *Oscillatoria Geminata* and *Spirulina Laxis-sima*) and Phaeophyceae (*Laminaria digitata* and *L. Japonica*; waste products from alginate production) at different pH. *J. Appl. Phycol*, 18 (2):135-143.

Prabha, Y., Soni, S. K., Gupta S., and Sonal. (2016). Potential of algae in bioremediation of wastewater: current research. *Int. J. Curr. Microbiol. App. Sci*, 5(2):693-700.

Priyadarshani, I., Sahu, D., and Rath, B. (2011). Microalgal bioremediation: Current practices and perspectives. *J.Biochem. Technol*, 3(3): 299-304.

Qdais, H.A., and Moussa, H. (2004). Removal of heavy metals from wastewater by membrane processes: a comparative study. *Desalination*, 164 (2):105-110.

Romera, E., Gonzalez, F., Ballester, A., Blazquez, M. L., and Mun, J. A. (2007). Comparative study of biosorption of heavy metals using different types of algae. *Bioresour. Technol*, 98 (17): 3344-3353.

Rylott, E. L., Lorenz, A., and Bruce, N. C. (2011).Biodegradation and biotransformation of explosives. *Curr. Opin. Biotechnol*, 22, (3): 434–440.

Shannon, M. J., and Unterman, R. (1993). Evaluating bioremediation: distinguishing fact from fiction. *Annual. Rev.Microbiol*, 47, 715 (24).

Singh, N., and Gadi, R., J. (2012). Bioremediation of Ni (II) and Cu(II) from wastewater by the nonliving biomass of *Brevundimonas vesicular*. *Environ. Chem. Ecotoxicol*, 4,137–142.

Sridhara Chary, N. C. T., Kamala, R. D., and Samuel, S. (2008). Assessing risk of heavy metals from consuming food grown on sewage irrigated soils and food chain transfer. *Ecotoxicol Environ Saf,* 69: 513-524.

Tabatabaei, M., Mohtashami, S. K., Tohidfar, M., and Moradi, F. (2013). Comparative salt stress study on intracellular ion concentration in marine and salt-adapted freshwater strains of microalgae. *Not. Sci. Biol*, 5 (3): 309-315.

Takagi, M., and Yoshida, T. (2006). Effect of salt concentration on intracellular accumulation of lipids and triacylglyceride in marine microalgae *Dunaliella* cells. *J. Biosci. Bioeng*, 101: 223-226.

Van, Ho, A., Ward, D. M., Kaplan, J. (2002). Transition metal transport in yeast. *Ann. Rev. Microbiol,* 56, 237–261.

Wang, J., and Chen, C. (2009). Biosorbents for heavy metals removal and their future, *Biotechnol. Adv,* 27, 195–226.

Wang, L. K., Tay, J. H., and Hung, T. U. (2010). *Handbook of environmental bioengineering,* Vol (11), Humana Press, USA.

Yuan, C., and Weng, C. H. (2006). Electrokinetic enhancement removal of heavy metals from industrial wastewater sludge. *Chemosphere,* 65 (1): 88-96.

Zalups, R. K., and Ahmad, S. (2003). Molecular handling of cadmium in transporting epithelia. *Toxicol. App. Pharmacol*, 186, 163–188.

In: The Role of Photosynthetic Microbes…
Editors: K. Tripathi, N. Kumar et al.
ISBN: 978-1-53614-032-3

Chapter 10

Recent Advances in the Production and Purification Technology of Phycobiliproteins: A Sustainable Approach

***Vinod K. Kannaujiya*[1], *Md. Akhlaqur Rahman*[2,3], *Adi Nath*[2], *Shailendra K. Singh*[2], *Shanthy Sundaram*[2] *and Rajeshwar P. Sinha*[4,*]**

[1]Department of Botany, MMV,
Banaras Hindu University, Varanasi, India
[2]Centre of Biotechnology, Nehru Science Centre,
University of Allahabad, Allahabad, India
[3]Department of Biotechnology, S. S. Khanna Girl's Degree College, Allahabad, India
[4]Laboratory of Photobiology and Molecular Microbiology, Centre of Advanced Study in Botany, Banaras Hindu University, Varanasi, India

Abstract

Phycobiliproteins (PBPs) are brilliantly colored accessory light harvesting protein complexes around the periphery of thylakoid membrane. They can capture photonic energy over a broad region of solar spectrum which is essential for photosynthesis in cyanobacteria. Apart from photosynthesis, PBPs play an indispensable role in various biotechnological and pharmaceutical industries, and have increased demand in market worldwide. To fulfill the great demand, the utmost need is to have large-scale production

[*]Corresponding Author Email: rpsinhabhu@gmail.com.

of cyanobacterial biomass using cost effective technology. Several industries across the world have successfully developed cost effective strategies for large scale production of cyanobacterial biomass for the extraction of PBPs. The wide range of open and closed photobioreactor is an effective tool for large-scale production of biomass. Recently, the development of a rotating algal biofilm photobioreactor has simplified the production strategy of PBPs. Apart from quantity, quality is also extremely important for commercial for the cost effective extraction and purification of PBPs. Several physical and chemical protocols including aqueous two phase system, hydrophobic interaction/ion exchange chromatography have been developed in this context. This chapter primarily focuses to find out the gaps and current developments in the biomass production and purification technology for PBPs.

Keywords: aqueous two-phase system, biomass, cyanobacteria, photobioreactor, phycobiliproteins purity index

INTRODUCTION

Cyanobacteria are a primitive group of Gram-negative photosynthetic prokaryotes that originated around 3.5 billion years ago (Guan et al., 2007). They are ubiquitously distributed throughout the Earth surface and play a distinguished role in natural ecosystems for the production of primary biomass (Häder et al., 2015). They are predominantly dependent on solar photonic energy for survival and growth, and to procure solar energy, cyanobacteria have a group of accessory light harvesting protein complex that play a prominent role in harvesting photonic energy for carbon storage during photosynthesis (Sinha et al., 1995). These accessory light harvesting proteins are known as phycobiliproteins (PBPs) having various types of chromophores for absorption of light energy (Kannaujiya and Sinha, 2015; Kannaujiya and Sinha, 2017). PBPs are water soluble, vividly colored macromolecules of proteins organized in array of subunits on the photosynthetic apparatus in cyanobacteria, red algae and cryptomonads. Photosynthesis is driven by the continuous transfer of photonic energy from PBPs complex to Chl*a* of photosystem (Adir et al., 2006). PBPs are elementary composition of phycocyanin (PC), phycoerythrin (PE), allophycocyanin (APC) interlinked by linker polypeptides (Figure 1) (Grossman et al., 1993; Cai et al., 2001). Generally, the PBPs constitute about 24% of the dry cell mass of cyanobacteria but it enhances up to 40-50% after growth in low light in certain cyanobacteria (Muramatsu and Hihara, 2012). The percentage composition of phycobiliproteins varies from 6-35% in different cyanobacteria (Figure 2). According to Sidler (1994) the color spectrum various subunit of PBPs are identified at different wavelengths such as PE (Pink; 540-570 nm), PC (Blue; 610-620 nm), and APC (Blue-Green; 650-655 nm).

Each hetero-dimer subunits of PBPs are associated with linear tetrapyrrolechromophores (ring A-D) that covalently link with apo-proteins of cysteine residue by thioether bond C-3 position (A) and C-18 (D) position (Figure 3). The hetero-dimer is made up of α and β subunits which are structurally aggregated to form a functional group such as αβ, $(\alpha\beta)_3$ and $(\alpha\beta)_6$. Apart from α and β subunits, a γ subunits have also been found in R-phycoerythrin (R-PE) in red algae (Isailovic et al., 2004). The range of molecular weight for α subunits (12-20 kDa) and β subunits (15-22 kDa) have been found while γ subunits are almost fixed at 30 kDa (Galland-Irmouli et al., 2000). The brilliantly colored PBPs have a major role in the development of bio products that are widely used in commercial foods, dairy products, colorants, cosmetics and beverages industries (Richa et al., 2011).Apart from food products, they are commonly used for the purpose of fluorescence detection in molecular biology, pharmaceutics and clinical therapeutics (Eriksen, 2008).

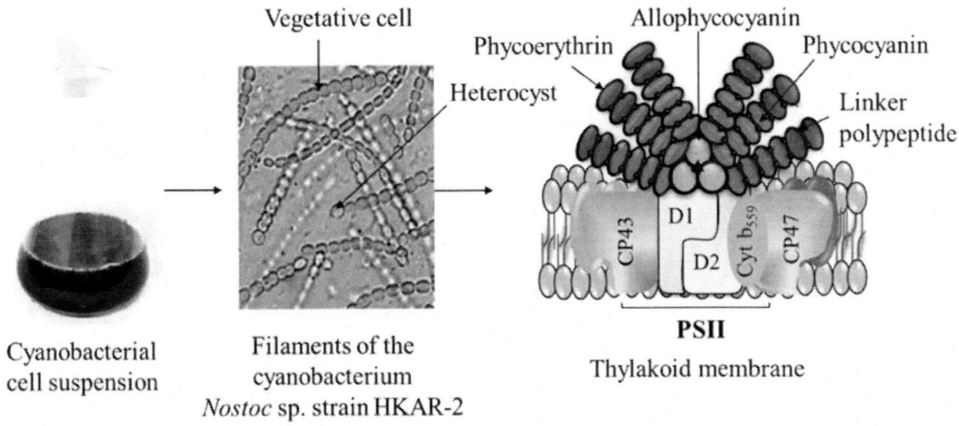

Figure 1. Physiological and structural location of phycobliproteins on thylakoid membrane of cyanobacterial filaments exhibit tri cylindrical core, six peripheral rods and linker polypeptides. PSII: Photosystem II.

The commercial and non-commercial applications of PBPs have received several patents in the field of production, food and molecular biology (Sekar and Chandramohan, 2008). The global utilization of PBPs has always been limited. In the recent years, global attention has always been focused on the large scale cyanobacterial biomass cultivation for efficient production of PBPs. Therefore, the aim of this chapter is to outline the recent developments in biomass production and purification technology for large scale harvesting of pure PBPs which could be used for quality products in the field of biotechnology and food industry.

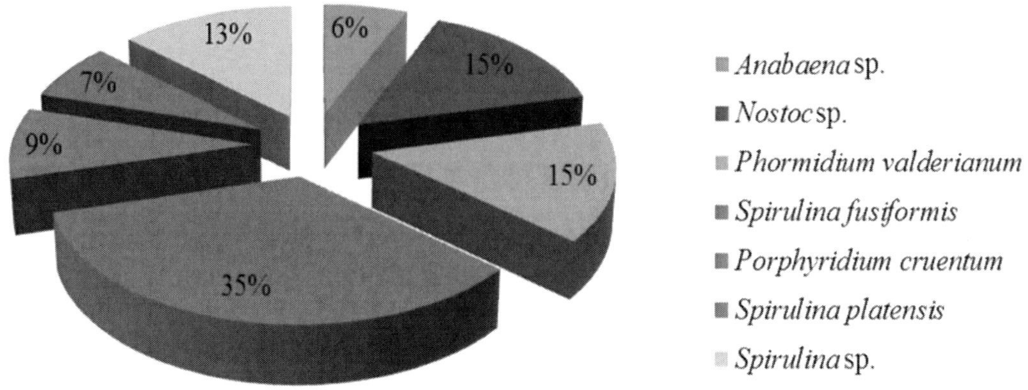

Figure 2. Composition of phycobiliproteins in different cyanobacteria.

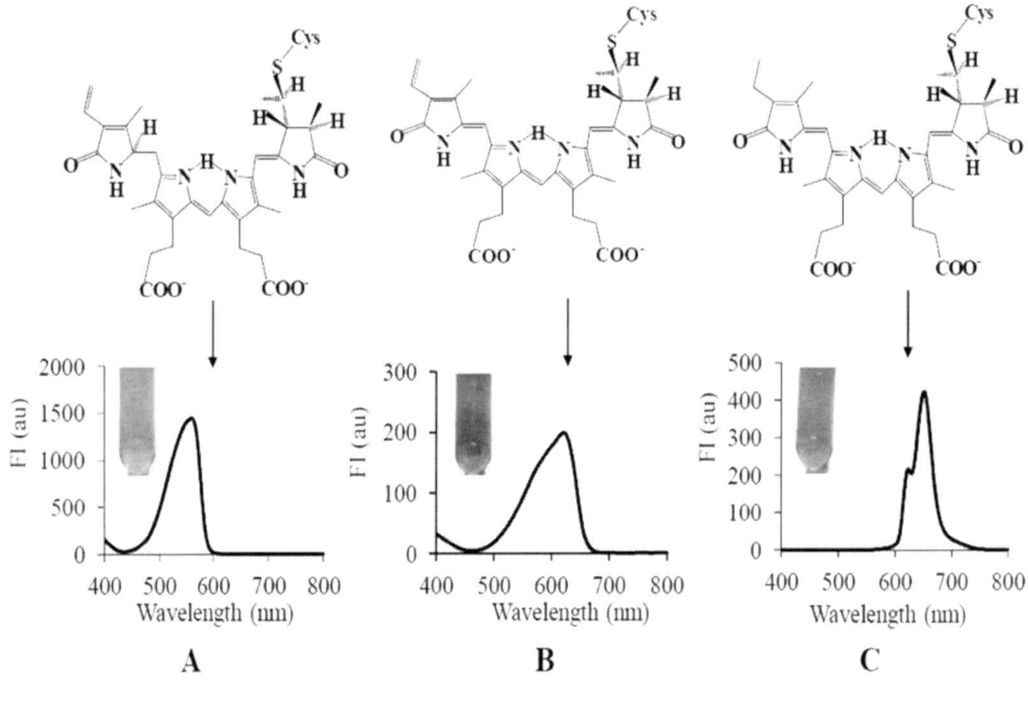

Figure 3. Molecular structure of chromophore of native bili proteins with corresponding fluorescence intensity (FI) of phycoerythrin (A), phycocyanin (B) and allophycocyanin (C) of *Nostoc*sp. strain HKAR-11. *Inset* vividly colored phycoerythrin (Pink), phycocyanin (Blue) and allophycocyanin (Bluish-Green).

CULTIVATION OF CYANOBACTERIA AND PRODUCTION OF PHYCOBILIPROTEINS

The overflowing world population increases the demand of carbon biomass which would increase more than 50% in the coming decades (Foley et al., 2011). The large scale cultivation of cyanobacteria is important for the production of biomass for its utilization in biotechnological industries. Cyanobacterial biomass can be luxuriantly grown in agricultural fields, thus there is no need to spend on costly technology for their production (Vandamme et al., 2013). Downstream processing is another tool for reduction in the cost effective production of biomass (Greenwell et al., 2010; Chen et al., 2011).

Table 1. Physiochemical parameters for the cultivation of microalgae (adapted from Chen et al., 2011)

Methods of cultivation	Energy source	Carbon Source	Cell density	Photobio-reactor
Phototrophy	Light	Inorganic	Low	Open
Heterotrophy	Organic substrate	Organic	High	Closed
Mixotrophy	Light + organic substrate	Inorganic and organic	Medium	Closed
Photo heterotrophy	Light	Organic	Medium	Closed

In the era of modern research and development, use of unique photobioreactor (Morweiser et al., 2010), fertile productive strains (Larkum et al., 2012) and gene construction for metabolic pathway (Georgianna and Mayfield, 2012) have contributed to enhance the protection of microalgae biomass. Cyanobacterial species such as *Spirulina* and *Nostoc* have been extensively utilized for the commercial production of PBPs for health and food products (Dillon et al., 1995).Various methodologies of cultivation of biomass and production of PC/PE/APC have been addressed by several workers (Eriksen, 2008; Kuddus et al., 2013). Table 1 depicts the various methods and energy sources for the successful cultivation of microalgae.

Photoautotrophic method is basically the outdoor culture of cyanobacteria under obligate dependence on photonic energy from sunlight (Jiménez et al., 2003). The cyanobacterium, *Arthrospira platensis* is luxuriantly grown in alkaline pH in open environment or outdoor culture. The growth of *A. platensis* reaches up to 3000 metric tons (dry biomass) annually. This is critical in the production of value added products. The heterotrophic growth of cyanobacteria is partially dependent on photonic energy of solar spectrum with higher growth rate as compared to photoautotrophic growth (Kuddus et al., 2013). Graverholt and Eriksen (2007) have found the growth of *Galdieria sulphuraria* considerably lower as compared to non-heterotrophic cyanobacterium *A. platensis* with a significantly higher (0.86 g L^{-1} day^{-1}) PC yield. The specific growth rate of mixotrophic culture on glucose medium is almost similar to photoautotrophic growth of cyanobacteria (Marquez et al., 1993). However, the productivity of PC was found to be more in a mixotrophic cultures in closed photobioreactor (Eriksen, 2008).

Open System of Cultivation

Numerous cyanobacteria exist in natural open ponds. Most of the cyanobacteria are exposed to harsh environmental conditions with remarkable hindrance in their survivability. Among hundreds of species of microalgae, *A. platensis* is a remarkable cyanobacterium found in natural soda lake and has unique adaptive mechanisms under environmental stresses. Therefore, production of cyanobacteria is being limited in open environmental conditions. Davis et al., (2011) have observed productivity up to 25 g^{-2} day^{-1} $year^{-1}$ in semi-continuous mode with cell concentration (0.5 g/L) in the cyanobacterium *A. platensis*. To enhance the productivity of biomass, and the production of cyanobacterium, several natural and man-made ponds have been maintained world-wide by leading industries (Table 2).

Table 2. Leading companies equipped with open photobiorector for the large scale cultivation of microalgae (adapted from Singh and Gu, 2010)

Industry	Locality and Country
Kelco	San Diego, USA
Neptune Industries	Boca Raton, USA
Blue Marble Energy	Seattle, USA
Aquaflow Binomics	Nelson, New Zealand
LiveFuels, Menlo Park	California, USA
OriginOil Inc.	Los-Angeles, USA
PetroSun, Scottsdale	Arizona, USA
Neste Oil	Helsinki, Finland
Ingrepo	Heure, Netherlands
Seambiotic	Israel, Israel

Closed System of Cultivation

Closed photobioreactors have several advantages over the open photobioreactor, because it will be able to maintain aseptic culture conditions, monoculture and prevention of water evaporation and utilization of energy which enable higher productivity of biomass (Chrismadha and Borowitzka, 1994; Barbosa et al., 2003). Several industries have been established closed system for optimum environmental conditions (Table 3).

Recently, several advancements have taken place in terms of design and physical operation of closed photobioreactor to obtain large scale biomass for commercial biological products. Davis et al., (2011) found high productivity (2-6 g/L) in closed photobioreactors as compared to open ponds (0.5 g/L). Similarly, PC productivity of outdoor culture in *Arthrospira platensis* and *Anabaena* sp. have been found to be about 14-23 and 0.82-1.32 g m^{-2} day^{-1}, respectively (Moreno et al., 2003). However, PC, PE and APC productivity of the same culture may enhance up to 10 fold in close photobioreactor as compared to open cultivation methods. Tubular photobioreactors are common type of closed bioreactor for large scale production at industrial level of cyanobacterial biomass (Pulzet al., 2013).

Table 3. Industries equipped with closed photobiorector for large scale cultivation of microalgae (adapted from Singh and Gu, 2010)

Company	Locality and Country
A2BE Carbon Capture	Boulder Colorado, USA
Green Fuel Technologies	Cambridge, Massachusetts, USA
Solazyme, Inc.	San Francisco, USA
Algeneol Biofuels	Fort Meyers, Florida, USA
Sapphire Energy	San Diego, USA
Inventure Chemical Technology	Seattle, USA
Solena	Washington State, USA
Solix Biofuels	Fort Collins, Colorado, USA
XL Renewables	Phoenix Arizona, USA
Bionavitas	Snoqualmie, Washington, USA
Cellena	Hawaii, USA

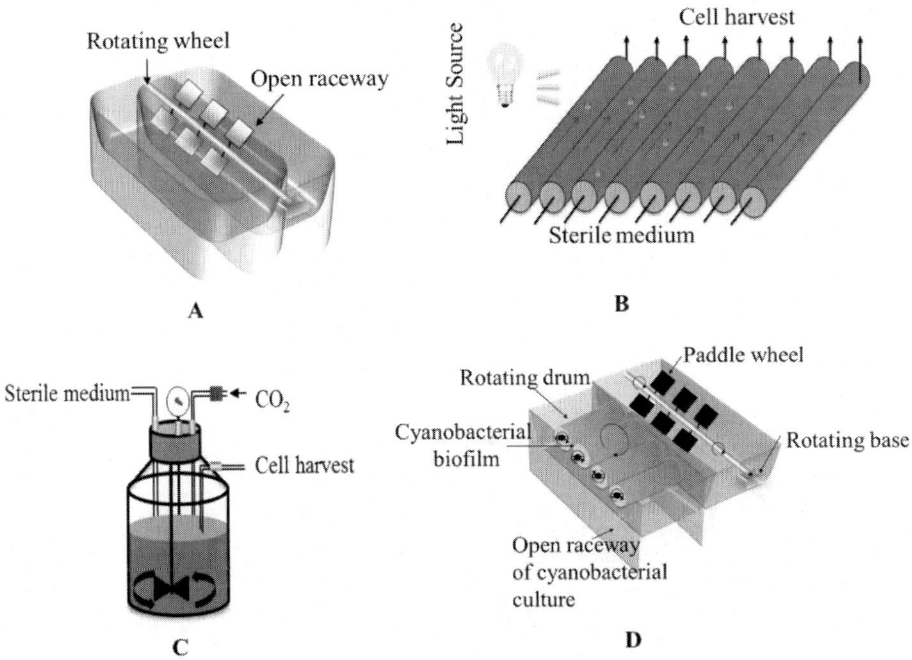

Figure 4. 3D architecture of open raceway (A), tubular 9B), fed-batch (C) and rotating algal biofilm photobioreactor (D) for large scale cultivation of cyanobacteria (adapted from Kannaujiya et al., 2017).

The advantages and disadvantage of tubular photobioreactor have been discussed (Torzillo, 1997; Tredici et al., 2010; Chini Zittelli et al., 2013; Torzillo and Zittelli, 2015). Similarly, fed-batch photobioreactor is influenced the volumetric productivity and final biomass concentration. Relative production of PC may enhance up to 16.1% with higher productivity (94.8 mg/L/d) by incorporation of 5 mM growth medium (Xieet al.,

2015). Flat panel are bioreactors is unique for wide volume culture (3.4-fold) and significant yield of biomass (1.7 g DW L^{-1} d^{-1}) as compared to other photobioreactors (0.5 g DW L^{-1} d^{-1}) (Bergmann and Trösch, 2016). Recently, rotating algal biofilm bioreactor has been developed for efficient and low cost production of biomass (Christenson and Sims, 2012; Gross et al., 2013). This photobioreactor is equipped with growth platform coupled with water mobility system for addition of growth medium which show a marked reduction in the cost of production of microalgae(Gross et al., 2015; Wood et al., 2015). Although, net productivity of PC was found to be lower (820-850 mg/m^2-day) as compared to open photobioreactor (1350 ± 173 mg/m^2-day) (Pushparaj et al., 1997). However, purity of PC has been increased as compared to other closed photobioreactors.

Global Scenario of Cyanobacteria Cultivation

Several industries have focused on mass cultivation of microalgae in controlled as well as uncontrolled environmental conditions. To obtain high end products at low cost, industries are looking for cost effective production. Most of these industries have successfully employed tubular closed system for obtaining high value products. A few industries have also focused on open pond large scale production of biomass by using their unique architecture (Schlipalius, 1991) for obtain high cell density cyanobacteria, paddle-wheel driven raceway ponds are used in Israel (Ben-Amotz, 1995). Figure 4 depicts the different cultivation system used for large scale biomass production in cyanobacteria.

Purification and Extraction of PBPs

PBPs have wide range of commercial and biotechnological applications. However, the purity of PBPs is a quality symbol for most of the applications. In order to exploit the PBPs, it is necessary to develop an efficient and cost effective method for the extraction and purification of PBPs. The thickness of cell wall and size of cyanobacteria is primary barrier for the efficient extraction of PBPs. To disrupt the primary barrier, several physical and chemical methods have been developed. Among physical methods, sonication (20 KHz) is the most common method for breakage of the strong cell wall in cyanobacteria (Benedetti et al., 2006). In addition, sonication with sand particles is more advantageous over simple sonication (Wiltshire et al., 2000). The consequence of freeze-thaw cycles of cyanobacterial biomass under 4 to -20°C is another cost effective method for easy cell disruption for obtaining pure PBPs (Viskari and Colyer, 2003, Stewart and Farmer, 1984). Viskari and Colyer, (2003) have reported nitrogen cavitation method for fast cell disruption and release of PBPs. Recently, a unique automated microwave-assisted faster extraction reactor (MAE) has been introduced for 8-10 time faster extraction of PBPs from freeze dried cells of *Porphyridium purpureum* (Juin et al., 2015). However, for easy extraction of PBPs from cyanobacteria aqueous two phase system is more advantageous over other conventional physical methods. Details of certain physical methods of PBPs extraction has been depicted in Table 4.

The total recovery of PBPs from cyanobacteria is not possible by the incorporation of only physical methods. Several chemicals have been reported to disrupt cell wall as well as inhibit protease degrading enzyme activity and protect them from degradation. Chemicals compositions such as of EDTA/lysozyme, sucrose (10%), PMSF (2.5 mM) and sodium azide are being used for the extraction PBPs in of the any physical methods (Sinha et al., 1995; Wang et al., 2014; Kannaujiya and Sinha, 2015; Kannaujiya and Sinha, 2016 a,b). Hydrochloric acid (2 to 10 N) is another reagent used for the extraction using of PBPs in phosphate buffer (50 mM, pH 6.8) from sheathed cyanobacteria (Sarada et al., 1999). Lysozyme induced disintegration/degradation of cell wall followed by cellular fractionation could be more helpful as compared to crude extraction (Boussiba and Richmond, 1979). The composition of phosphate buffer (1M) and potent surfactant (NP-40) are used for the rapid extraction of PBPs from certain microalgae (Zhao et al., 2015). Wood et al., (2015) have found an easy process for the PC extraction from lyophilized and powdered biomass subjected to repeated freeze/thaw cycles (Table 5).

Recently, PBPs were extracted from cyanobacteria at room temperature by using different solvents such as distilled water (pH 7.0), phosphate buffer (0.1 M, pH 6.8), and seawater (pH 8.13) (Sudhakar et al., 2015). Su et al., (2014) have found specific pH (6.8-7.5) and temperature (30-60°C) range for rapid extraction of PBPs in phosphate buffer (10 mM).

Table 4. Various physical methods for the extraction of phycobiliproteins from cyanobacteria and red algae

Method	References
Ultrasonic homogenizer (20-50 kHz) at 4°C	Bermejo et al., (2006); Horváth et al., (2013); Kannaujiya and Sinha, (2016 a,b)
Polytron homogenizer (50 Hz) at 4°C	Horváth et al., (2013)
French press cell	Bryant et al., (1976); Sinha et al., (1995)
Liquid nitrogen	Stewart and Farmer, (1984)
Repeated freeze thawing at 0 to 4°C/4 to -20°C	Sinha et al., 1995; Kannaujiya and Sinha (2016 a,b)
Sonication with fine sand	Wiltshire et al., (2000)
Nitrogen cavitation	Viskari and Colyer, (2003)
Aqueous two-phase systems (ATPS)	Patil et al., (2006); Zhao et al., (2014); Luo et al., (2016)
Microwave-assisted extraction (MAE)	Juin et al., (2015)

Table 5. Chemical methods for the extraction of phycobiliproteins from cyanobacteria and red algae

Chemicals	References
Phosphate buffer (25 mM) (pH 7), Lysozyme 10 mg/ml	Kilpatrick, (1985)
Cellulase (1%), Pectolyase (0.1%)	Viskari and Colyer, (2003)
Phosphate buffer (0.75 M) (pH 7) with (0.1-1.0%) Triton X-100	Sinha et al., (2003)
Phosphate buffer (50 mM),) phenylmethanesulfonylfluoride (PMSF) (1 mM), EDTA (10% w/v) and sucrose (5% w/v)	Sinha et al., (1995) Kannaujiya and Sinha, (2016a,b)
Acetate buffer and (2 mg/ml) Lysozyme (Triton X-100)	Viskari and Colyer, (2003)
Trizma (250 mM), EDTA (10 mM) in phosphate buffer	Viskari and Colyer, (2003)
Rivanol (1%)	Minkova et al., (2003)
Tris (1 M) (pH 8.0), 0.5 M EDTA, Sucrose (20% w/v) with Lysozyme (5 mg/ml)	Bhaskar et al., (2005)
Phosphate buffer (1 M) and NP-40	Zhao et al., (2015)

Purification and Purity Analysis of PBPs

A number of techniques including precipitation, gel permeation and column chromatography have been found to be effective tools for the rapid enrichment and purification of PBPs from cyanobacteria (Table 6). Column chromatography has been applied as effective methods for purification of PBPs (Sonani et al., 2016). An anion exchange column chromatography (DEAE cellulose) was used for the purification of PBPs from the cyanobacterium *Microcystis aeruginosa* (Padgett and Krogmann, 1987). Herrera et al., (1989) formulated a series of chromatography techniques that includes ultra-filtration, charcoal adsorption and spray drying technique to obtain pure PC (purity index 3.91) with a yield of (9%) from total proteins in biomass. Similarly, the purity index of PC (5.06) and APC (5.34) have been obtained by using series of fractional precipitation, ion exchange chromatography (DEAE-Sepharose) and gel permeation chromatography (Sephadex G-100) (Zhang and Chen, 1999). Rivanol reagent has been introduced for efficient extraction of PC from *Spirulina fusiformis* with purity up to 4.3, coupled with high recovery (Minkova et al., 2003). Recently, used two-phase aqueous extraction method has been followed by ion exchange chromatography which showed remarkable increment in purity index of PC (6.69) (Soni et al., 2008). The advantage of cell disruption in liquid nitrogen followed by cell lysis using lysozyme yielded of a purity index of 4.98 after sepharose column chromatography (Bhaskar et al., 2005). The recovery of pure PC is crucial task for optimum yield of pure PBPs. Recently scientists have developed a process by expanding bead adsorption chromatography using streamline-column DEAE anion for better recovery and optimum yield of PBPs (Babu et al., 2006; Bermejo et al., 2006). Apart from anion exchanger, hydrophobic interaction chromatography (HIC) has also been carried out separately (Niu et al., 2006) or a combination with ion-exchange chromatography (Wang et al., 2002) for the purification of R-PE from *Palmaria palmata* and *Polysiphoni aurceolata.* A single-step chromatographic technique is more advantageous over conventional chromatography for purification of R-PE (Soni et al., 2008) and PC/PE (Kannaujiya and Sinha, 2016 b) from *Polysiphoni aurceolata* and *Nostoc* sp. strain HKAR-11 respectively. The combination of DEAE Sepharose Fast Flow and hydrophobic interaction column chromatography under pH gradient has achieved the highest purity ratio (11.53) for PE (Kannaujiya and Sinha 2016 b). Chakdar and Pabbi (2012) obtained purity index (4.95) of PE in *Anabaena variabilis* using anion exchange chromatography. Tripathi et al., (2007) isolated and purified PE from desiccation-tolerant cyanobacterium *Lyngbya arboricola* using acetone precipitation, gel filtration and ion-exchange chromato-graphy and obtained high purity >5 after purification. Although, most of these methods have been non-commercialized and non-scalable due to high cost, series of stages, low recovery and low yield.

Recently, aqueous two-phase extraction (ATPS) with PEG 4000 and potassium phosphate buffer have been introduced as an attractive and alternative method for better optimization of purification of PC from cyanobacteria (Zhao et al., 2014).ATPS consists of two units which include a mixing component and phase separation that interacts by the mechanism of droplet-droplet collision and coalescence (Hatti-Kaul, 2000). The phase separation in PC of PBPs is created by gravity and centrifugation (Chethana et al., 2015). Luo et al., (2016) have updated ATPS method with interconnection of vortex fluidic device for enhancing phase demixing which accelerated more recovery of intact PC isolated from *Spirulina maxima*. The involvement of multiple ATPS-VFD units further increased the purity of PC up to 2.3 as compared to conventional ATPS system with higher recovery 78-93% (Luo et al., 2016). Concerns about two-step membrane process have been addressed for filtration of B-phycoerythrin using polyethersulfone flat membrane (Marcati et al., 2014). Senthilkumar (2013) obtained purity index of 5.2 using Q-Sepharose column chromatography. The purity index of PC was also enhanced with affinity chromatography (Munier et al., 2015). Various strategies for the extraction and purification of PBP from cyanobacteria have already been discussed (Table 6) (Cucllar-Bermudez et al., 2015, Sonani et al., 2016).

Conclusion and Future Perspectives

PBPs are commercially produced and are highly potent pigments found in cyanobacteria. In the last 20-25 years, the potential applications of PBPs have been elaborated in various sectors of science throughout the world. Although, the cyanobacteria are potential source of PC, PE and APC, they have not been exploited

Table 6. List of selected and relevant protocols employed for the purification of phycobiliproetins from cyanobacteria and red algae. (ATPS: Aqueous Two Phase System)

Species	Steps involved in PBPs purification	PC		PE		APC		References
		P^I	$Y^{\%}$	P^I	$Y^{\%}$	P^I	$Y^{\%}$	
Spirulina maxima	ATPS; Ultrafiltration; Ammonium sulphate precipitation	3.8	29.0	-	-	-	-	Rito-Palomares et al., (2001)
Spirulina fusiformis	Rivanol treatment; Ammonium sulphate precipitation; Gel filtration chromatography	4.3	45.7	-	-	-	-	Minkova et al., (2003)
O.quadri-punctulata	Ammonium sulphate precipitation; Gel filtration chromatography; Anion exchange chromatography	3.3	44.2	-	-	-	-	Soni et al., (2006)
Anabaena flos-aquae	Ammonium sulphate precipitation; Hydroxyapatite interaction chromatography	4.7	-	-	-	-	-	Benedetti et al., (2006)
Spirulina platensis	Chitosan adsorption; ATPS	5.1	66.0	-	-	-	-	Patil et al., (2006)
Spirulina platensis	Expanded bed adsorption chromatography; Ion exchange Chromatography	3.6	8.7	-	-	-	-	Niu et al., (2007)
P. fragile	Ammonium sulphate fractionation; Hydrophobic interaction chromatography	4.5	62.0	-	-	-	-	Soni et al., (2008)

Table 6. (Continued)

Species	Steps involved in PBPs purification	PC		PE		APC		References
		P^I	$Y^{\%}$	P^I	$Y^{\%}$	P^I	$Y^{\%}$	
Heterosiphonia japonica	Ammonium sulphate precipitation; Gel filtration chromatography; Ion exchange chromatography	-	-	4.8	-	-	-	Sun et al., (2009)
Geitlerinema sp. A28DM	Ethodin mediated precipitation; Gel filtration chromatography	-	-	-	-	3.2	62	Parmar et al., (2010)
S. platensis	Ammonium sulphate precipitation; Ion-exchange chromatography	5.5	67.4	-	-	5.1	80	Yan et al., (2011)
Pseudanabaena sp.	Ammonium sulphate precipitation; Gel filtration chromatography; Ion exchange chromatography	-	-	6.8	47.0	-	-	Mishra et al., (2011)
Anabaena variabilis	Ammonium sulphate precipitation; Ion exchange chromatography	-	-	4.9	62.5	-	-	Chakdar and Pabbi, (2012)
Lyngbya sp. A09DM	Triton X-10; ammonium sulphate precipitation; Ion exchange chromatography; Gel filtration chromatography	5.5	62.2	6.7	76.1	5.4	71	Sonani et al., (2014)
Nostoc sp. strain HKAR-11	Ammonium sulphate precipitation; Repeated gel filtration chromatography, Hydrophobic interaction chromatography	5.7	73	11	83	-	-	Kannaujiya et al., (2016b)

properly for novel applications. Production of high value PBPs is currently being done by outdoor culture. Recently, developed techniques such as Rotating Algal Biofilm Reactor (RABR) and fed-batch cultivation might reduce the cost of PBPs production. All enzymes responsible for synthesis of holo-PBPs are now known and it could be used for the recombinant production of PBPs using appropriate bacterial strains. Several optimized protocols have been developed for the extraction and purification of PBPs from cyanobacteria. However, there is a pressing need for specific procedures that would be applicable to all the cell types. Further, extensive research in the field of production and purification technology is needed for the development of quality products and to meet the ever increasing demand of such novel molecules.

ACKNOWLEDGMENTS

V. K. Kannaujiyais thankful to UGC, New Delhi for financial assistance.

REFERENCES

Adir, N, (2005). Elucidation of the molecular structures of components of the phycobilisome: reconstructing a giant. *Photosynth. Res.* 85: 15-32.

Adir, N., Dines, M., Klartag, M., McGregor, A., and Melamed-Frank, M. (2006). Assembly and disassembly of phycobilisomes. In: Shively J. M. (ed), *Microbiology Monographs*. Springer, Verlag-Berlin, Heidelberg, 47-77.

Babu, B. R., Rastogi, N. K., and Raghavarao, M. S. (2006). Mass transfer in osmotic membrane distillation of phycocyanin colourant and sweet lime juice. *J. Mem. Sci.* 212: 58-69.

Barbosa, M. J, Albrecht, M., Wijffels, R. H. (2003). Hydrodynamic stress and lethal events in sparged microalgae cultures. *Biotechnol. Bioeng.* 83:112-120.

Ben-Amotz, A. (1995). New mode of *Dunaliella* biotechnology: two-phase growth for b-carotene production. *J. Appl. Phycol.* 7: 65-68.

Benedetti, S., Rinalducci, S., Benvenuti, F., Francogli, S., Pagliarani, S., Giorgi, L., Micheloni M, D'Amici, G. M., Zolla, L., and Canestrari, F. (2006). Purification and characterization of phycocyanin from the blue green alga *Aphanizomenon flos-aquae*. *J. Chromatogr. B Analyt. Technol. Biomed. Life. Sci.* 833: 12-18.

Benedetti, S., Rinalducci, S., Benvenuti, F., Francogli, S., Pagliarani, S., Giorgi, L., Micheloni M, D'Amici, G. M., Zolla, L., and Canestrari, F. (2006). Purification and characterization of phycocyanin from blue-green alga *Aphanizomenon flos-aquae*. *J. Chromatogr. B*, 833: 12-18.

Bergmann, P., and Trösch, W. (2016). Repeated fed-batch cultivation of *Thermosynechococcus elongatus* BP-1 in flat-panel airlift photobioreactors with static mixers for improved light utilization: Influence of nitrate, carbon supply and photobioreactor design. *Algal Res.* 17: 79-86.

Bermejo, R., Felipe, M. A., and Talaver, E. M. (2006). Expanded bed adsorption chromatography for recovery of phycocyanin from the microalga *Spirulina platensis*. *Chromatographia* 63: 59-66.

Bhaskar, U. S., Gopalswamy, G., and Raghu, R. (2005). A simple method for efficient extraction and purification of C-Phycocyanin from *Spirulina platensis* Geitler. *Ind. J. Exp. Biol.* 43: 277-279.

Boussiba, S., and Richmond, A. E. (1979). Isolation and characterization of phycocyanins from the blue-green alga *Spirulina platensis*. *Arch. Microbiol.* 120: 155-159.

Bryant, D. A., Glazer, A. N., and Eiserling, F. A. (1976). Characterization and structural properties of the major biliproteins of *Anabaena* sp. *Arch. Microbiol.*, 110: 61-75.

Cai, Y. A., Murphy, J. T., Wedemayer, G. J., and Glazer, A. N. (2001). Recombinant phycobiliproteins. *Anal. Biochem.* 290: 186-204.

Chakdar, H., and Pabbi, S., (2012). Extraction and purification of phycoerythrin from *Anabaena variabilis* (CCC421). *Phykos* 42: 25-31.

Chen, C-Y., Yeh, K-L., Aisyah, R., Lee, D-J., and Chang, J-S. (2011). Cultivation, photobioreactor design and harvesting of microalgae for biodiesel production: A critical review. *Bioresour. Technol.* 102: 71-81.

Chethana, S., Nayak, C., Madhusudhan, M. C., and Raghavarao, K. S. M. S. (2015). Single step aqueous two-phase extraction for downstream processing of C-phycocyanin from *Spirulina platensis*. *J. Food. Sci. Technol.* 52: 2415-2421.

Chini, Zittelli, G., Biondi, N., Rodolfi, L., and Tredici, M. R. (2013). Photobioreactors for mass production of microalgae. In: Richmond, A., Hu, Q., (eds) *Handbook of microalgal culture: applied phycology and biotechnology*, Wiley, Oxford, 225-266.

Chrismadha, T., and Borowitzka, M. A. (1994). Effect of cell density and irradiance on growth, proximate composition and eicosapentaenoic acid production of *Phaeodactylum tricornutum* grown in a tubular photobioreactor. *J. Appl. Phycol.* 6: 67-74.

Christenson, L., and Sims, R. (2012). Rotating algal biofilm reactor and spool harvester for wastewater treatment with biofuels by-products. *Biotechnol. Bioeng.* 109: 1674-1684.

Cuellar-Bermudez, S. P., Aguilar-Hernandez, I., Cardenas-Chavez, D. L., Ornelas-Soto, N., Romero-Ogawa, M. A. and Parra-Saldivar, R. (2015). Extraction and purification of high-value metabolites from microalgae: essential lipids, astaxanthin and phycobiliproteins. *Microbiol. Biotechnol.* 8: 190-209.

Davis, R., Aden, A, and Pienkos, P. (2011). Techno-economic analysis of autotrophic microalgae for fuel production. *Appl. Energy* 88: 3524-3531.

Dillon, J. C., Phuc, A. P., and Dubacq, J. P. (1995). Nutritional value of the alga *Spirulina. World Rev. Nut. Diet* 77: 32-46.

Eriksen, N. T. (2008). Production of phycocyanin–a pigment with applications in biology, biotechnology, foods and medicine. *Appl. Microbiol. Biotechnol.* 80: 1-14.

Foley, J. A., Ramankutty, N., et al. (2011). Solutions for a cultivated planet. *Nature* 478: 337-342.

Galland-Irmouli, A. V., Pons, L., Lucon, M., Villaume, C., Mrabet, N. T., Gueant, J. L., and Fleurence, J. (2000). One-step purification of R-phycoerythrin from the red macroalga *Palmaria palmata* using preparative polyacrylamide gel electrophoresis. *J. Chromatogr. B. Biomed. Sci. Appl.* 739: 117-123.

Georgianna, D. R., and Mayfield, S. P. (2012). Exploiting diversity and synthetic biology for the production of algal biofuels. *Nature* 488: 329-335.

Graverholt, O. S., and Eriksen, N. T. (2007). Heterotrophic high cell density fed batch and continuous -flow cultures of *Galderia sulphuraria* and production of phycocyanin. *Appl. Microbial. Biotech.* 77: 69-75.

Greenwell, H. C., Laurens, L. M., Shields, R. J., Lovitt, R. W., and Flynn, K. J. (2010). Placing microalgae on the biofuels priority list: a review of the technological challenges. *J. Royl. Soc. Interface* 7:703-726.

Gross, M., Henry, W., Michael, C. and Wen, Z. (2013). Development of a rotating algal biofilm growth system for attached microalgae growth with in situ biomass harvest. *Bioresour. Technol.* 150: 195-201.

Gross, M., Mascarenhas, V., and Wen, Z. (2015). Evaluating algal growth performance and water use efficiency of pilot-scale revolving algal biofilm (RAB) culture systems. *Biotechnol. Bioeng.* 112: 2040-2050.

Grossman, A. R., Schaefer, M. R., Chiang, G. G., and Collier, J. L. (1993). The phycobilisomes a light harvesting complex responsive to environmental conditions. *Microbiol. Rev.* 57: 725-749.

Guan, X., Qin, S., Zhao, F., Zhang, X., and Tang, X. (2007). Phycobilisomes linker family in cyanobacterial genomes: divergence and evolution. *Int. J. Biol. Sci.* 3: 434-445.

Häder, D-P, Williamson, C. E.,Wängberg S, Rautio M., Rose K. C., Gao K., Helbling E. W., Sinha R. P., and Worrest R. (2015). Effects of UV radiation on aquatic ecosystems and interactions with other environmental factors. *Photochem. Photobiol. Sci.* 14: 108-126.

Hatti-Kaul R (2000). *Aqueous two-phase systems.* Methods and Protocols, Humana Press.

Herrera, A., Boussiba, S., Napoleone, V., and Hohlberg, A. (1989). Recovery of C-phycocyanin from the cyanobacterium *Spirulina maxima. J. Appl. Phycol.* 1: 325-331.

Horváth, H., Kovács, A. W., Riddick, C., Présing, M. (2013). Extraction methods for phycocyanin determination in freshwater filamentous cyanobacteria and their application in a shallow lake. *European. J. Phycol.* 48: 278-286.

Isailovic, D., Li, H. W., and Yeung, E. S. (2004). Isolation and characterization of R-phycoerythrin subunits and enzymatic digests. *J. Chromatogr. A.* 1051: 119-130.

Jiménez, C., Cossıo, B. R., Labella, D., and Niell, F. X. (2003). The feasibility of industrial production of *Spirulina* (*Arthrospira*) in Southern Spain. *Aquaculture* 217: 179-190.

Juin, C., Chérouvrier, J. R., Thiéry, V., Gagez, A. L., Bérard, J. B., Joguet, N., Kaas, R., Cadoret, J. P. and Picot, L. (2015). Microwave-assisted extraction of phycobiliproteins from *Porphyridium purpureum. Appl. Biochem. Biotechnol.* 175: 1-15.

Kannaujiya, V. K., and Sinha, R. P. (2015). Impacts of varying light regimes on phycobiliproteins of *Nostoc* sp. HKAR-2 and *Nostoc* sp. HKAR-11 isolated from diverse habitats. *Protoplasma* 252: 1551-1561.

Kannaujiya, V. K., and Sinha, R. P. (2016a). Thermokinetic stability of phycocyanin and phycoerythrin in food grade preservatives. *J. Appl. Phycol.* 28: 1063-1070.

Kannaujiya, V. K. and Sinha, R. P. (2016b). An efficient method for separation and purification of phycobiliproteins from rice-field cyanobacterium *Nostoc* sp. strain HKAR-11. *Chromatographia* 79: 335-343.

Kannaujiya, V.K., and Sinha, R.P. (2017). Impacts of diurnal variation of ultraviolet-B and photosynthetically active radiation on phycobili-proteins of the hot-spring cyanobacterium *Nostoc* sp. strain HKAR-2. *Protoplasma* 254: 423-433.

Kannaujiya, V. K., Sundaram, S., Sinha, R. P. (2017). *Phycobiliproteins: recent developments and future applications.* Springer Nature, Singapore.

Kilpatrick, K. A. (1985). *The development of a method for measure marine cyanobacterial phycoerythrin extracted in solvents.* Dissertation, Texas AM University.

Kuddus, M., Singh, P., Thomas, G., and Al-Hazimi, A., (2013). Recent developments in production and botechnological applications of c-phycocyanin. *Bio. Med. Res. Int.* 74: 2859.

Larkum, A. W., Ross, I. L., Kruse, O., and Hankamer, B. (2012). Selection, breeding and engineering of microalgae for bioenergy and biofuel production. *Trends Biotechnol.* 30: 198-205.

Luo, X., Smith, P., Raston, C. L., and Zhang, W. (2016). Vortex fluidic device-intensified aqueous two phase extraction of c-phycocyanin from *Spirulina maxima. ACS Sustainable Chem. Eng.* 4: 3905-3911.

Marcati, A., Ursu, A. V., Laroche, C., Soanen, N., Marchal, L., Jubeau, S., Djelveh, G., and Michaud, P. (2014). Extraction and fractionation of polysaccharides and B-phycoerythrin from the microalga *Porphyridium cruentum* by membrane technology. *Algal Res*. 5: 258-263.

Marquez, F. J., Sasaki, K., Kakizono, T., Nishio, N., and Nagai, S. (1993). Growth characteristics of *Spirulina platensis* in mixotrophic and heterotrophic conditions. *J. Ferment. Bioeng*. 76: 408-410.

Minkova, K. M., Tchernov, A. A., Tchorbadjieva, M. I., Fournadjieva, S. T., Antova, R. E., and Busheva, M. Ch. (2003). Purification of C-phycocyanin from *Spirulina (Arthrospira) fusiformis*. *J. Bio-technol.*102: 55-59.

Mishra, S. K., Shrivastav, A., and Mishra, S. (2011). Preparation of highly purified C-phycoerythrin from marine cyanobacterium *Pseudanabaena* sp. *Protein Expr.Purif.* 80: 234-238.

Moreno, J., Vargas, M. A., Rodriguez, H., Rivas, J. and Guerrero, M. G. (2003). Outdoor cultivation of a nitrogen-fixing marine cyano-bacterium, *Anabaena* sp. ATCC 33047. *Biomol. Eng*. 20: 191-197.

Morweiser, M., Kruse, O., Hankamer, B., and Posten, C. (2010). Developments and perspectives of photobioreactors for biofuel production. *Appl. Microbiol. Biotechnol*. 87: 1291-1301.

Munier, M., Moranc, M., Dumay, J., Jaouen, P., and Fleurence, J. (2015). One-step purification of R-phycoerythrin from the red edible seaweed *Grateloupiaturuturu*. *J. Chromatogr. B*992: 23-29.

Muramatsu, M., and Hihara, Y. (2012). Acclimation to high-light conditions in cyanobacteria: from gene expression to physiological responses. *J. Plant. Res*. 125: 11-39.

Niu J. F., Wang G. C., Lin X. Z. and Zhou B. C. (2007). Large-scale recovery of C-phycocyanin from *Spirulina platensis* using expanded bed adsorption chromatography. *J. Chromatogr. B. Analyt. Technol. Biomed. Life Sci.* 850: 267-276.

Niu, J. F., Wang, G. C., and Tseng, C. K. (2006). Method for large-scale isolation and purification of R-phycoerythrin from red alga *Polysiphoni aurceolata* Grev. *Protein Expr. Purif.* 49: 23-31.

Padgett, M. P., and Krogmann, D. W. (1987). Large scale preparation of pure phycobiliproteins. *Photosynth. Res*. 11: 225-235.

Parmar, A., Singh, N. K and Madamwar, D. (2010). Allophycocyanin from a local isolate *Geitlerinema* sp. A28DM (cyanobacteria): A simple and efficient purification process. *J. Phycol.* 46: 285-289.

Patil, G, Chethana, S., Sridevi, A. S., and Raghavarao, K. S. (2006). Method to obtain C-phycocyanin of high purity. *J. Chromatogr*. 1127: 76-81.

Pulz, O., Broneske, J., Waldeck, P. (2013). IGV GmbH experience report, industrial production of microalgae under controlled conditions: innovative prospects. In: Richmond A, Hu Q (eds) *Handbook of microalgal culture: applied phycology and biotechnology,* Wiley, Oxford, 445-460.

Pushparaj, B., Pelosi, E., Tredici, M. R., Pinzani, E., Materassi, R., and Autotrofi, M. (1997). An integrated culture system for outdoor production of microalgae and cyanobacteria. *J. Appl. Phycol.* 9: 113-119.

Richa, Kannaujiya, V. K., Kesheri, M., Singh, G., and Sinha, R. P. (2011). Biotechnological potentials of phycobiliproteins. *Int. J. Pharma Bio. Sci.* 2: 446-454.

Rito-Palomares, M., Nunez, L., and Amador, D. (2001). Practical application of aqueous two-phase systems for the development of a prototype process for c-phycocyanin recovery from *Spirulina maxima. J. Chem. Technol. Biotechnol* 76: 1273-1280.

Sarada, R., Pillai, M. G. and Ravishankar, A. (1999). Phycocyanin from *Spirulina* sp: influence of processing of biomass on phycocyanin yield, analysis of efficacy of extraction methods and stability studies on phycocyanin. *Process Biochem.* 24: 795-801.

Schlipalius, L. (1991). The extensive commercial cultivation of *Dunaliella salina. Bioresour. Technol.* 38: 241-243.

Sekar, S. and Chandramohan, M. (2008). Phycobiliproteins as a commo-dity: trends in applied research, patents and commercialization. *J. Appl. Phycol.* 20: 113-136.

Senthilkumar, N., Suresh, V., Thangam, R., Kurinjimalar, C., Kavitha, G, Murugan, P., Kannan, S., and Rengasamy, R. (2013). Isolation and characterization of macromolecular protein R-Phycoerythrin from *Portieria hornemannii. Int. J. Biol. Macromol.* 55: 150-160.

Shrivastav, A., Mishra, S. K., and Mishra, S. (2010). Polyhydroxyal-kanoate (PHA) synthesis by *Spirulina subsalsa* from Gujarat coast of India. *Int. J. Biol. Macromol.* 46: 255-260.

Sidler, W. A. (1994). Phycobilisome and phycobiliprotein structures. In: Bryant D. A. (ed), *Molecular Biology of Cyanobacteria,* The Kluwer Academic Publishers, Dordrecht, Netherlands, 139-216.

Singh, J., and Gu. (2010). Commercialization potential of microalgae for biofuels production. *Renew. Sustainable Energy Rev.* 14: 2596-2610.

Sinha, R. P., Lebert, M., Kumar, A., Kumar, H. D., Häder, D. P. (1995). Spectroscopic and biochemical analyses of UV effects of phycobilisomes of *Anabaena* sp. and *Nostoccarmium. Bot. Acta* 108: 87-92.

Sonani, R. R., Rastogi, R. P., Patel, R., and Madamwar, D. (2016). Recent advances in production, purification and applications of phyco-biliproteins. *World J. Biol. Chem.* 26: 100-109.

Sonani, R. R., Singh, N. K., Kumar, J., Thakar, D., and Madamwar, D. (2014). Concurrent purification and antioxidant activity of phycobiliproteins from *Lyngbya* sp. A09DM: An antioxidant and anti-aging potential of phycoerythrin in *Caenorhabditiselegans. Process Biochem.* 49: 1757-1766.

Soni, B., Kalavadia, B., Trivedi, U. and Madamwar, D. (2006). Extraction, purification and characterization of phycocyanin from *Oscillatoria quadri punctulata*-isolated from the rocky shores of Bet-Dwarka, Gujarat, India. *Process Biochem.* 41: 2017-2023.

Soni, B., Trivedi, U., and Madamwar, D. (2008). A novel method of single step hydrophobic interaction chromatography for the purification of phycocyanin from *Phormidium fragile* and its characterization for antioxidant property. *Bioresour. Technol.* 99: 188-194.

Stewart, D. E., and Farmer, F. H. (1984). Extraction, identification and quantitation of phycobiliproteins pigments from phototropic biomass. *Limnol. Oceanogr.* 29: 392-397.

Su, C-H., Liu, C. S., Yang, P. C., Syu, K. S., and Chiuh, C. C. (2014). Solid–liquid extraction of phycocyanin from *Spirulina platensis*: kinetic modeling of influential factors. *Sep. Purif. Technol.* 123: 64-68.

Sudhakar M. P., Jagatheesan A., Perumal K. and Arunkumar K. (2015). Methods of phycobiliprotein extraction from *Gracilaria crassa* and its applications in food colourants. *Algal Res.* 8: 115-120.

Sun, L., Wang, S., Gong, X., Zhao, M., Fu, X., and Wang, L. (2009). Isolation, purification and characteristics of R-phycoerythrin from a marine macroalga *Heterosiphonia japonica*. *Protein Expr. Purif.* 64: 146-154.

Torzillo, G. (1997). Tubular reactors. In: Vonshak A (ed) *Spirulina platensis (Arthrospira) physiology cell biology and biotechnology,* Taylor and Francis, London, 101-115.

Torzillo, G., and Zittelli, G. C. (2015). Tubular photobioreactors. In: Prokop et al. (eds) *Algal Biorefineries,* Springer, Switzerland, 187-212.

Tredici, M. R., Chini-Zittelli, G., Rodolfi, L, (2010). Photobioreactors. In: Flickinger M. C., Anderson S (eds) *Encyclopedia of industrial biotechnology: bioprocess, bioseparation, and cell technology,* Wiley, Hoboken, 3821-3838.

Tripathi, S. N., Kapoor, S., and Shrivastava, A. (2007). Extraction and purification of an unusual phycoerythrin in a terrestrial desiccation tolerant cyanobacterium *Lyngbya arboricola*. *J. App. Phycol.* 19: 441-447.

Vandamme, D., Foubert, I., and Muylaert, K. (2013). Flocculation as a low-cost method for harvesting microalgae for bulk biomass production. *Trends. Biotechnol*, 31: 233-239.

Viskari, P. J., and Colyer, C. C. (2003). Rapid extraction of phyco-biliproteins from cultured cyanobacteria samples. *Anal. Biochem.* 319: 263-271.

Wang, G. C., Sun, H. B., Fan, X., and Tseng, C. K. (2002). Functional ingredients from algae for foods and nutraceuticals. *Acta Bot. Sinica* 44:541-546.

Wang, L., Qu, Y., Fu, X., Zhao, M., Wang, S., and Sun, L. (2014). Isolation, purification and properties of an R-phycocyanin from the phycobilisomes of a marine red macroalga *Polysiphoni aurceolata*. *PLoS ONE* 9: e87833.

Wiltshire, K. H., Boersma, M., and Moller, A. (2000). Extraction of pigments and fatty acids from the green alga *Scenadesmus oblique* (cholorophycae). *Aquat. Ecol.* 34: 119-126.

Wood, J. L., Miller, C. D., Sims, R. C., and Takemoto, J. Y. (2015). Biomass and phycocyanin production from cyanobacteria dominated biofilm reactors cultured using oilfield and natural gas extraction produced water. *Algal Res.* 11: 165-168.

Xie, Y., Jin, Y., Zeng, X., Chen, J., Lu, Y., and Jing, K. (2015). Fed-batch strategy for enhancing cell growth and C-phycocyanin production of *Arthrospira (Spirulina) platensis* under phototrophic cultivation. *Bioresour. Technol.* 180: 281-287.

Yan, S. G., Zhu, L.P., Su, H. N., Zhang, X. Y., Chen, X. L., Zhou, B. C., and Zhang, Y. Z. (2011). Single-step chromatography for simultaneous purification of C-phycocyanin and allophycocyanin with high purity and recovery from *Spirulina* (*Arthrospira*) *platensis*. *J. Appl. Phycol.* 23: 1-6.

Zhang, Y. M., and Chen, F. (1999). A simple method for efficient separation and purification of C-phycocyanin and allophycocyanin from *Spirulina platensis. Biotechnol. Technol.* 13: 601-603.

Zhao, L., Peng, Y., Gao, J., and Cai, W. (2014). Bioprocess intensification: an aqueous two-phase process for the purification of C-phycocyanin from dry *Spirulina platensis. Eur. Food. Res. Technol.* 238: 451-457.

Zhao, M., Sun, L., Sun, S., Gong, X., Fu, X. and Chen, M. (2015). Phycoerythrins in phycobilisomes from the marine red alga *Polysiphonia urceolata. Int. J. Biol. Macromol.* 73: 58-64.

In: The Role of Photosynthetic Microbes…
Editors: K. Tripathi, N. Kumar et al.
ISBN: 978-1-53614-032-3

Chapter 11

Genetic Modification of Cyanobacteria for Sustainable Agriculture

Meenakshi Singh[1], Manoj Kumar Tripathi[1], Smita Shukla[3], Himanshu Dubey[4] and Keshawanand Tripathi[2,*]

[1]Institute of Bioengineering and Biological Sciences, Varanasi, Uttar Pradesh, India
[2]Centre for Conservation and Utilization of Blue Green Algae, IARI, New Delhi, India
[3]Deen Dayal Upadhyaya Gorakhpur University, Gorakhpur, India
[4]National Research Centre on Plant Biotechnology, New Delhi, India

Abstract

Cyanobacteria are one of the most primitive organisms which can adjust to various extremes of the environment. The growing concern for meeting demand for sustainability of agriculture food has diverted the attention of the scientific community towards natural resources to combat this problem. Prolonged use of chemical fertilizers along with using irrigation poor water has deteriorated the quality of soil and increased its salinity. Cyanobacteria with their inherent characteristics of nitrogen fixation and oxygenic photosynthesis prove a boon for enhancing crop yield. Manipulating these photosynthetic diazotrophs using different biotechnological techniques can be a potential way to resolving agricultural problems. However, application of these transgenics is a challenge due to several biotic and abiotic factors. The biotechnological regulatory issue is also a major constraint in their field application at present. The scientific community is to

* Corresponding Author Email: tripathikn009@gmail.com.

exploit cyanobacteria as a natural source for increasing crop yield and reclaiming non usable agricultural land.

Keywords: agricultural, cyanobacteria, genetic modification, sustainable, transgenic cyanobacteria

INTRODUCTION

In this era of exploding population and urbanization, the biggest challenge before scientific community is to increase the crop yield. It is anticipated that world's population will be reaching to nine billion in 2050 and to feed, we will need to increase food production up to 70 Percent Increase in temperature, shortage of potable water, rise in greenhouse gases from livestock and cattle are one of the many factors disturbing climate which in turn is exerting stress on agriculture. To meet this demanding situation, we need to address the issue of climate change and become more productive and sustainable. In light of this prospect, The Food and Agriculture Organization (FAO) of the United Nations observe World Food Day every year on October 16 and global message for 2016 was 'Climate is changing, Food and Agriculture must too'.

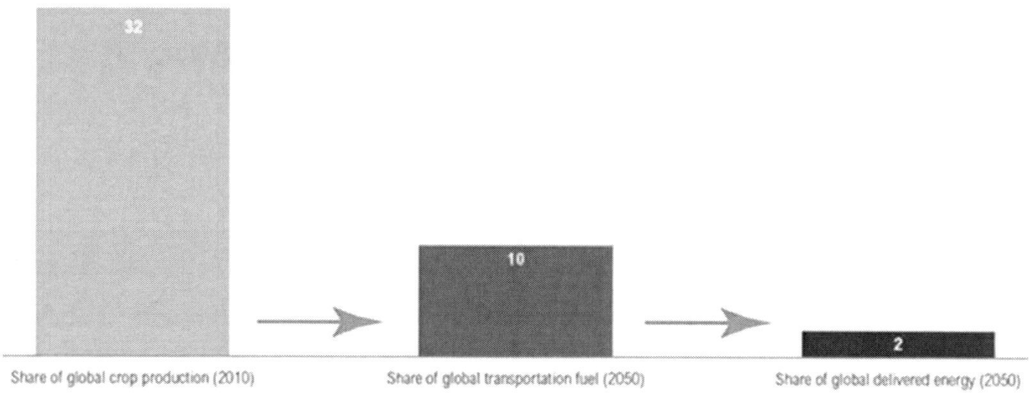

Figure 1. Profitability of biofuel crops. Biofuel production is also giving a tough competition to the food crops and the land acquired by them. Production of two percent of energy would require 32 percent of total crop production (reproduced with permission from World Resources Institute).

The term, sustainable literally means using a resource so that it is not permanently damaged or depleted. According to 'Farm bill' (1990), sustainable agriculture is stated to be "an integrated system of plant and animal production practices having a site-specific application that will, over the long term:

- satisfy human food and fiber needs
- enhance environmental quality and the natural resource base upon which the agricultural economy depends
- make the most efficient use of nonrenewable resources and on-farm resources and integrate, where appropriate, natural biological cycles and controls
- sustain the economic viability of farm operations
- enhance the quality of life for farmers and society as a whole

FACTORS AFFECTING AGRICULTURE

For solving this problem, we should first focus on the causes of stress that decrease food yield. Urbanization and infrastructure development in the last century have rapidly swallowed land meant for the agricultural purpose. This has enormously reduced the overall production of crops. Construction of roads, dams, and buildings have made the surrounding soil infertile, hence making it unavailable for farming. In a bid to save non-renewable fuels, the emphasis is given to increasing the production of bio-fuel based crops. Statically, till 2050 for producing 10% of all transport fuel, we need 32% of global crop production which in turn will produce only 2% of global energy (Figure 1). Global water deficit is continuously increasing due to rise in the temperature and demand of water supply (Figure 2). Poor irrigation practices and heavy use of fertilizers have slowly degraded the quality of the soil and rendered it saline. Salinity will be our prime concern in the discussion below.

On a worldwide scenario, the total irrigated land is roughly 310 million hectares of which around 62 million hectares (20 percent) is salt-affected. Since more than 20 years, an average of 2000 hectares of irrigated land in arid and semi-arid regions have been degraded by salt, according to a recent study, 'Economics of Salt-induced Land Degradation and Restoration' published by UNU Institute for Water, Environment, and Health (UNU-INWEH). Increasing content of salt is degrading the alluvial soil of the productive lands like Indo-Gangetic Basin, Euphrates River Basin and many more. Above study quotes about an estimated loss of US$ 441 per hectare due to salinity in 2013 which comes to around 27.3 billion USD per year (Qadir et al., 2014). Extensive use of chemical fertilizers for higher crop yield have gradually rendered these soils saline. We need to look for an eco-friendly and greener alternative to solve the problem. Cyanobacteria (blue-green algae) can be used as natural biofertilizer fertilizer without compromising on the fertility of the soil.

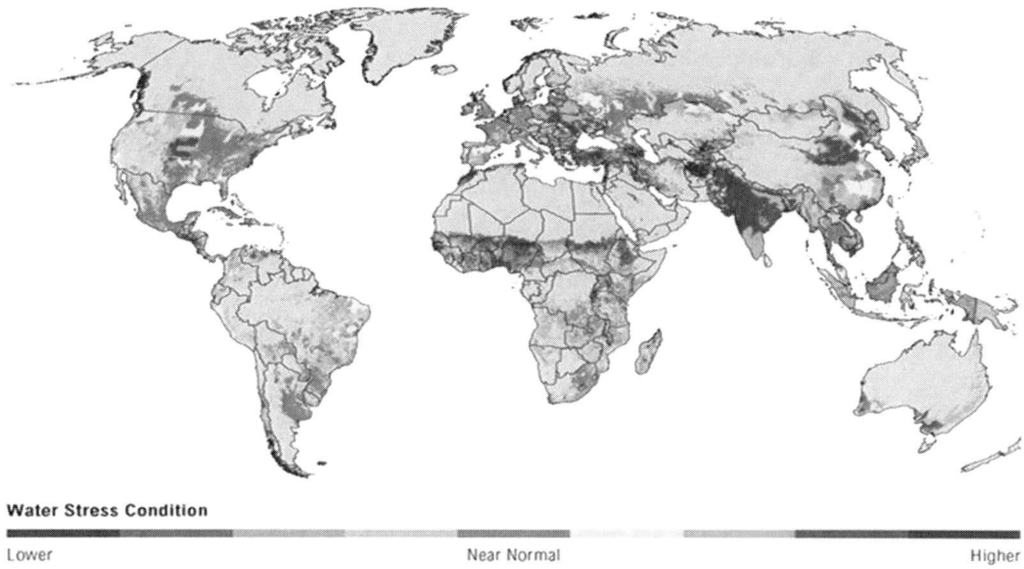

Figure 2. Water stress in future. Increased consumption of water and rise in temperature would adversely go to affect the world in coming years. Major northern parts of India will be deeply truck by water crisis (reproduced with permission from World Resources Institute).

Cyanobacteria were the organisms that changed primitive earth's atmosphere from reducing into oxidizing one. They have a unique characteristic of combining oxygenic photosynthesis with nitrogen fixation and thus, play a vital role in the nitrogen and carbon biogeochemical cycle. Due to their great evolutionary age, adaptability, and similarity to eukaryotic organisms, cyanobacteria have emerged as interesting objects of basic research. In a typical prokaryotic cell organization, they accomplish water-oxidizing photosynthesis of higher plants. Cyanobacteria have a wide range of morphologies containing about 150 genera and 2,000 species (Sharma et al., 2011). Based on taxonomy, cyanobacteria have been classified into five principal groups (Boone and Castenholz, 2001). The agronomic significance of filamentous N_2-fixing cyanobacteria (or diazotrophs) lies in their ability to fix atmospheric nitrogen. Although nitrogen accounts for almost 78% of the earth's atmosphere, its inert nature mostly keeps its unutilized except for some free-living and symbiotic bacteria (Figure 3). Free-living N_2-fixing species such as *Anabaena* and *Nostoc* often account half or more of the species present in paddy field soils deficient in nitrogen or with low nitrogen input, and play a significant role in sustaining and improving the rice field productivity (Roger et al., 1993). Rice fields in southeastern part of the world, which account for more than half of the world total rice production, are mainly dependent on these diazotrophs for fixing nitrogen (Singh et al., 2014; Tripathi et al., 2013a). But the question still remains the same. Are these conventional methods of farming are enough to fulfill the exploding and hungry population? Excessive use of these chemical fertilizers is not only polluting our environment but also changing physio-chemical properties of soil. Eutrophication of water bodies and increase the salinity of soil are the main problems which sprouted up in due course of time. Reports indicate that cyanobacteria provide nitrogen to surrounding plants and organisms (Rana et al., 2012). Thus, cyanobacteria can provide an effective answer to industrial inorganic fertilizers used in heavy doses in the fields. Recent advances in the field of biotechnology and molecular biology have provided an important breakthrough for designing living organisms for our own benefit. Genetically manipulating the cyanobacteria can help us in combating other issues such as increased salinity, removal of toxic waste etc.

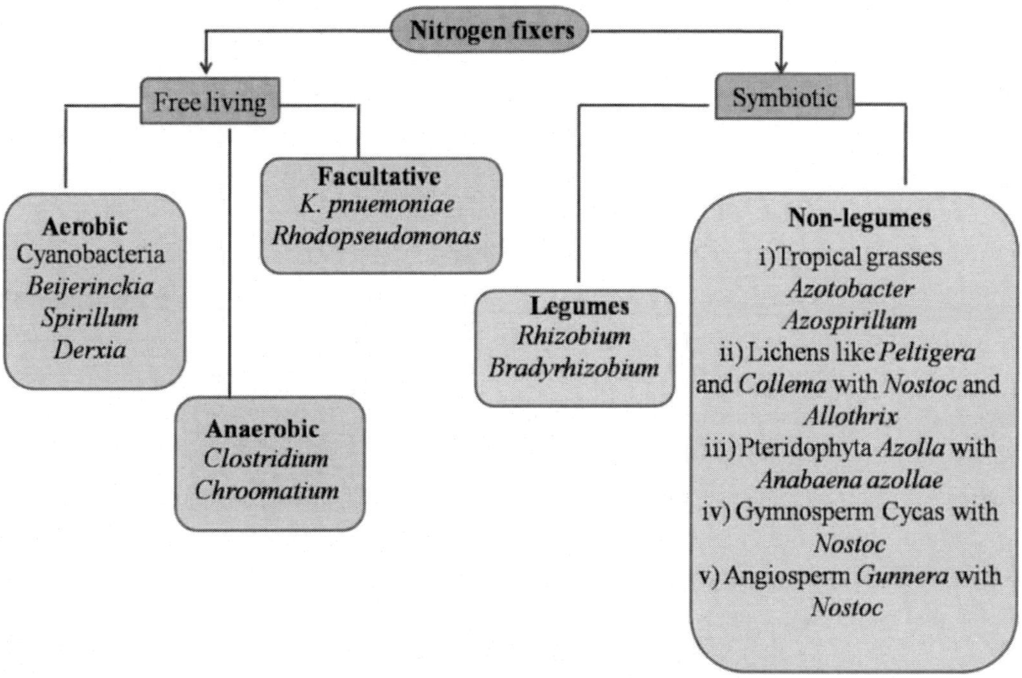

Figure 3. Natural nitrogen fixers. Many bacteria and cyanobacteria naturally fix nitrogen from atmosphere and help plants in assimilating them. These organisms either live freely or in symbiotic association with plants.

PLANT PERSPECTIVE

Salt-loving or halophilic cyanobacteria can be found in high salt environments such as hypersaline lakes, hypersaline lagoons, and saltern ponds. Cyanobacteria inhabiting such environments are generally unicellular (*Aphanothece halophytica, Euhalothece, and Halothece*) (Tripathi et al., 2013b) found in the salt concentration approaching NaCl saturation or non-heterocystous filamentous species (*Phormidium* sp, *Halospirulina tapeticola, Halomicronema excentricum)* which dominate in intermediate salinities

(Oren, 2015). At high salinities, life becomes energetically expensive. Basically, there are two stratagems for these organisms to survive under high saline conditions. First is "salt-in", which adapts the intracellular enzymes and proteins to remain stable under high concentration of K^+ and Cl^-. Another one is biosynthesis/accumulation of "compatible solutes", which are small, polar, highly soluble organic molecules with a neutral charge at physiological pH and do not interfere with the cellular metabolism (Roberts 2000, 2005; Roebler and Müller 2001). With the exception of proline, they are characterized as amino acid derivatives: betaines, ectoines, N-acetylated diamino acids and N-derivative carboxamides of glutamine. Species which are slightly tolerant to salt synthesize/accumulate sucrose or trehalose; glucosylglycerol, ectoine, and proline with intermediate tolerance; glycine betaine being synthesized by the salt requiring ones. Genetically modifying these extreme halophiles can be of great help in reinstating saline agricultural fields.

For constructing efficient genetically engineered cyanobacteria, the strong expression vector is needed which could work in the lab as well in field conditions. Many ways have been developed to transform cyanobacteria *viz.,* transformation, electroporation, and conjugation. Some cyanobacteria have been known to transform naturally. Waditee et al., (2005) transformed *Synechococcus* with N-methyltransferase genes naturally and found it to coexpress the enzymes conferring salt tolerance and capacity to grow at 0.5 M NaCl and in seawater. Studies showed that presence of extracellular nucleases hinders the process of transformation in heterocyst-forming, filamentous cyanobacteria (Wolk and Kraus, 1982). Yoshihara et al., (2001) suggested that a product of *Synechocystis* ORF *slr*0197 is required for the uptake of DNA. Nowadays, electroporation is used for transforming many organisms, which cannot be made competent. Thiel and Poo (1989) transformed filamentous cyanobacterium *Anabaena* sp. strain M131 with the shuttle vector pRL6 through electroporation and also showed that presence of restriction enzymes significantly reduced the efficiency of transformation.

Toyomizu et al., (2001) and Ravindran et al., (2006) successfully transformed *Spirulina platensis* with chloramphenicol acetyltransferase (*CAT*) genes and *Oscillatoria* MKU 277 with shuttle plasmid pRL489 by electroporation. *Arabidopsis* has been transformed with N-methyl-transferase gene through electroporation (Waditee et al., 2005), which became salt tolerant. Chaurasia et al. (2008) constructed a stable integrative expression vector pFPN which can be easily transferred in desired strain by electroporation. This is reported to be suitable for *Anabaena* strains and for other nonconjugal strains.

Conjugation is generally used to introduce DNA from *Escherichia coli* into cyanobacteria and is advantageous in transferring large segment of foreign DNA. Elhai (1993) wanted to access strong promoters which can be used in cyanobacteria. He fused *luxAB,* encoding bacterial luciferase to several promoters such as P_{tac} and P_{psbA}, with sequences nearly identical to consensus *E. coli* σ^{70} promoters. Both of them gave higher expression than that of the strong *Anabaena* PCC 7120 promoter. Xiaoqiang et al., (1997) introduced various combinations of *cry* gene isolated from *Bacillus thuringiensis* subsp. *israelensis* along with p20 in *Anabaena* sp. 7120 using triparental conjugation to produce mosquitocidal cyanobacteria.

Photosynthetic diazotrophic cyanobacteria such as *Nostoc* and *Anabaena* play an important role in the carbon and nitrogen cycles in tropical paddy fields and are salt sensitive. Therefore, genetically modifying them to make salt tolerant is interesting to study. There are few reports describing the genetic transformation of cyanobacteria for improving salinity tolerance of which many are on transforming with compatible solute synthesizing genes. Chaurasia and Apte (2009) overexpressed *groESL* operon encoding heat shock proteins in *Anabaena* sp. PCC 7120 which conferred heat and salinity tolerance.

There are two known pathways for synthesis of glycine betaine; first involved oxidation of choline and the other one is by the action of methyltransferases. Choline oxidation pathway involves enzyme choline monooxygenase (CMO) found in plants; membrane-bound choline dehydrogenase (CDH) in animals and bacteria, and choline oxidase (COD) in some other bacteria for converting choline to betaine aldehyde. NAD^+-dependent betaine aldehyde dehydrogenase (BADH) further transforms betaine aldehyde to glycine betaine in all the organisms (Takabe et al., 2006) (Figure 4). Studies have been going on to develop transgenic plant species expressing CMO, CDH or COD for improving the stress tolerance. Genes responsible for choline dehydrogenase (*gbsB*) and glycine betaine aldehyde dehydrogenase (*gbsA*) together constitute an operon. Nomura et al., (1995) transformed freshwater *Synechococcus* sp. PCC 7942 with *E. coli bet* gene cluster (encoding for glycine betaine); *betA* (choline dehydrogenase), *betB* (betaine aldehyde dehydrogenase), *betl* (a putative regulatory protein), and *betT* (the choline transport system). The transformant showed increased salt tolerance up to 0.3 M NaCl with an accumulation of glycine betaine.

In the second pathway, enzyme glycine sarcosine methyltransferase (GSMT) catalyzes the methylation of glycine to sarcosine (N-monomethylglycine) and subsequently to dimethylglycine, and sarcosine dimethylglycine methyltransferase (SDMT), which catalyzes the methylation of sarcosine to dimethylglycine, and then to glycine betaine (Nyyssola et al., 2000, 2001) (Figure 5). Increasing salinization of paddy fields and their association with rice field cyanobacteria opens an exciting field for developing salt-tolerant transgenic cyanobacteria. Two N-methyltransferases encoding ApGSMT (glycine sarcosine methyl-transferase) and ApDMT (dimethylglycine transferase) from *A. halophytica* were transferred and expressed in *A. doliolum* and

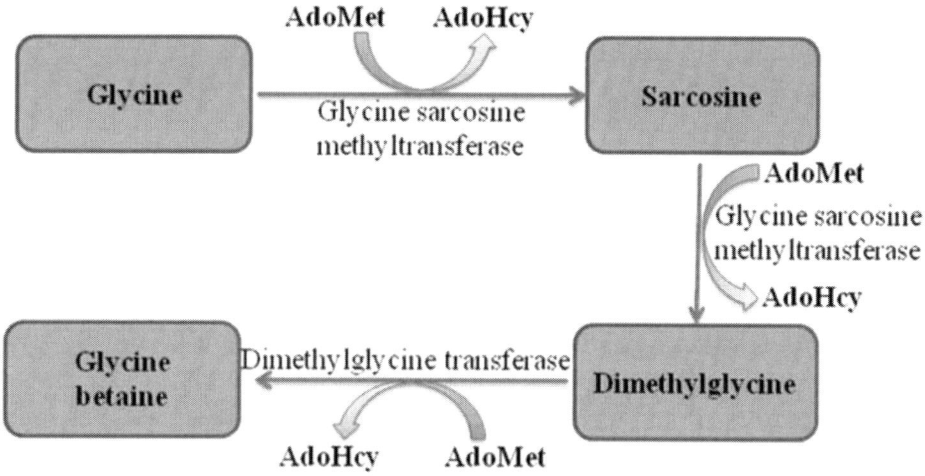

Figure 4. Glycine betaine biosynthetic pathway from choline. It involves choline monooxygenase (CMO) in plants; membrane bound choline dehydrogenase (CDH) in animals and bacteria, and choline oxidase (COD) in other bacteria. (Source: Tripathi et al., 2017.)

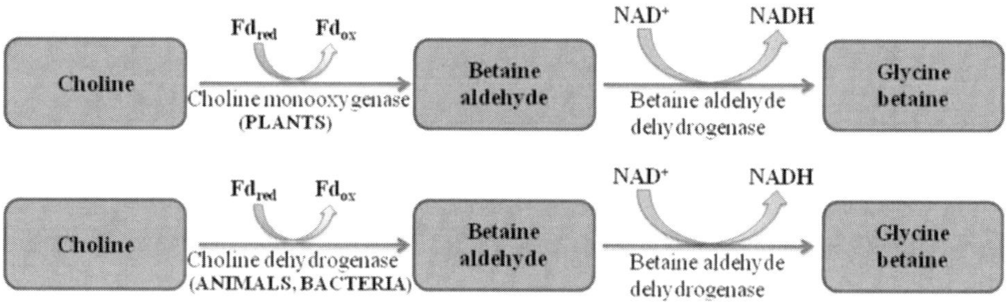

Figure 5. Biosynthetic pathway of glycine betaine by glycine methylation in *A. halophytica*. The pathway requires action of glycine sarcosine methyltransferases to convert glycine to sarcosine and them to dimethylglycine which is further converted to glycine betaine by action of dimethyltransferase. (Source: Tripathi et al., 2017.)

Anabaena 7120 using triparental conjugation. The *ApGSMT-DMT* transformed *Anabaena* 7120 and *A. doliolum* cells were able to tolerate high salt concentrations, i.e., 0.2 and 0.3 M NaCl, respectively, which were otherwise lethal to the WT cells. Also, nature of *ApGSMT-DMT* transformed *Anabaena* 7120 and *A. doliolum* changed from freshwater to halophilic. There seemed to be an obligate requirement of NaCl to survive by these transformants. Observations either with *Anabaena* 7120 or *A. doliolum* indicated that introduction of N-methyltransferase genes in cyanobacterial cells increased the activity of the enzymes involved in nitrogen metabolism (Singh et al., 2013). Further studies on physiological mechanisms like photosynthesis showed high activity whereas respiration decreased depicting efficient photosynthates (Swapnil et al., 2015).

CONCLUSION

Salinity is taking a toll over the gross yield of crops worldwide. Degradation of already shrinking agricultural lands due to salinity is having a deleterious impact on the food needs. The biotechnological implication of the salinity tolerant genes from various halotolerant organisms could be a boon in this going to be the pandemic situation. Studying these genes and transforming them into suitable species such as diazotrophic cyanobacteria can be of great help in increasing nitrogen fixation in salt-affected lands. These can act as natural biofertilizers which will enhance the soil health and help in reducing the use of chemical fertilizers. Despite the advances made on the transgenic cyanobacteria, the regulatory issues constrain their application in the field. The hard work of the scientist may prove fruitful in these years to come with these interventions of regulatory authority and the government of India. Studies are underway to develop commercially important transgenic crops such as tobacco, carrot, potato, tomato, soyabean and rice with salt tolerant genes (Khan et al., 2015). For the very first time, Sugino et al. in 1999 introduced *dnaK*1 gene from *Aphanothece halophytica* in tobacco plant which exhibited increased stress tolerance. Ectoine genes from *Halomonas elongate* were transferred to tobacco plants using an *Agrobacterium*-mediated gene delivery system and the plant showed photosynthesis at increased salinity (Moghaieb et al., 2006). Two proline expressing genes *OsP5CS1 and OsP5CS2* were expressed in tobacco plants that increased the osmolyte accumulation and also reduced the damage to cells under abiotic stress conditions (Zhang et al., 2014). Similarly, workers like Zhu et al., 1998; Sawahel and Hassan, 2002; and Han and Hwang, 2003 studied the overexpression of various proline synthesis genes conferred stress resistance to rice, wheat and carrot plants, respectively. The transgenic wheat crop was also developed by introducing glycine betaine synthesizing *betA* genes (He et al., 2010). Ahmad et al., (2008) developed salt and drought resistant potato plants by expressing glycine betaine synthesizing *codA* genes. Similarly, Goel et al., (2011) transformed the same *codA* gene in tomato and rendered them resistant to salt and water. Recently, transgenic soybean was produced using a salinity tolerant gene *Ncl,* and salt tolerant lines of the plant were produced through fine mapping using DNA marker-assisted selection (MAS). Under salt stress field conditions, these lines showed 3-5 times more grain yield (Do et al., 2016).

ACKNOWLEDGMENTS

KT is thankful to Department of Science and Technology (DST), New Delhi for financial assistance in form of DST-Young Scientist.

Conflict of Interest

Authors declare no conflict of interest.

References

Ahmad, R., Kim, M. D., Back, K. H., Kim, H. S., Lee, H. S., Kwon, S. Y., Murata, N., Chung, W. and Kwak, S. (2008). Stress-induced expression of choline oxidase in potato plant chloroplasts confers enhanced tolerance to oxidative, salt, and drought stresses. *Plant. Cell. Rep*, 27: 687–698.

Boone, D. R., and Castenholz, R. W. (2001). The *Archaea* and the deeply branching and phototrophic bacteria. In: Garrity, G. M. (ed) *Bergey's manual of systematic bacteriology,* 2nd Edn. Springer-Verlag, New York: 33-38.

Chaurasia, A. K., and Apte, S. K. (2009). Overexpression of the *groESL* operon enhances the heat and salinity stress tolerance of the nitrogen fixing cyanobacterium *Anabaena* sp. strain PCC7120. *Appl. Environ. Microbiol*, 75, 6008–6012.

Chaurasia, A. K., Parasnis, A., and Apte, S. K. (2008). An integrative expression vector for strain improvement and environmental applications of the nitrogen fixing cyanobacterium, *Anabaena* sp. strain PCC7120. *J. Microbiol. Meth*, 73: 133–141.

Do, T. D., Chen, H., Hien, V. T., Hamwieh, A., Yamada, T., Sato, T., Yan, Y., Cong, H., Shono, M., Suenaga, K. and Xu, D. (2016). *Ncl* synchronously regulates Na^+, K^+, and Cl^- in Soybean and greatly increases the grain yield in saline field conditions. *Sci. Rep*, 6: 19147.

Elhai, J. (1993). Strong and regulated promoters in the cyanobacterium *Anabaena* PCC 7120. *FEMS. Microbiol. Lett*, 114: 179-184.

Goel, D., Singh, A. K., Yadav, V., Babbar, S. B., Murata, N. and Bansal, K. C. (2011). Transformation of tomato with a bacterial *codA* gene enhances tolerance to salt and water stresses. *J. Plant. Physiol*, 168:1286–1294.

Han, K. H. and Hwang, C. H. (2003). Salt tolerance enhanced by transformation of a *P5CS* gene in carrot. *J. Plant. Biotechnol*, 5: 149–153.

He, C., Yang, A., Zhang, W., Gao, Q. and Zhang, J. (2010). Improved salt tolerance of transgenic wheat by introducing *betA* gene for glycine betaine synthesis. *Plant. Cell. Tiss. Organ. Cult*, 101: 65–78.

Khan, M. S., Ahmad, D. and Khan, M. A. (2015). Utilization of genes encoding osmoprotectants in transgenic plants for enhanced abiotic stress tolerance. *Electron. J. Biotechnol*, 18: 257–266.

Moghaieb, R. E. A., Tanaka, N., Saneoka, H., Murooka, Y., Ono, H., Morikawa, H., Nakamura, A., Nguyen, N. T., Suwa, R. and Fujita, K. (2006). Characterization of salt tolerance in ectoine-transformed tobacco plants (*Nicotiana tabaccum*): photosynthesis, osmotic adjustment, and nitrogen partitioning. *Plant. Cell. Environ.* 29: 173–182.

Nomura, M., Ishitani, M., Takabe, T., Rai, A. K. and Takabe, T. (1995). *Synechococcus* sp. PCC 7942 transformed with *Escherichia coli bet* genes produces betaine from choline and acquires resistance to salt stress. *Plant. Physiol*, 107: 703-708.

Nyyssola, A., Kerovuo, J., Kaukinen, P., von Weymarrn, N. and Reinikainen, T. (2000). Extreme halophiles synthesize betaine from glycine by methylation. *J. Biol. Chem*, 275: 22196–22201.

Nyyssola, A., Reinikainen, T. and Leisola, M. (2001). Characterization of glycine sarcosine N-methyltransferase and sarcosine dimethylglycine N-methyltransferase. *Appl. Environ. Microbiol*, 67: 2044–2050.

Oren, A. (2015). Cyanobacteria in hypersaline environments: biodiversity and physiological properties. *Biodivers. Conserv*, 24: 781–798.

Qadir, M., Quillérou, E., Nangia, V., Murtaza, G., Singh, M., Thomas, R. J., Drechsel, P. and Noble, A. D. (2014). Economics of salt-induced land degradation and restoration. *Nat. Resour. Forum*, 38: 282–295.

Rana, A., Joshi, M., Prasanna, R., Shivay, Y. S. and Nain, L. (2012). Biofortification of wheat through inoculation of plant growth promoting rhizobacteria and cyanobacteria. *Eur. J. Soil. Biol*, 50: 118-126.

Ravindran, C. R. M., Suguna, S. and Shanmugasundaram, S. (2006). Electroporation as a tool to transfer the plasmid pRL489 in *Oscillatoria* MKU 277. *J. Microbiol. Meth*, 66: 174-176.

Roberts, M. F. (2000). Osmoadaptation and osmoregulation in archaea. *Front. Biosci*, 5: 796-812.

Roberts, M. F. (2005). Organic compatible solutes of halotolerant and halophilic microorganisms. *Saline. Systems*, 1: 5.

Roeßler, M. and Müller, V. (2001). Osmoadaptation in bacteria and archaea: common principles and differences. *Environ. Microbiol*, 3: 743-754.

Roger, P. A., Zimmerman, W. J. and Lumpkin, T. A. (1993). Microbiological management of wetland rice fields. In: Metting, B. (eds), Soil *Microbial Ecology: Applications in Agricultural and Environmental Management,* New York Dekker: 417–455.

Sawahel, W. A., and Hassan, A. H. (2002). Generation of transgenic wheat plants producing high levels of the osmoprotectant proline. *Biotechnol. Lett*, 24: 721–725.

Singh, H., Khattar, J. S. and Ahluwalia, A. S. (2014). Cyanobacteria and agricultural crops. *Vegetos*, 27 (1): 37-44.

Singh, M., Sharma, N. K., Prasad, S. B., Yadav, S. S., Narayan, G. and Rai, A. K. (2013). Freshwater cyanobacterium *Anabaena doliolum* transformed with *ApGSMT-DMT* exhibited enhanced salt tolerance and protection to nitrogenase activity but became halophilic, *Microbiol, SGM,* 159(3): 640-647.

Sharma, N. K., Tiwari, S. P., Tripathi, K. (2011). Sustainability of cyanobacteria (blue-green algae): facts and challenges. *J. App. Phycol* 23:1059-1081.

Sugino, M., Hibino, T., Tanaka, Y., Nii, N. and Takabe, T. (1999). Overexpression of *DnaK* from a halotolerant cyanobacterium *Aphanothece halophytica* acquires resistance to salt stress in transgenic tobacco plants. *Plant. Sci*, 146: 81–88.

Swapnil, P., Singh, M., Singh, S., Sharma, N. K. and Rai, A. K. (2015). Recombinant glycinebetaine improves metabolic activities, ionic balance and salt tolerance in diazotrophic freshwater cyanobacteria. *Algal. Res,* 11: 194–203.

Takabe, T., Rai, V. and Hibino, T. (2006). Metabolic engineering of glycine-betaine. In: Rai, A. K. and Takabe, T. (eds.), *Abiotic Stress tolerance in plants; towards the improvement of global environment and food*, Netherlands, Springer, pp. 137-151.

Thiel, T. and Poo, H. (1989). Transformation of a filamentous cyanobacterium by electroporation, *J. Bact*, 171: 5743-5746.

Toyomizu, M., Suzuki, K., Kawata, Y., Kojima, H., and Akiba, Y. (2001). Effective transformation of the cyanobacterium *Spirulina platensis* using electroporation, *J. Appl. Phycol*, 13: 209-214.

Tripathi, K., Abraham, G., Rai, V., Dhar, D. W., Rashi, V., Yadav, R. K. (2017). Compatible solute in relation to microbes and their biotechnological significance. In: Prashad R., Kumar N. (eds) *Microbes and Sustainable Agriculture,* IK publication, New Delhi, pp. 208-221.

Tripathi, K., Sharma, N. K., and Rai, V. and Rai, A. K. (2013a). Low cellular P-quota and poor metabolic adaptations of the freshwater cyanobacterium *Anabaena fertilissima* Rao during Pi-limitation. *Antonie Leeuwenhoek*, 103:277–291.

Tripathi, K., Sharma, N. K., Kageyama, H., Takabe, T., and Rai, A. K. (2013b). Physiological, biochemical and molecular responses of the halophilic cyanobacterium *Aphanothece halophytica* to Pi-deficiency. *Eur. J. Phycol*. 48 (4): 461–473.

Waditee, R., Bhuiyan, M. N. H., Rai, V., Aoki, K., Tanaka, Y., Hibino, T., Suzuki, S., Takano, J., Jagendorf, A. T., Takabe, T. and Takabe, T. (2005). Genes for direct methylation of glycine provide high levels of glycine betaine and abiotic stress tolerance in *Synechococcus* and *Arabidopsis*. *Proc. Natl. Acad. Sci. USA*, 102: 1318-1323.

Wolk, C. P., and Kraus, J. (1982). Two approaches to obtaining low, extracellular deoxyribonuclease activity in cultures of heterocyst forming cyanobacteria. *Arch. Microbiol*, 131: 302–307.

Xiaoqiang, W., Vennison, S. J., Huirong, L., Ben-dov, E., Zaritsky, A. and Boussiba, S. (1997). Mosquito larvicidal activity of transgenic *Anabaena* Strain PCC 7120 expressing combinations of genes from *Bacillus thuringiensis* subsp. Israelensis. *Appl. Environ. Microbiol*, 63(12): 4971–4975.

Yoshihara, S., Geng, X. X., Okamoto, S., Yura, K., Murata, T., Go, M., Ohmori, M. and Ikeuchi, M. (2001). Mutational analysis of genes involved in pilus structure, motility and transformation competency in the unicellular motile cyanobacterium *Synechocystis* sp. PCC 6803. *Plant. Cell. Physiol*, 42: 63–73.

Zhang, X., Tang, W., Liu, J. and Liu, Y. (2014) Co-expression of rice *OsP5CS1* and *OsP5CS2* genes in transgenic tobacco resulted in elevated proline biosynthesis and enhanced abiotic stress tolerance. *Chin. J. Appl. Environ. Biol*, 20(4): 717–722.

Zhu, B., Su, J., Chang, M., Verma, D. P. S., Fan, Y. L. and Wu, R. (1998). Overexpression of a 1-pyrroline- 5-carboxylate synthase gene and analysis of tolerance to water and salt stress in transgenic rice. *Plant. Sci*, 139: 41–48.

In: The Role of Photosynthetic Microbes…
Editors: K. Tripathi, N. Kumar et al.
ISBN: 978-1-53614-032-3

Chapter 12

BIOREMEDIATION OF PESTICIDES RESIDUES: A PSYCHOLOGICAL APPROACH

Savita Singh[1], Ashutosh Kumar[2,3*], S. P. Jeevan Kumar[2], Mohd. Imran[3], Madan Kumar[2], Arvind Nath Singh[2], and Manoj Kumar Tripathi[4]

[1]Department of Botany, Banaras Hindu University, Varanasi, India
[2]Indian Institute of Seed Science, Mau, India
[3]Integral University, Lucknow, India
[4]Institute of Bioengineering and Biological Sciences, Varanasi, Uttar Pradesh

ABSTRACT

To increase the food production, pesticides have been used excessively; as a result, they have become a threat to the environment. In the present era, microbial degradation is one of the important techniques for degradation of pesticides from agricultural lands. Studies have been conducted on the use of different microbes such as bacteria, fungi and genetically modified microorganisms for degradation of pesticides. Case studies and researches have revealed that microbial consortia of naturally occurring microbes isolated from particular contaminated environments have potential to degrade pesticides at faster rate rather than individual microbes. Microalgae and cyanobacteria exhibiting high growth rate and biodegradation potential are yet to be explored. Microalgae and cyanobacteria are cosmopolitan in nature ranging from unicellular to filamentous forms that inhabit several environmental conditions. In order to re-evaluate the current scenario of pesticide contamination and the role of micro algae and cyanobacteria have been

* Corresponding Author Email: ashu.nbaim@gmail.com.

discussed. The chapter highlights the use of microalgae and cyanobacteria in bioremediation of pesticide contamination and remedies using psychological approaches.

Keywords: bioremediation, cyanobacteria, pesticides residue

INTRODUCTION

India ranked second after China in terms of population growth and agriculture productivity is required to feed such huge population (Reshmita et. al., 2015). Introduction of pesticides initially helped in controlling the pest to boost agriculture production. Many pesticides are used worldwide to control pests in agricultural practices. However, continuous application of pesticides is serious threat to the environment. The term "pesticide" includes substances intended for use as plant growth regulators, defoliants, desiccants, fruit thinning agent or an agent to prevent the premature fall of fruit and substances applied to crops either before or after harvest and to prevent deterioration during storage or transport (Subramanyam et al., 2012). Pesticides kill pests, thereby protecting and indirectly aid in increasing the production. All these compounds are of particular concern because of the deleterious effects of used pesticides on different beneficial microorganisms and their recalcitrant behavior on the environment. Consequently, they persist in the environment because they are not susceptible to normal rate of biological transformation. Excessive use of pesticides is highly detrimental to the environment and affects the soil health and indirectly affects the production in the long run.

Pesticide consumption has crossed the mark of 2 million tones and among them Europe utilizes 45% followed by USA (Abhilash and Nandita, 2009). India is the largest producer of pesticides in the Asia and ranks 12th in the world. In India, use of pesticide is about 0.5 kg/hectare and major contribution is from organochlorine pesticides due to wide spectrum of actions and apt for usage in warm humid climatic conditions (Bhat and Padmaja, 2014). Due to excessive use of pesticides and insecticides in agriculture, the concept of sustainable agriculture has been abolished and causes severe ill effect to existing agriculture system and environment. In this regard bioremediation achieved by means of microorganisms and plant or their enzymes for bringing the polluted environment back to its original state has gained an up thrust. As chemical methods of remediation which includes, the precipitation and adsorption for heavy metals present in pesticides have disadvantage of high economical and energy input (Monteiro et al., 2012). Thus, there is a need for the sustainable remediation which should be nontoxic to nature, reusable and reproducible. The ability of microorgansims to degrade and/or detoxify chemical substances such as petroleum products, pesticides including Polycyclic aromatic hydrocarbons (PAHs), Polychlorinated biphenyls (PCBs), Polybrominated biphenyl ethers (PBDEs), heavy metals make microorganisms very prominent for bioremediation and cleaning environmental pollution (Biswas et al., 2015). A number of reviews dealing with the remediation are available but there are few reports of microorganisms assisted bioremediation, particularly microalgae and cyanobacteria. Bioremediation brought about by microalgae and cyanobacteria have been given the term "phycoremediation" (Rao et al., 2011). Bioremediation of pesticides by microalgae and cyanobacteria have been discussed in detail in the section below.

India and Worldwide Scenario of Pesticides Consumption

Pesticides play an important role in the Indian agriculture to meet increasing demand for food. India is the 12th largest producer of chemical pesticides in the world. Global pesticides consumption has increased 50 fold since 1950, with the drastic growth in human population (Vendan, 2016). India is the country where maximum human population is involved in agriculture and is expected to reach 1.5 billion by 2050 (Figure 1) (Carter and Honmann, 1991; Nanda and Carl Haub, 2007). The main challenge is to increase food production with the consistent reduction of pests (plant pathogens and insects) and infestation.

The worldwide consumption of pesticides is about two million tonnes per year. Out of which 45% is used by Europe alone, 25% is consumed by the USA, and 25% in the rest of the world. India's share is just 3.75%. The usage of pesticides in South Korea and Japan is 6.6 and 12.0 Kg/ha, respectively whereas in India, it is only 0.5 Kg/ha. Globally, the pesticides cover only 25% of the cultivated land area (Subramanayam et al., 2012). The three most commonly used pesticides are Hexacholorocyclohexane (HCH) (only gamma-HCH is allowed), dichlorodiphenyltrichloroethane (DDT) and Malathion that account for about 70% of the total pesticide consumption. Despite development of newer pesticide, these pesticides still remain the choice of small farmers because they are cost-effective, easily available and display a wide spectrum of bioactivity (Aktar et al., 2009). The total consumption of pesticides in India is in mainly the form of insecticides (80%) herbicides (15%), fungicides (2%) and others less than (3%). While comparing the worldwide consumption of pesticide, 47.5% is the share of herbicides, 29.5% is the share of insecticides, 17.5% is of fungicides and others account for 5.5% only (Figure 1). On the contrary, the consumption of herbicides in India is probably low because weed control is mainly done manually. In addition to public health and agricultural use, pesticides also find their use in other sectors too (De A. et al., 2014). Among the widely used pesticides that are being used in India are organophosphorus pesticides. Some organophosphates include monorotophos, phorate, methyl parathion, malathion, chlorpyrifos, diazinon, phorate, quinalphos, ethion, etc., which are moderate to high bio-accumulating hazards according to WHO (Gallo and Lawryk, 1991) The episodic application of these pesticides makes the situation tormenting and this repetition in the extended period essentially leads to bioaccumulation of pesticides and their residues in environment, endangering the living populations by their versatile toxicity (Gupta, 2004). Recently, there has been increasing interest on using algae to remove, degrade or render these pesticides harmless residues due to CO_2 reducer from atmosphere and short doubling period.

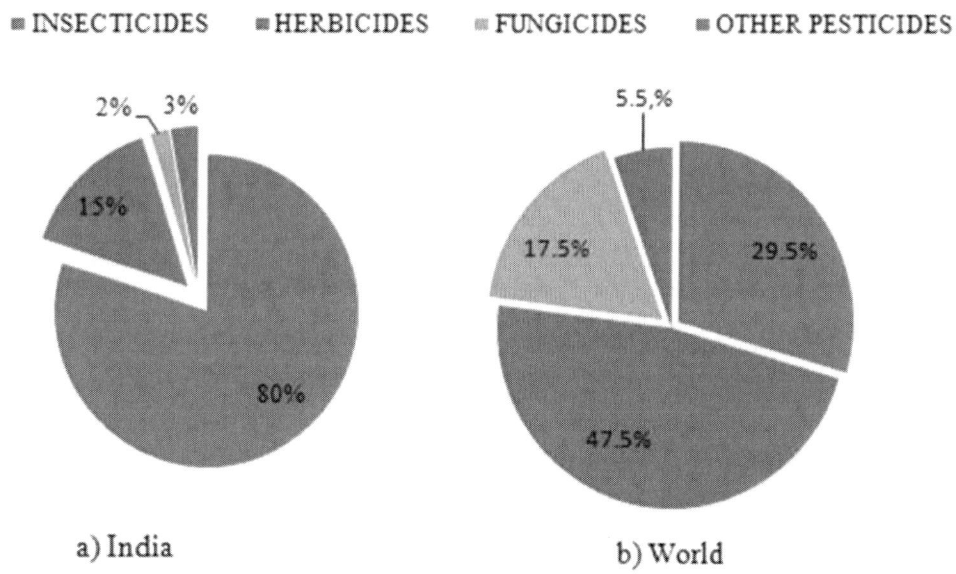

Source: H. neotia, 2016.

Figure 1. Application (%) of pesticides in agricultural sectors.

Micro-Algae and Cyanobacteria "Promising Bioremediation Agents"

Bioremediation refers to biological treatment to destroy or reduce the concentration of hazardous waste from a contaminated site which could be land, air or water (Biswas et al., 2015). Indigenous microorganisms are the important agents used for bioremediation. Microalgae along with other microbes such as bacteria and fungi have served as bioremediating agents. The first patent for a biological remediation substance was *Pseudomonas putida* capable of degrading petroleum long back in 1974 (Prescott et al., 2002). Investigation on pesticides bioaccumulation/ biodegradation in green algae and cyanobacteria is of great importance from environmental point of view because widespread distributions of these compounds in agricultural areas have become one of the major concerns (Jin et al., 2012). Globally, trees, grasses, herbs, and associated fungi and microorganisms are being used increasingly for cleaning polluted sites. Phytoremediation is "on the brink of commercialization" (Watanabe, 1997), and is given a rapid increasing market potential (Flathman and Lanza, 1998). The phytoremediation market is still emerging in the Europe, while in the US revenues are likely to exceed $300 million in 2007 (Campos et al., 2008). At present, the most widely used pesticides belong to the organophosphorus group. In the USA alone over 40 million kg of organophosphorus are applied annually (Mulchandani, 1999a; EPA, 2004, Tiwari, 2015).

Biodegradation of pesticides depends mainly on two factors; the first abiotic factors relates to the chemical structure of the pesticides while the second factor is microbial consortia and the optimum conditions for their survival and activity (Priyadarshani et al., 2011). Abiotic factors can be extended to pH, temperature, salinity, nutrients, light quality and intensity, water availability, oxygen tension and redox potential, surface binding, presence of alternative carbon substrates and electron acceptors, chemical structure, molecular weight and functional groups of the applied pesticides, their concentration, toxicity and solubility in water. On the other hand the biotic factors related to microorganisms including the presence and number of appropriate microorganisms, the contact between microalgae and the substrate (pesticide) that transform to toxic pesticides into nontoxic forms by metabolizing it through different metabolic pathways. A schematic representation of mode of action of pesticides and role of microalgae and cyanobacteria in bioremediation of pesticides is depicted in Figure 2.

Again, study by Ibrahim et al., (2016) mentioned the use of green algae i.e., *Chlorella vulgaris* and *Scenedesmus quadricuda* to degrade Malathion from the contaminated water. The ability of *A. azotica* to biodegrade lindane has potential use in the bioremediation of organochlorine pesticide. Many cyanobacteria exhibit a wide range of tolerance towards pesticides (Ahmad and Venkatraman, 1973; Kaushik and Venkatraman, 1983; Pabbi and Vaishya, 1992). Many of gaseous, liquid and solid pollutants namely carbon dioxide, nitrogen, phosphorus, phenolic, pesticides and detergents are detoxified/ metabolized by cyanobacteria (Subramanian and Uma, 1999). Both microalgae and cyanobacteria offer a great potential for bioaccumulation and transformation of pesticides (Table 1). Their biodegradability, non-toxicity, environmental compatibility, low cost features makes them efficient and promising bioremediating agent for pesticide removal from environment.

Malathion is a highly toxic organophosphorus insecticide to aquatic organisms. A mixed cyanobacterial population consisting of *Anabaena oryzae*, *Nosto cellipsosporum*, *Calothrix castellii*, *Tolypothrix ceytonica* and *Synechococcus* sp. were used to study the effect of Malathion on the growth and the diversity of the algae (Richa, 2015). This study

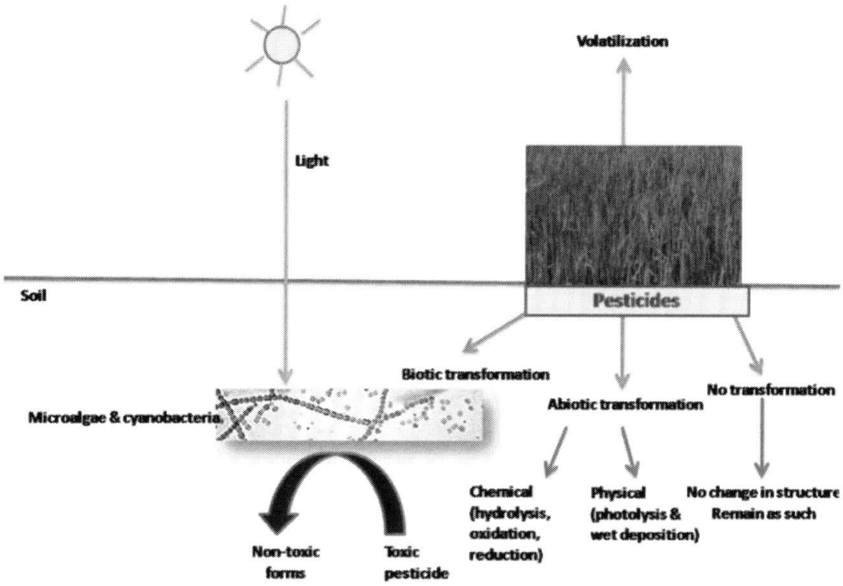

Figure 2. Mode of action of pesticides and role of microalgae and cyanobacteria in bioremediation of pesticides.

Table 1. Microalgae and cyanobacteria in bioaccumulation and biotransformation of pesticides

Microalgae/ Cyanobacteria	Bioaccumulation /Biotransformation	References
Anabaena azotica	Lindane	Zhang et. al., (2012)
Anabaena sp.	Butachlor	Agrawal et al., (2013)
Chlamydomonas sp.	Mirex, Lindane, naphthalene, Phenol	Jin et. al., (2012)
Chlorella sp.	Toxaphene, methoxychlor Lindane, chlordimeform	Muñoz et. al., (2003) Kobayashi and Rittman, (1982)
Chroococ custurgidus	Chlorpyrifos	Kumar et. al., (2014)
Cylindrotheca sp.	DDT	Kobayashi and Rittman, (1982)
Dunaliella sp.	Mirex, DDT, naphthalene	Kobayashi and Rittman, (1982)
Euglena gracilis	DDT, parathion, Phenol	Kobayashi and Rittman, (1982)
Scenedesmus Obliquus	DDT, parathion, Naphthalene sulfonic Acid	Tang et al., (2010)
Selenastrum capricornutum	Benzene, toluene, Benzo[a]pyrene chlorobenzene, 1,2-dichlorobenzene, Nitrobenzene, naphthalene, 2,6-dinitrotoluene, phenanthrene, di-nbutylphthalate, pyrene	Gavrilescu, (2010)

revealed that *A. oryzae* with its capability to utilize Malathion as a sole phosphorus source under limited or sufficient nitrogen conditions. This mode of phytoremediation might be considered as an inexpensive and efficient tool for the bioremediation of different organophosphorus insecticides from contaminated waste water (Ibrahim and Essa, 2010). The degradation experiments by Zhang (2012) showed that the percentage of lindane™ (*gamma*-hexachlorocyclohexane) removal efficiency by *Anabaena azotica* was 48.8% after 5 days, at an initial lindane concentration of 0.2 mg L^{-1}. The degradation of an organophosphorus pesticide, fenamiphos by different species of green algae and cyanobacteria were studied by Caceres, (2008). All the species tested were able to transform fenamiphos to its primary oxidation production, fenamiphos sulfoxide (FSO), while the majority of these cultures were able to hydrolyze FSO to fenamiphos sulfoxide phenol (FSOP). Fenamiphos sulfone phenol, FSOP and FSO were detected in the culture extracts of these algae and cyanobacteria.

FUTURE PROSPECTS AND BIOTECHNOLOGICAL ADVANCEMENTS

Microalgae and cyanobacteria have been reported to play an important role on bioremediation processes for such the removal of the heavy metals from polluted water, biotransformation of mercury (Hg^{2+}) and degradation of methyl parathion, crude oil and aromatic hydrocarbons. Recombinant filamentous cyanobacteria having ability to degrade organic pollutants have been reported by several workers (Kuritz and Wolk, 1995; Wipa and Fa-Aroonsawat, 2008). Such advancements are likely to be exploited. However, rigorous and careful development relating to its pros and cons in the environmental conditions are to be monitored for their success in field conditions.

CONCLUSION

Extensive applications of pesticides in agriculture have increased the issue of soil and water contamination. A chemical mode of pesticides remediation is expensive but biodegradation is becoming an ideal method. For the removal of hazardous pesticides residues from environment the use of microalgae is very efficient due to their small size, short doubling period, and ecofriendly and cost-effectiveness. These biological agents especially microalgae have potential to detoxify pesticides but there is a need to conduct extensive research and studies for exploration of degradation mechanisms of pesticides using of microalgae.

ACKNOWLEDGMENTS

The authors wish to thank director, ICAR-Indian Institute of Seed Science, Mau for his valuable support and guidance.

CONFLICT OF INTEREST

No conflict of interest among the authors.

REFERENCES

Abhilash, P. C., and Singh, N. (2009). Pesticide use and application: An Indian scenario. *J. Hazard. Mater.*, 165: 1-12.

Ahmad, M. H., and Venkataraman, G. S. (1973). Tolerance of *Aulosira fertilissima* to pesticides. *Curr. Sci.*, 42:108.

Aktar, M. W., Sengupta, D., and Chowdhury. A. (2009). Impact of pesticides use in agriculture their benefits and hazards. *Interdisc. Toxicol.*, 2: 1-12.

Bhat, D., and Padmaja, P. (2014). Assessment of organic pesticides in ground and surface water in Bhopal India. *IOSR- JESTFT*, 8: 51-52.

Biswas, K., Paul., D., and Sinha, S.N. (2015). Biological agents of bioremediation: A concise Review. *Front. Environ. Microbiol.,* 1: 39-43.

Carter, B., and Honmann, K. (1991). *Valuing environmental costs in India: The economy wide impact of environment degradation*, World Bank, Mimeo.

Campos, V., M., Merino, I., Casado, R., Pacios L. F., and Gómez, L. (2008). Review. Phytoremediation of organic pollutants. *Span. J. Agric. Res.,* 6:38-47.

Caceres, T. P., Megharaj, M., and Naidu, R. (2008). Biodegradation of the pesticide Fenamiphos by ten different species of green algae and cyanobacteria. *Curr. Microbiol.*, 57:643-646.

De, A. Bose, R., Kumar, A., and Mojumdar, S. (2014).*Targeted delivery of pesticides using biodegradable polymeric nanoparticles*, Springer Briefs in Molecular Science.

EPA (2004). http://www.epa.gov/oppsrrd1/REDS/factsheets.

Flathman, P. E, and Lanza, G. R. (1998). Phytoremediation: current views on an emerging green technology. *J. Soil. Contam.* 7:415-432.

Gallo, M. A., and Lawryk, N. J. (1991). Organic phosphorus pesticides. In: Haves W. J., Laws E. R., (eds). *Hand book of pesticide toxicology*. San Diego, Academic Press Inc.

Gupta, P. K. (2004). Pesticide exposure-Indian scene. *Toxicology*, 198 (1–3) 83-90.

Jin, Z. P., Luo, K., Zhang, S., Zheng. Q., and Yang, H. (2012). Bio-accumulation and catabolism of prometryne in green algae. *Chemosphere*, 87:278-84.

Ibrahim, W. M., Adawy, A. A, and Karam, M. A. (2016). *5th International Conference on Agriculture, Environment and Biological Sciences* (ICAEBS-16), April 28-29, Pattaya (Thailand).

Ibrahim, W. M., and Essa, A. M. M. (2010). Tolerance, biodegradation and utilization of Malathion, an organophosphorus pesticide, by some cyanobacterial isolates. *Egypt. J. Bot.* 27 (29): 225-240.

Kaushik, B. D., and Venkatraman, G. S., (1983). Response of cyano-bacterial nitrogen fixation of pesticides. *Curr. Sci.*, 52: 321-323.

Kuritz. T., and Wolk, C. P (1995). Use of filamentous cyanobacteria for biodegradation of organic pollutants. *Appl. Environ. Microbiol.* 61: 234-238.

Monteiro, C. M., Castro, P. M. L., and Malcata, F. W. (2012). Metal uptake by microalgae: underlying mechanisms and practical applications. *Biotechnol. Prog* 28(2): 299-311.

Mulchandani, A., Kaneva, I., and Chen, W. (1999a). Detoxification of organophosphate nerve agents by immobilized *Escherichia coli* with surface-expressed organophosphorus hydrolase. *Biotechnol. Bioeng*, 63: 216-223.

Nanda, A. R., and Carl, H. (2007). *The Future population of India-a long-range demographic view, presented population foundation of India and population reference bureau.*

Notia H., (2016) Next Generation Indian Agriculture - Role of Crop Protection Solutions. *Agriculture knowledge Report.*

Pabbi, S., and Vaishya. A. K. (1992). *Effect of insecticides on cyanobacterial growth and nitrogen fixation.* In: BD Kaushik (ed). Indian Agriculture Research Institute, New Delhi pp. 389-493.

Prescott, L.M., Harley, J.P., and Klein, D.A. (2002). *Microbiology*, 5th Edition, McGraw Hill, New York, pp. 1014.

Priyadarshani, I., Sahu, D., and Rath, B. (2011). Microalgal bio-remediation: Current practices and perspectives. *J. Biochem. Tech*, 3(3): 299-304.

Rao, P. H. R., Kumar, R. R., Raghavan, B. G., Subramanian, V., and Sivasubramanian, V. (2011). Application of phycoremediation technology in the treatment of wastewater from a leather-processing chemical manufacturing facility. *Water, SA* 37: 7-14.

Reshmita, N., Yibo, L., Wangming, Y., Chen, Y., Wen, Ch., Qian Li and Xuefeng, C. (2015). *Changes in arable land demand for food in India and China: A Potential Threat to Food Security Sustainability,* 7, 5371-5397.

Singh, R., Singh, R. P., and Singh, D. V. (2015). Distribution of Cyanobacteria (blue-green algae) in rice fields of Varanasi. *Int. J. Adv. Res.* 3(8): 1055-1060.

Subramanian, G., and Uma, L. (1999). The role of cyanobacteria in environmental management. Bulletin de l' Institut Oceanographique. *Marine Cyanobacteria* 19:599-606.

Subramanyam, D., Lokanatha, V., Rajendra, W., and Vijayalakshmi, B. (2012). Biomanagement of pesticides for sustainable environment: An Indian scenario. *IJBPR* (7): 826-834.

Tiwari, B., Singh, S., Kaushik, M. S., and Mishra, A. K. (2015). Regulation of organophosphate metabolism in cyanobacteria. A review. *Micro-biology,* 84(3)291-302.

Vendan, S. E. (2016) Current scenario of biopesticides and ecofriendly insect pest management in India. *South. Indian. J. Biol. Sci*. 2016 2(2); 268-271.

Watanabe, M. E. (1997). Phytoremediation on the brink of commercialisation. *Env. Sci. Technol*. 31:182A-186A.

Wipa, C. and Fa-Aroonsawat, S. (2008). Biodegradation of organo-phosphate pesticide using recombinant cyanobacteria with surface- and intracellular-expressed organophosphorus hydrolase. *J. Microbiol. Biotechnol.,* 18(5): 946-951.

Zhang, H., Hu, C., Jia, X., Xu, Y., Wu, C., Chen, L., and Wang, F. (2012). Characteristics of γ-hexachlorocyclohexane biodegradation by a nitrogen-fixing cyanobacterium, *Anabaena azotica*. *J. Appl. Phycol*. 24:221-225.

In: The Role of Photosynthetic Microbes…
Editors: K. Tripathi, N. Kumar et al.

ISBN: 978-1-53614-032-3

Chapter 13

BIOSYNTHESIS AND BIOTECHNOLOGICAL POTENTIAL OF UV-ABSORBING COMPOUNDS IN MICROALGAE

Md. Akhlaqur Rahman[1,2], Shailendra K. Singh[1], Vinod K. Kannaujiya[3], Kritika Dixit[1], Adi Nath[1,4] and Shanthy Sundaram[1,*]

[1]Centre of Biotechnology, Nehru Science Centre,
University of Allahabad, Allahabad, India
[2]Department of Biotechnology, S. S. Khanna Girl's Degree College, Allahabad, India
[3]Department of Botany, MMV, Banaras Hindu University, Varanasi, India
[4]Department of Botany, Nehru Gram Bharti University, Allahabad, India

ABSTRACT

The drastic depletion of stratospheric ozone layer are due to continuously released atmospheric pollutants such as organo bromide (OBs), chlorocarbons (CCs) and chlorofluorocarbons (CFCs) producing significant levels of ultraviolet-B radiation (UV-B 280–315 nm) which directly comes to the Earth surface. UV radiations (UVR) are abiotic stress factors and causes damaging effects on crop production and living organisms. UV-B (280-315) and UV-A (315–400 nm) radiations have excessive energy that may damage natural aquatic systems by deep penetration in water. A number of photo-protecting compounds have been reported from various organisms. Photosynthetic organisms as well as cyanobacteria have the capacity to counteract the negative effect of ultraviolet radiation by producing UV protecting compounds such as mycosporine-like amino acids (MAAs) and scytonemin. The syntheses of photo protecting compounds in cyanobacteria

[*]Corresponding Author Email: shanthy.cbt@gmail.com.

are of particular interest as promising biomolecules. They have applications in biomedical science, pharmaceutical and cosmetic industries.

Keywords: cyanobacteria, mycosporine-like amino acids (MAAs), photosynthesis, phytoplankton, scytonemin, ultraviolet

INTRODUCTION

Cyanobacteria are photosynthetic, aerobic and Gram negative, bacteria which vary in their size and shapes (single cell, colonial and filamentous cell). They are also called as Cyanophyta, as it comes under phylum of bacteria which obtain their energy by photosynthesis. The name "cyanobacteria" actually represents the color of bacteria. They are often known as blue-green algae, as we know cyanobacteria are prokaryotic and algae should be under eukaryotic group (Allaby, 1992). Cyanobacteria have also capability of biomass production both in aquatic and terrestrial ecosystems. Some cyanobacteria have capacity to fix atmospheric nitrogen with the help of nitrogenase enzyme which is important for paddy fields as biofertilizer. It has been reported that the cyanobacteria fix more than 35 million tons of nitrogen yearly and play an important role in the biological cycles of carbon, nitrogen, and oxygen (Tomitani et al., 2006). All photosynthetic organisms including cyanobacteria depend on solar radiation as the primary source of energy. Aquatic cyanobacteria employ several strategies known as a "carbon concentrating mechanism" to aid in the acquisition of bicarbonate (Kerfeld et al., 2010).

Ultraviolet radiations (UVR) are highly energetic radiations coming from the sun light; having wavelengths ranging from 100 nm to 400 nm (Figure 1). UV radiation is an effective abiotic stressor which causes damaging effect on crop production and living organisms. The UV radiations are categorized into three forms and include least dangerous radiation UV-A (315-400 nm), more dangerous UV-B (280-315 nm) and highly dangerous UV-C (100-280 nm). However, the UV-C does not reach earth's surface due to absorption by stratospheric ozone's layer (Figure 2). UV-B and UV-A radiation coming from solar radiation and reaches deep in to the water level and affects aquatic environment (Hader et al., 2007). It is reported that a number of active functions such as growth, survival, motility, photosynthesis, photosynthetic oxygen production, phycobiliprotein composition, protein profiling, cell differentiation, CO_2 uptake and nitrogen uptake affected by both UV-B and UV-A radiation (Gao et al., 2007; Lesser, 2008; Sinha et al., 2008). UV-B radiation damage different physiological and biological activities of cyanobacteria which can modify the structure and function of entire ecosystem (Häider et al., 2011).

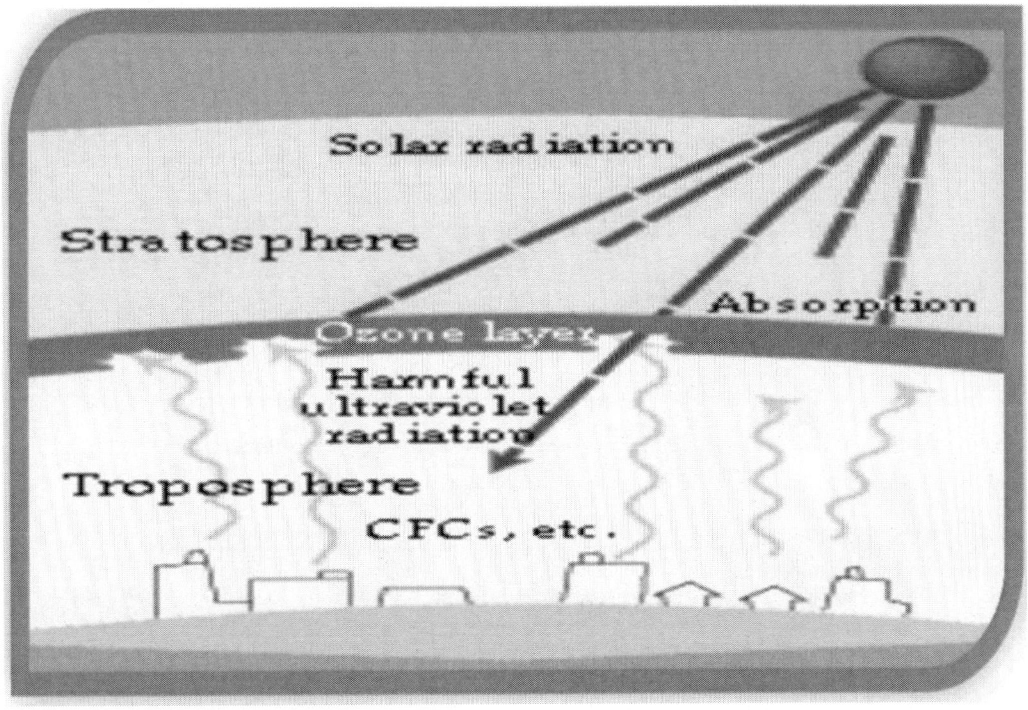

Figure 1. Diagrammatic representations for transmittance of harmful solar radiation.

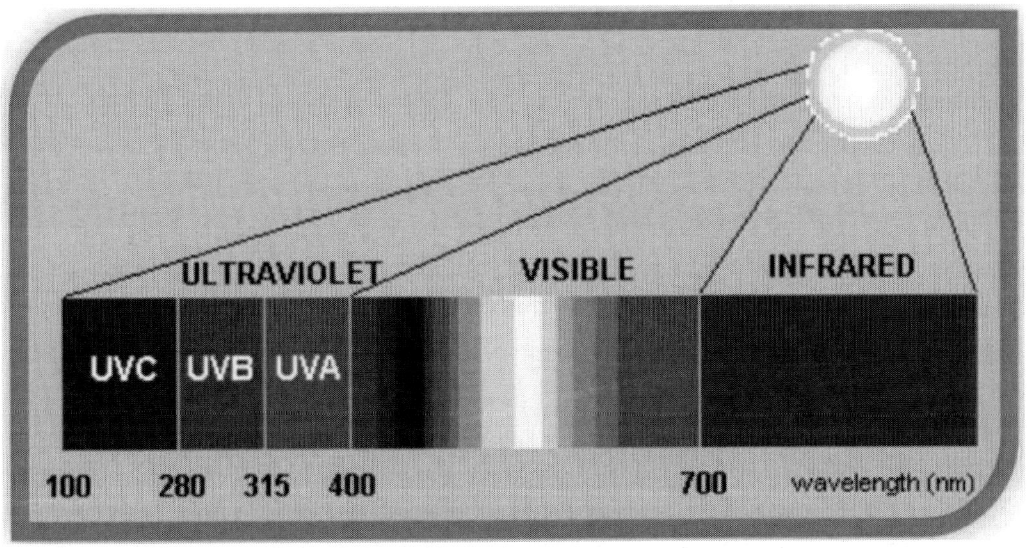

Figure 2. Distribution of wavelengths of solar spectrum including UV radiation.

Solar visible light (400-700 nm) is important for routine functioning of the physiology and biochemistry of the cell. It is shown that during last decade, UV-B radiation has great damaging effect on photosynthetic organelles in plants (Teramura and Sullivan, 1994). Chlorophyll (Chl) biosynthesis is severely affected by UV-B radiation

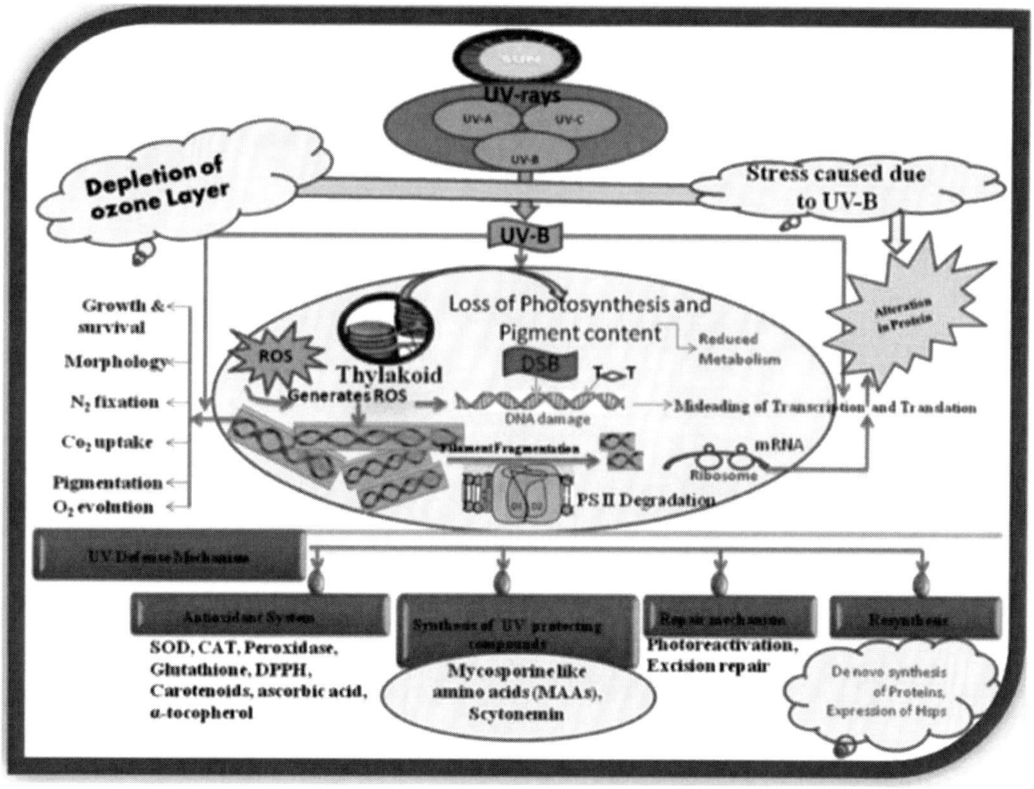

Figure 3. A proposed diagram for ultraviolet-B stress response in cyanobacteria.

(Subbaiah et al., 1987). It has potency to inhibit electron transport system (Eichhorn et al., 1993) and also inhibition of net photosynthesis (Brandle et al., 1977). It was also shown that UV-B radiation has a greater damaging effect on phycocyanin in comparison to Chl a (Rahman et al., 2015). Cyanobacteria have inherent ability to repair the damaging effects of UV radiation. The UV-B radiation generates high level of reactive oxygen species (ROS) through photosynthetic pigments and by redox components Rahman et al., 2015). Several cyanobacterial species have shown variation in their tolerance to UV radiation. These include defense strategies such as avoidance, scavenging of ROS by different enzymatic or non-enzymatic antioxidant molecules, UV protecting compounds (MAAs and Scytonemin). They play a role in providing defense strategy to adapt and grow in environment with high doses of UV-B radiation (Figure 3).

Biosynthesis of UV Protecting Compounds

Some cyanobacteria either fresh water or marine water has overcome UV-B stress by the synthesis of UV protecting compounds which protect them from deleterious effect of

UV radiation (Singh et al., 2010). Besides this, cyanobacteria synthesize some extracellular polysaccharides which provide matrix for mycosporine-like amino acids (MAAs) and scytonemin after long duration of UV-B exposure (Ehlin Schulz et al., 1997; Rahman et al., 2016). A number of MAAs have been reported from different organisms, including cyanobacteria, micro/macroalgae, fungi, lichen, heterotrophic bacteria, and some marine animals such as arthropods, fishes, rotifers, cnidarians, mollusks, echinodermates etc. (Helbling et al., 2002). It has also been reported that UV protecting compounds have the capability of dissipating absorbed radiation as heat without producing reactive oxygen species (Conde et al., 2000). Hence, it is shown that MAAs provides protection from UV radiation but also to primary and secondary consumers through the food chain (Whitehead et al., 2001).

Mycosporine-Like Amino Acids and Its Structure

Shibata (1969) observed the biosynthesis of MAAs in large amounts in cyanobacteria for the first time. Mycosporine-like amino acids (MAAs) are present in a number of organisms. They are secondary metabolite that absorbs solar radiation directly or indirectly and protects organisms from high energy solar radiation (Häder et al., 2007). MAAs is a colorless compound with small molecular weight (<400 Da) and water soluble. It is composed of cyclohexenone or a cyclohexenimine chromophore carrying nitrogen substituent of an amino acid or an amino alcohol. The ring system of MAAs structure contains a glycine subunit on the third position of carbon atom, although several MAAs contain sulfate esters or glycoside linkages through its imine substituents. The MAAs occur at great UV absorption maxima (310-362 nm), which have high molar extinction coefficients (e=28, 100–50,000 $M^{-1}cm^{-1}$) and is photo stabile and antioxidant in nature (Conde et al., 2000; Whitehead and Hedges 2005). MAAs are UV protective/screening capable of dissipating absorbed radiation efficiently resistance to some abiotic stressors (Conde et al., 2000; Coba, et al., 2009). It was also found that MAAs not only protect the cell from UV-B radiation by dissipating the absorbed energy as heat but also by scavenging ROS such as superoxide anions, singlet oxygen, hydroxyl radicals, hydroperoxyl radicals and also inhibit lipid peroxidation (Cobaet et al., 2007; Suh et al., 2003; Oren et al., 2007). Different cyanobacteria have been reported having the potential for the production of different types of MAAs in diverse habitats (Table 1). The variation observed regarding the absorption spectra of various MAAs are due to the alteration in the side groups and nitrogen substituent's.

Table 1. Different MAAs, its absorption maxima and molecular structure

Mycosporine-like amino acid	λmax (nm)	Molecular structure
Palythine	320	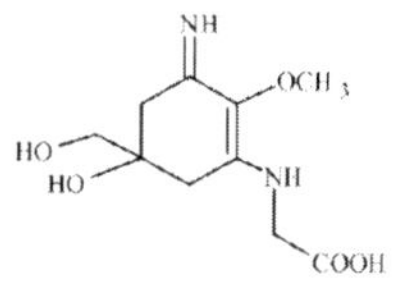
Mycosporine-glycine	310	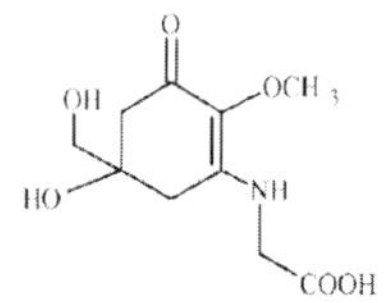
Palythine-serine-sulfate	320	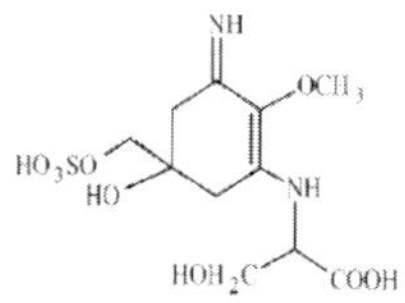
Mycosporine-methylamine-serine	327	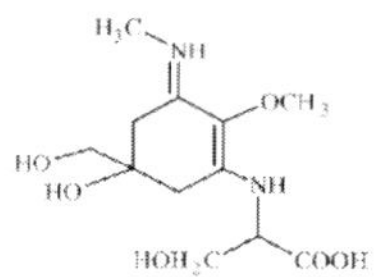
Palythine-serine	320	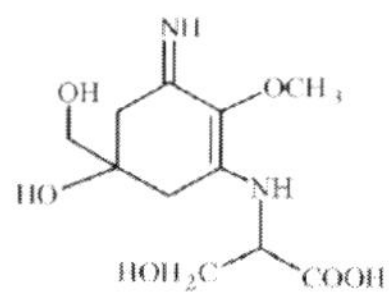
Mycosporine-methylamine-threonine	327	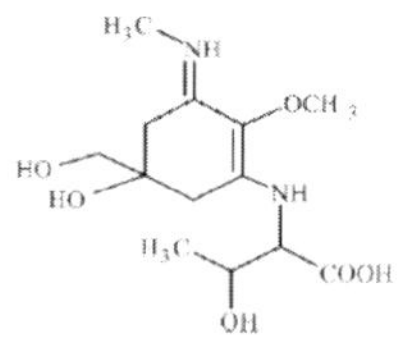

Table 1. (Continued)

Mycosporine-like amino acid	λmax (nm)	Molecular structure
Mycosporine-glutamic acid-glycine	330	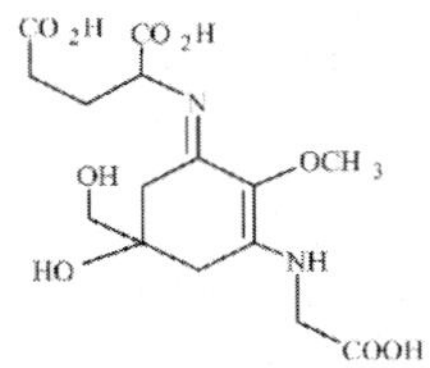
Asterina-330	330	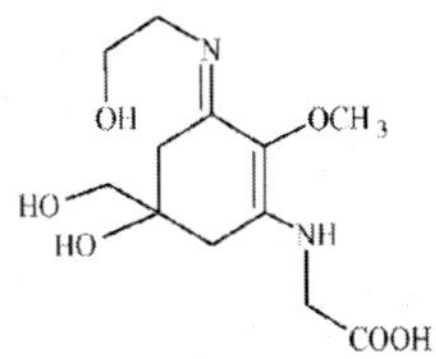
Porphyra-334	334	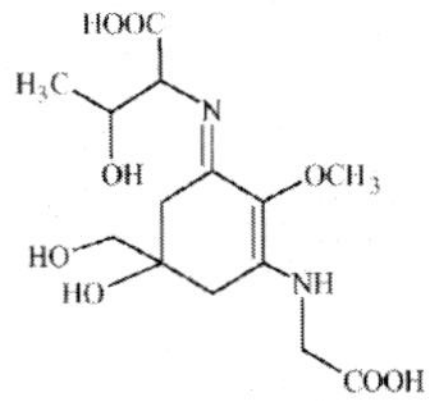
Mycosporine-2-glycine	334	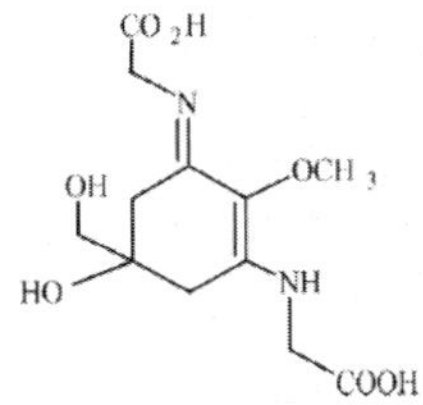
Palythinol	332	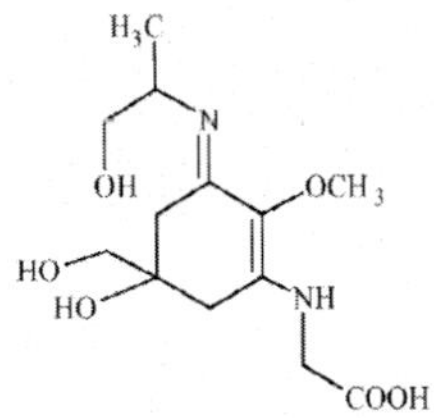
Shinorine	334	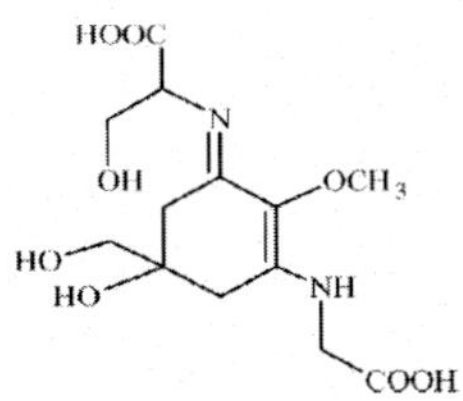

Mycosporine-like amino acid	λmax (nm)	Molecular structure
Mycosporine-glycine-veline	335	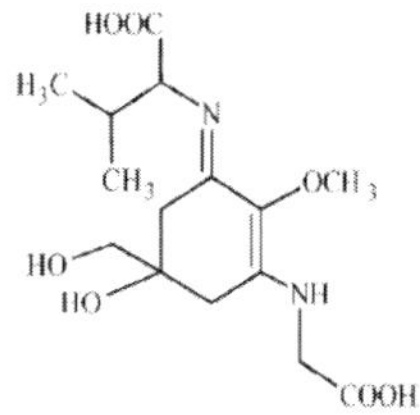
Usujirene	357	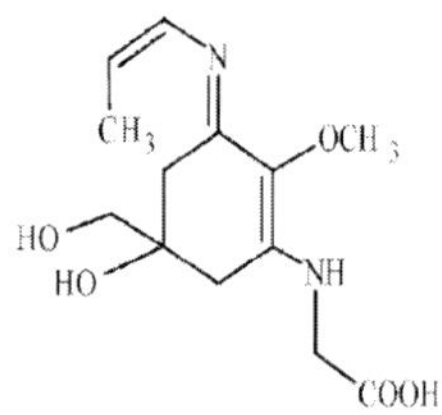
Palythenic-acid	337	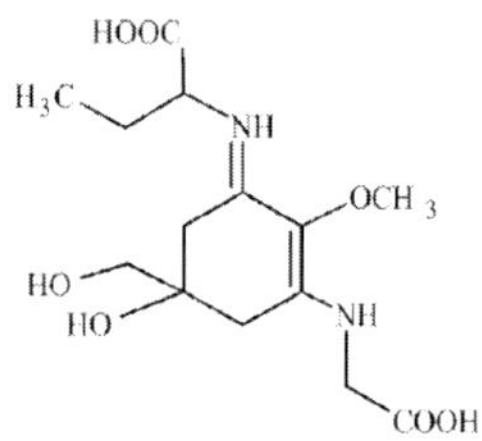
Palythene	360	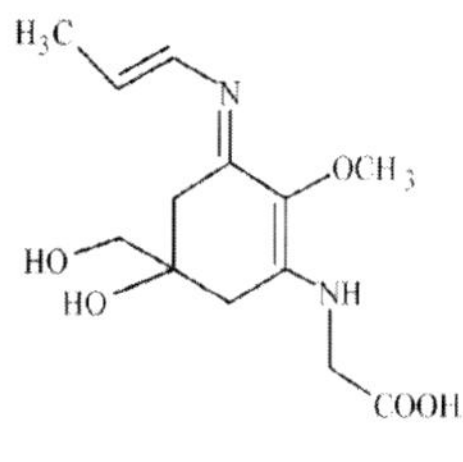
Euhalothece-362	362	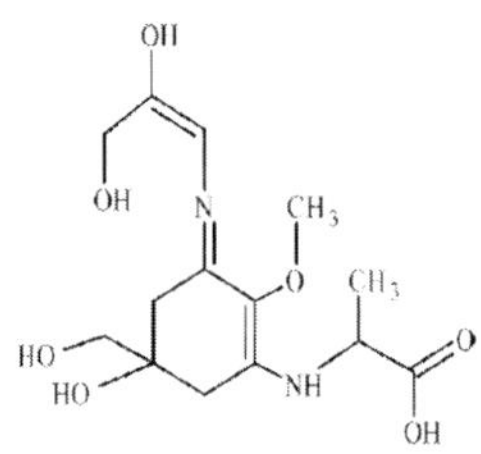

Table 1. (Continued)

Mycosporine-like amino acid	λmax (nm)	Molecular structure
Mycosporine-taurine	309	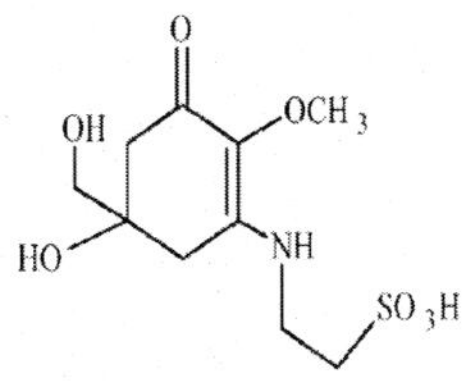

Biosynthesis of MAAs

It was shown that the biosynthesis of MAAs occur via the first step of the shikimate pathway in different groups of organisms such as cyanobacteria and fungi. In the first step of the shikimate pathway, the glycolytic intermediate pentose phosphate pathway intermediate erythrose-4-phosphate and the phosphoenol pyruvate are condensed in to heterocyclic compound, 3-deoxy-D-arabinoheptulosonate 7-phosphate (DAHP), which may formally be considered as 2-deoxy-glucose-6-phosphate derivative (Garner and Herrmann, 1984).

In the next stage, the ring oxygen is replaced for the exocyclic C7 of DAHP to form a highly substituted cyclohexane derivative, 3-dehydroquinate. The 3-dehydoquinate, which is produced during the earlier part of the shikimate pathway, this function as a pioneer for the biosynthesis of fungal mycosporine and MAAs via gadusols (Sinhaet al., 2007; Favre-Bonvinet al., 1987). Issue regarding the long stability and whether MAAs biosynthesis contains an intermediate of the shikimate pathway are great challenges in the present (Balskus and Walsh 2010). From a previous study it was reported that the protein sequence 3-phosphoshikimate 1-carboxyvinyl transferse has an important role in production of mycosporine glycine in *Synechocystis* PCC 6803 and 3-dehydroquinate synthase for prodution of shinorine in *Anabaena cylindrica* (Rahman et al., 2014). It was also shown that some genes are responsible for MAA biosynthesis in cyanobacteria and gene such as YP_324358 and YP_324357 are involved in the synthesis of the common core (deoxygadusol) of all MAAs (Singh et al., 2010).

The cyanobacteria which have capacity to synthesize MAAs are mostly present in hypersaline environments. Garcia-Pichel et al., (1998) studied morphology, physiology and gene sequence (16S rRNA of unicellular halophilic cyanobacteria. The unicellular cyanobacterium (*Euhalothece* sp.) isolated from the upper layer of a gypsum crust of a hypersaline water system having two different MAAs with absorption maxima at 331 and 362 nm were retrieved when it was grown at high light intensities. The compound obtained was purified, characterized and identified as mycosporine-2-glycine (Kedar et al., 2002). Beside this MAAs such as asterina-330, palythinol, shinorine and mycosporine–glycine have also been reported having photoprotective role in epilithic (growing on stone) cyanobacteria derived from fresh water lake (Sommaruga and Garcia-Pichel, 1999). The bloom-forming cyanobacterium *Microcystis aeruginosa* isolated from Lake Taihuhad have the capacity to biosynthesize MAAs (shinorine and porphyra-334) and has the potency to absorb UV radiations (Liu et al., 2004).

Increased UV-B radiation was found to enhance the capacity of synthesis of two Mycosporine-Like amino acids, shinorine and porphyra-334 (both compound having absorption maxima at 334 nm) in three cyanobacterial strains of *Nodularia* i.e., *N. baltica, N. spumigenaand N. harveyana* (Sinha et al., 2003). The cyanobacterium *Anabaena doliolum* when radiated with PAR and UV radiation showed to converting of mycosporine–glycine into two MAAs, porphyra-334 and shinorine (Singh et al., 2008). In another study it was shown that sulfur deficiency is helped in the synthesis and bioconversion of primary MAA into secondary MAA in the cyanobacterium *Anabaena variabilis* PCC 7937 (Singh et al., 2010). Abiotic factors such as different UV radiation, salt, nutrient, desiccation, and photoperiods are responsible for damaging the production of photoprotective compounds in cyanobacteria (Bhatia et al., 2011). The biosynthesis of MAA is a light (energy) dependent process.

ROLE OF MAAS IN BIOTECHNOLOGY

MAAs Acts as Photo Protecting Compound

Because of their UV-absorption capacity, MAAs act as strong photo-protecting compounds against harmful ultraviolet radiation in cyanobacteria. Mycosporine-like amino acids (MAAs) have the capability to offersite-dependent protection to whole cells by capturing damaging doses of ultraviolet radiations and after that converting this energy in the form of normal radiation in environment (Oren and Gunde-Cimerman, 2007). Wavelengths lower than 300 nm coming from solar radiation are more harmful to cellular components and genetic material (Callone et al., 2006).) Studies by Hoyer et al., (2001) confirmed that out of 10 photons coming from light energy, MAAs to prevent only 3 from striking cytoplasmic targets in different cyanobacteria.The concentration of MAAs within cyanobacterial cell has the ability to absorb 10-26% photon energy coming from UV-B radiation, which gives more resistance capacity to UV-B radiation against damage (Ehling-Shulz and Scherer, 1999).

MAAs as Natural Sunscreening Compounds

MAAs is actively excreted and accumulated in the epidermal layer of skin and at this location MAAs show great sunscreen activities (Oren and Cimerman, 2007). Several studies on the photophysical characteristics and photodegradation of MAAs have showed that they are highly stable and very effective sunscreen compounds. In a current study it was shown that porphyra 334 which is isolated from an Indian species *Porphyra vietnamensis* has great sun protecting effects against similar to the efficient of *Aloe vera* gel as photo protecting compound (Bhatia et al., 2010). It was also reported that MAAs contains less photodynamic reactivity in comparison to other commercially available sunscreen compounds in the market. In the recent times several synthetic analogues of MAAs have been used for production of MAAs due to the economic angle. One of the best examples of analogues of mycosporine glycine (3-alkyl amino 2-methoxy cyclohex 2-enones) and tetra hydro pyridine (1-alkyl 3-alkanoyl 1, 4, 5, 6-tetra hydro pyridines) were synthesized although they have problems related to stability. A product available in the market known as Helioguard® 365 isolated from the red algal *Porphyra umbilicalis*. Bhatia et al., (2010) also showed that the some protection capacity in different phytoplanktons and micro organisms. These MAAs also play an important role in human beings by blocking the thymine dimer formation caused by ultra violet radiation *in vitro* and recently they were also found to provide growth stimulation in human cells (Bandaranayake, 1998; Schulz and Scherer, 1999; Cockell and Knowland, 1999). MAAs can block the synthesis of both 6-4 photoproducts and cyclobutane pyrimidine dimer (CPD) formation and thus they may be used as possible candidates for natural sunscreen compounds (Sinha et al., 1998). Recently Coba et al., (2009) reported that MAAs have capacity to prevent UV-induced skin damage *in vivo*.

MAAs Acts as Antioxidant Molecules

The importance of MAAs is in that it acts as antioxidant molecules. For example mycosporine glycine, it acts as a biological antioxidant molecules and was shown that these biological compounds effectively inhibit the various detrimental effects of the type II photosensitization and decline the level of singlet oxygen radicals produced by a stressor eosin Y or methylene blue (Whitehead and Hedges, 2005). A number of mycosporines having high antioxidant activity of scavenging reactive oxygen species produced of supersaturated oxygen in the depth of the water are known. One of the best examples is mycosporine glycine which was found in *Platygyra ryukyuensis* (Bhatia et al., 2010). Mycosporine glycine has moderate antioxidant activity, while 4-deoxygadusol acts as a precursor of MAAs, with powerful antioxidant potency and its retro biosynthesis with the help of bacterial conversion of algal MAAs has been performed for commercial purpose (Bhatia et al., 2011). Some imino mycosporine-like amino acids such as example asterina 330, palythine, porphyra-334, shinorine and palythinol do not have the capabilities of oxidization and hence exhibit less or no activity (Dunlap and Yamamoto, 1995, Tao et al., 2008). Therefore, it was concluded that some MAAs may play an important role in protection from scavenging $^{1}O_2$ produced by several endogenous photosensitizers (Oren and Cimerman, 2007; Dunlap and Yamamoto, 1995; Tao et al., 2008).

SCYTONEMIN

Scytonemin is commonly found as yellow-brown, lipid soluble photoprotective compounds which are deposited in the exopolysaccharide sheath of different cyanobacterial species commonly found on rock surface, bark of trees and marine intertidal mats (Garcia-Pichel and Belnap, 1996). Scytonemin is chemically a dimeric compound which consists of indolic and phenolic subunits carrying molecular mass of 544 Da. This is also a kind of important photoprotective compound which was firstly reported in some terrestrial cyanobacteria (Nägeli and Gattungene inzelliger Algen, 1949). Recently this photo-protecting compound scytonemin have been obtained from cyanobacteria the *Lyngbya* sp. inhabiting rock on the shores near Sao Fransisco do Sul, Brasil (Richter et al., 2006). The photoprotecting compound Scytonemin shows *in vivo* absorption maximum at 370 nm, while the purified scytonemin occurs at absorption maximum at 386 nm and it also reported that absorbs significantly at 252, 278, and 300 nm. This important photoprotecting compound Scytonemin allows cyanobacteria to survive at lethal doses of UV-A or UV-B radiation in environments with high intensity of solar radiation (Jones et al., 2011). Scytonemin is a highly stable compound and it has ability to screen without any metabolic investment even after physiological inactivity. Hence, it can be concluded that, it is a valuable compound to protect cyanobacteria from UV radiation (Soule et al., 2007). This UV protective compound scytonemin is mostly found in the green oxidized form; while, two other forms of scytonemin i.e., oxidized (yellow; e.g., fuscochlorin) and the reduced (red; e.g., fuscorhodin) have been found that are synthesized depending on the redox and acid-base conditions during the extraction and purification process (Garcia-Picheland Castenholz, 1991). Beside these, some other UV protecting compounds with absorption maxima at 312 and 330 nm have been also found in the terrestrial cyanobacterium, *Nostoc commune*, which also has the capability to synthesize Scytonemin (Scherer et al., 1988). Scytonemin has great pharmacological application and also contains anti-proliferative activities which may be applicable for synthesizing different therapeutically important drugs (Stevenson et al., 2002). The UV protecting compound Scytonemin has great therapeutic application in acute and chronic disorders of inflammation and proliferation. It is reported that in the case of human T-cell leukemia Jurkat cells, scytonemin repressed cell proliferation (IC50 = 7.5 μM) and it has the capability to enhance apoptosis rate in 24% of cells.

CONCLUSION

The biosynthesis and accumulation of different UV protecting/ screening compounds have been recognized from different group of photosynthetic organisms having the capacity to counteract some structural and physiological damages generated by ultra violet radiation. Mycosporine-like amino acids and Scytonemin are low cost UV protecting compounds found in photosynthetic or marine organisms such as different kind of microalgae/cyanobacteria. These UV protecting compounds have tremendous potential as photo protecting compounds, antioxidative molecules, natural sunscreen compounds and therapeutic agents. It is concluded that these UV protecting compounds have valuable role in industry and health care and their commercial potential can be exploited through biotechnology.

ACKNOWLEDGMENTS

Md. Akhlaqur Rahman is thankful to University Grants Commission, New Delhi, India for financial assistance under Maulana Azad National Junior and Senior Research Fellowship to carry out this work.

CONFLICT OF INTEREST

No conflict of interest among the authors.

REFERENCES

Allaby, M., (ed.) (1992). *The concise Oxford Dictionary of Botany*. Oxford University Press.

Balskus, E. P., and Walsh, C. T., (2010). The genetic and molecular basis for sunscreen biosynthesis in cyanobacteria, *Science*, 329: 1653–1656.

Bandaranayake, W. M., (1998). Mycosporines: Are they nature's sunscreens? *Nat. Pro. Rep.*, 15: 159-72.

Bhatia, S., Garg, A., Sharma, K., Kumar, S., Sharma, A., Purohit, A. P., (2011). Mycosporine and mycosporinelike amino acids: A paramount tool against ultra violet irradiation. *Pharmacognosy Rev*. 5: 138-146.

Bhatia, S., Sharma, K., Namdeo, A. G., Chaugule, B. B., Kavale, M. and Nanda, S. (2010). Broad-spectrum sun protective action of Porphyra 334 derived from *Porphyra vietnamensis*. *Pharmacog. Res*, 2: 45-9.

Brandle, J. R., Campbell, W. F., Sission, W. B., and Cadwell, M. M., (1977). Net photosynthetis, electron transport capacity, and ultra structure of *Pisum sativum* L. exposed to ultraviolet-B radiation. *Plant. Physiol.* 60: 165-169.

Callone, A. I., Carignan, M., Montoya, N. G., and Carreto, J. I., (2006). Biotransformation of mycosporine-like amino acids (MAAs) in the toxic dinoflagellate *Alexandrium tamarense*. *J. Photochem. Photobiol. Biol*. 84: 204-12.

Coba F. De La., Aguilera J., Figueroa, F. L., Gálvez, de M. V., and Herrera, E., (2009). Antioxidant activity of mycosporine-like amino acids isolated from three red macro algae and one marine lichen. *J. Appl. Phycol,* 21:161–169.

Coba, F. De La., Aguilera, J., and Figueroa, F. L., (2007). Use of mycosporine-type amino acid Porphyra-334 as an antioxidant. *US patent WO2007/026035 A2*.

Cockell, C. S., and Knowland, J. (1999). Ultraviolet radiation screening compounds. *Biol. Rev*. 74: 311-45.

Conde, F. R., Churio, M. S., and Previtali, C. M., (2000). The photo-protector mechanism of mycosporine-like amino acids. Excited-state properties and photo- stability of porphyra-334 in aqueous solution. *J. Photochem. Photobiol. B. Biol.* 56: 139–144.

Dunlap, W. C., and Yamamoto, Y. (1995). Small-molecules antioxidants in marine organisms: antioxidant activity of mycosporine-glycine. *Comp. Biochem. Physiol.* 112: 105-114.

Ehlin-Schulz, M., Bilger, W., and Scherer, S., (1997). UV B induced synthesis of photoprotective pigments and extracellular poly-saccharides in the terrestrial cyanobacterium *Nostoc commune*. *J. Bacteriol*. 179: 1940-1945.

Ehling-Shulz, M. and Scherer, S., (1999). UV Protection in cyanobacteria. *E. J. Phycol,* 34: 329-338.

Eichhorn, M., Doehler, G., and Augsten, H., (1993). Impact of UV radiation on photosynthetic electron transport of *Wolffiaarrhiza* (L.) *Wimm*, *Photosynthetica*, 29: 613-618.

Favre-Bonvin, J., Bernillon, J., Salin, N., and Arpin, N., (1987). Biosynthesis of mycosporines: mycosporine glutaminol in *Tricho-thecium roseum*. *Phytochemistry*, 26: 2509–2514.

Gao, K., Yu, H., and Brown, M. T., (2007). Solar PAR and UV radiation affects the physiology and morphology of cyanobacterium *Anabaena* sp. PCC 7120. *J. Photochem. Photobiol. B. Biol*. 89; 117-124.

Garcia-Pichel, F., and Castenholz, R. W., (1991). Characterization and biological implications of scytonemin, a cyanobacterial sheath pigment. *J. Phycol.* 27: 395–409.

Garcia-Pichel, F., Nübel, U., and Muyzer, G., (1998). The phylogeny of unicellular, extremely halo tolerant cyanobacteria, *Arch. Microbiol.* 169: 469–482.

Garcia-Pichel, F., and Belnap, J. (1996). Micro environments and micro-scale productivity of cyanobacterial desert crusts. *J. Phycol.* 32: 774–782.

Garner, C. C., and Herrmann, K. M., (1984). Structural analysis of 3-deoxy-D-arabino-heptulosonate 7-phosphate by ^{1}H- and natural abundance 13C NMR spectroscopy. *Carbohydrate. Res.* 132: 317-322.

Häder, D. P., Helbling, E. W., Williamson, C. E., and Worrest, R. C., (2011). Effects of UV radiation on aquatic ecosystems and interactions with climate change. *Photochem Photobiol. Sci,* 10: 242–60.

Hader, D. P., Kumar, H. D., Smith, R. C., and Worrest, R. C., (2007). Effects of solar UV radiation on aquatic ecosystem and interaction with climate change. *Photochem. Photobiol. Sci,* 6: 267-285.

Helbling, E. W., Menchi, C. F., and Villafane, V. E., (2002). Bioaccumulation and role of UV-absorbing compounds in two marine crustacean species from Patagonia, Argentina. *Photochem. Photobiol. Sci.* 1: 820–825.

Hoyer, K., Karsten, U., Sawall, T., and Wiencke, C., (2001). Photo-protective substances in Antarctic macroalgae and their variation with respect to depth distribution, different tissues and developmental stages. *Marine. Ecol. Progr. Series.* 211: 117-29.

Jones, C. S., Esquenazi, E., Dorrestein, P. C., and Gerwick, W. H., (2011). Probing the *in vivo* biosynthesis of scytonemin, a cyanobacterial ultraviolet radiation sunscreen, through small scale stable isotope incubation studies and MALDI-OF mass spectrometry. *Bioorg. Med. Chem.*

Kedar, L., Kashman, Y., and Oren, A., (2002). Mycosporine-2-glycine is the major mycosporine-like amino acid in a unicellular cyano-bacterium (*Euhalothece* sp.) isolated from a gypsum crust in a hyper saline saltern pond. *FEMS. Microbiol. Lett.* 208: 233–237.

Kerfeld, C. A., Heinhorst, S. C., and Gordon, C., (2010). Bacterial micro compartments. *Annu Rev. Microbiol.* 64 (1): 391–408.

Lesser, M. P., (2008). Effects of ultraviolet radiation on productivity and nitrogen fixation on the cyanobacterium, *Anabaena* sp. (Newton's strain). *Hydrobiologia.* 598: 1-9.

Liu, Z., Hader, D. P., and Sommaruga, R., (2004). Occurrence of mycosporine-like amino acids (MAAs) in the bloom-forming cyanobacterium *Microcystis aeruginosa. J. Plankton Res.* 26: 963–966.

Nägeli, C., and Gattungen einzelliger, Algen., (1949). Physiologist und Systematics barbette, Neue Denkschrift, Allg. Schweiz. *Nature. Ges.* 10: 1–138.

Oren, A., and Gunde-Cimerman, N., (2007). Mycosporines and mycosporine-like amino acids: UV protectants or multipurpose secondary metabolites? *FEMS. Microbiol. Lett.* 269:1-10.

Rahman, M. A., Kannaujiya, V. K., Rajneesh, Adinath., Dixit, K., Sinha, R. P., and Sundaram, S., (2016). Impact of ultraviolet-B radiation on photoprotection and pigmentation of *Anabaena cylindrica* and *Synechocystis* PCC 6803: a comparative study. *I.J.O.N.S*, 6, 36

Rahman, M. A., Pandey, J. K., Sundaram, S., and Gopal, R., (2015). Response of Growth, Photosynthetic Pigments, Laser-Induced Pigment Fluorescence, Antioxidant Enzymes and Lipid Peroxidation to Ultraviolet-B radiation in two Cyanobacteria. *Indian. J. Plant. Physiol,* 20:240-248.

Rahman, M. A., Sinha, S., Sachan, S., Kumar, G., Singh, S. K., and Sundaram, S., (2014). Analysis of proteins involved in production of MAA's in two Cyanobacteria *Synechocystis* PCC 6803 and *Anabaena cylindrica. Bioinformation.* 10:449-453.

Richter, P. R., Sinha. R. P., and Hader, D. P., (2006). Scytonemin-rich epilithic cyanobacteria survive acetone treatment. *Curr. Trends, Microbiol.* 2: 13–19.

Scherer, S., Chen, T. W., and Böger, P., (1988). A new UV-A/B protecting pigment in the terrestrial cyanobacterium *Nostoc commune. Plant. Physiol.* 88: 1055–1057.

Schulz, M. E., and Scherer, S., (1999). UV protection in cyanobacteria. *Eur. J. Phycol,* 34: 329-38.

Shibata, K., (1969). Pigments and a UV-absorbing substance in coral and a blue-green alga living on the Great Barrier Reef. *Plant. Cell. Physiol,* 10: 325–335.

Singh, S. P., Sinha, R. P., Klisch, M., and Häder, D. P., (2008). Mycosporine-like amino acids (MAAs) profile of a rice-field cyanobacterium *Anabaena doliolum* as influenced by PAR and UVR. *Planta.* 229: 225–233.

Singh, S. P., Häder, D. P., and Sinha, R. P., (2010). Cyanobacteria and ultraviolet radiation (UVR) stress: mitigation strategies. *Age. Res. Rev.* 9: 79–90.

Singh, S. P., Klisch, M., Sinha, R. P., and Häder, D. P., (2010) Sulfur deficiency changes mycosporine-like amino acid (MAA) composition of *Anabaena variabilis* PCC7937: a possible role of sulfur in MAA bioconversion. *Photochem. Photobiol.* 86: 862–870.

Sinha, R. P., and Hader, D. P., (2008). UV-protectants in cyanobacteria. *Plant. Science.* 174: 278–289.

Sinha, R. P., Singh, S. P., and Haider, D.P., (2007). Database on mycosporines and mycosporine-like aminoacids (MAAs) in fungi, cyanobacteria, macro-algae, phytoplankton and animals. *J. Photochem. Photobiol. B Biol.* 89: 29–35.

Sinha, R. P., Ambasht, N. K., Sinha, J. P., Klisch, M., and Häder, D. P., (2003). UVB-induced synthesis of mycosporine-like amino acids in three strains of *Nodularia* (cyanobacteria), *J. Photochem. Photobiol. B. Biol,* 71: 51–58.

Sinha, R. P., Klisch, M., Groniger, A., and Hader, D. P., (1998). Ultraviolet absorbing/screening substances in cyanobacteria, phyto-plankton and macroalgae. *J. Photochem. Photobiol. B. Biol.* 47: 83-94.

Sommaruga, R., and Garcia-Pichel, F., (1999). UV-absorbing myco-sporine-like compounds in planktonic and benthic organisms from a high-mountain lake. *Arch .Hydrobiol.* 144: 255–269.

Soule, T., Stout, V., Swingley, W. D., Meeks, J. C., and Garcia-Pichel, F., (2007). Molecular genetics and genomic analysis of scytonemin biosynthesis in *Nostocpunctiforme* ATCC 29133, *J. Bacteriol*, 189: 4465–4472.

Stevenson, C. S., Capper, E. A., Roshak, A. K., Marquez, B., Grace, K., Gerwick, W. G., Jacob, R. S., and Marshall, L. A., (2002). Scytonemin, a marine natural product inhibitor of kinases key in hyper proliferative inflammatory diseases. *Inflamm. Res.* 51: 112-114.

Subbaiah, C. C., Kulandaivelu, G., and Gnamam, A., (1987). Effect of ethedium bromide and UV radiation on chloroplast development. Change in seedlings growth parameters, pigments and thylakoid membrane. *Curr. Trends. Life. Sci,* 13: 347–356.

Suh, H. J., Lee, H. W., and Jung, J., (2003). Mycosporine-glycine protects biological systems against photodynamic damage by quenching singlet oxygen with a high efficiency. *Photochem Photobiol.* 78: 109-113.

Tao, C., Sugawara, T., Maeda, S., Wang, X., and Hirata, T., (2008). Antioxidative activities of a mycosporine-like amino acid, Porphyra-334. *Fisheries. Sci.* 74: 1166-1172.

Teramura, A. H., and Sullivan, J. H., (1994). Effect of UV-B radiation on photosynthesis and growth of terrestrial plants, *Photosynth. Res.* 39: 463-473.

Tomitani, A., Knoll, A. H., Cavanaugh, C. M., and Ohno, T., (2006). The evolutionary diversification of cyanobacteria: molecular phylogenetic and paleontological perspectives. *Proc. Nat. Acad. Sci. USA.* 103: 5442–5447.

Whitehead, K., and Hedges, J. I., (2005). Photodegradation and photo-sensitization of mycosporine-like amino acids. *J. Photochem. Photobiol. Biol*, 80:115-21.

Whitehead, K., Karentz, J. I., and Hedges, D., (2001). Mycosporine-like amino acid (MAAs) in phytoplankton, a herbivorous pteropod (*Limacina helicina*), and its pteropod predator (*Clione antarctica*) in Mc Murdo Bay, Antarctica. *Mar. Biol.* 139: 1013–1019.

In: The Role of Photosynthetic Microbes…
Editors: K. Tripathi, N. Kumar et al.
ISBN: 978-1-53614-032-3

Chapter 14

ADAPTIVE STRATEGIES OF CYANOBACTERIA IN RESPONSE TO SALINITY

Keshawanand Tripathi[1,*], Ravindra Kumar Yadav[1], Yadurappa Prashad Reddy[1] and Devendra Kumar[2]

[1]Centre for Conservation and Utilization of Blue Green Algae,
ICAR-Indian Agricultural Research Institute, New Delhi, India
[2]Department of Microbiology, CCS University,
Meerut, Uttar Pradesh, India

ABSTRACT

Soil salinity is major abiotic factors that limit the growth of microorganisms including cyanobacteria. Salinity affects agricultural productivity worldwide. It decreases plant growth and productivity by inhibiting various physiological, biochemical and molecular processes. The intracellular accumulation of sodium ions under salt stress also changes the ionic ratio of sodium potassium and calcium, which affect the bioenergetics of vital metabolic processes. Cyanobacteria are very well adapted to aquatic habitats during the course of evolution and salinity regulation involves specific ion transporter and antiporter to generate the driving force for the influx and efflux ions. The basic mechanisms adopted by the cyanobacterial cells involve active transport of toxic sodium ions from cells to the external medium. However, synthesis or uptake of compatible solutes and status of antioxidant level also contribute to salinity tolerance. This chapter focuses on the different strategies involved in the mechanism of salinity tolerance of photosynthetic microbe cyanobacteria.

---

[*]Corresponding Author Email: tripathikn009@gmail.com.

Keywords: cyanobacteria, ions, osmolytes, salinity, transporters

INTRODUCTION

Problem of Salinity

Abiotic stresses such as salinity, heat, drought and cold adversely affect the growth and development of plants (Zhu, 2003; Golldack et al., 2014). It is estimated that more than 50% of the agricultural land will be salinized by 2050 (FAO, 2009). Salinity is one of the major abiotic stresses, which affect approximately 20% of the arable land and half of the irrigated lands of the world (Glick et al., 2007). Patel et al., (2010) reported that 7 million hectares of land in India are affected by salinization. Salinity increases the total concentration of inorganic ions in aquatic organisms especially in cyanobacteria (Hagemann, 2011; Oren, 2000, 2007; Allakhverdiev and Murata, 2008). Agricultural crops exhibit a wide range of physiological, biochemical and ecological responses under salinity stress (Hu and Schmidhalter 2002). It influences the biochemical process of cyanobacteria (Akbarimoghaddam et al., 2011). Rai and Sharma, (2006) found that salinity and phosphate is a limiting factor of growth and developments of cyanobacteria. Inorganic phosphates combined with calcium ions precipitated in the form of calcium phosphate which is unavailable to cyanobacteria (Tripathi et al., 2013). Salinity also affects the biodiversity of micro-flora (Wood 1999: Mittler and Blumwald, 2010; Yamaguchi and Blumwald, 2005). Stress responses are generally known as osmoregulation and responsible for maintaining the turgor presser and volume of the cells for the normal cellular physiology (Record et al., 1998a). The consequences of salt stress in cyanobacteria are an ionic imbalance, hyperosmotic stress, nutritional stress and oxidative stress (Zhu, 2003; Yadav et al., 2016).Salt toxicity comprises of osmotic and ionic stress which severely affect growth and metabolic activity of cyanobacteria (Patel et al., 2010; Gupta and Huang, 2014; Tripathi et al., 2013).

Cyanobacteria and Salinity

Cyanobacteria are Gram-negative photosynthetic microbes and distributed almost all habitats such as oceans, fresh-water, desert, fertile soil, polar region and hot springs (Thajuddin and Subramanian, 2005; Sharma et al., 2011). The cyanobacteria in general luxuriantly grow in paddy field (Dubey and Rai, 1995; Tripathi et al., 2013). Photosynthesis and nitrogen fixations are two important physiological and biochemical processes for the sustainable agriculture (Sharma et al., 2011; Tripathi et al., 2013). Fossil records show that cyanobacteria originated 2.8 billion years ago (Olson, 2006). Soo et al., (2014) reported that non-photosynthetic chemoheterotrophs are ancestors of cyanobacteria and photosystem are acquired after differentiation of the classes *Oxyphotobacteria* and *Melainabacteria*. Castonholdz, (2001) classified cyanobacteria into five sub-sections based on morphological characters. According to symbiogeneses, cyanobacteria are ancestors of the plant chloroplast cells (Cavalier-Smith, 2002). The photosynthetic machinery of the cyanobacteria is similar to the green plants and two distinct photosystems (PS I and PS II) are associated with series of electron carriers (Sudhir and Murthy, 2004). Light-harvesting phycobiliproteins (phycocyanin, and phycoerythrin) are responsible for the blue-green pigmentations (Patel et al., 2005; Singh et al., 2015). Phycobilisomes are attached to the thylakoid membrane and associated with the supra-molecular complex in a regular fashion (Glazer, 1984; Watanabae et al., 2014). Transfer of excitation energy takes place in a series of phycobiliproteins from phycoerythrin to phycocyanin to allophycocyanin and ultimately to chlorophyll *a*. Phycobiliproteins are water soluble and serve as reserve protein (Campbell et al., 1998; Singh et al., 2015).

Sodium Ion Acts as Nutrients and Toxicants

Sodium is the second most abundant solute in the ocean after chloride and sixth dominant elements in the earth crust (Lutgens and Tarbuck, 2003; Kronzucker et al., 2013). During the course of evolution, early life originated in salt-dominated environments and group such as Monera to Animalia are capable of tolerating salinity and require Na^+ for survival. Few groups of cyanobacteria known as halophilic have developed the ability to tolerate salt concentration up to saturation level (5.5 M NaCl) and survive at high NaCl environments. Development of the ability for utilization of Na^+ in cellular processes is osmotically challenging in cyanobacteria. The requirement of Na^+ for the growth of cyanobacteria up to certain level of salinity has been reported (Allen and Arnon, 1955; Tester and Davenport, 2003). It is important to emphasize that every substance has a threshold level for particular organisms and below is not toxic and dose decide if it is poison or a nutrient (Tester and Davenport, 2003). According to the Theophrastus Bombastus von Hohenheim, beneficial effects of Na^+ are seen in the range of concentrations and work as a nutrient ion (Flowers and Colmer, 2008).

The close similarity of hydrated ionic radii of Na^+ and K^+ due to discrimination between both cations by transporter proteins is the main reason of Na^+ toxicity (Hagemann, 2011; Almeida et al., 2017). Cyanobacterial cells maintain elevated Na^+ in the cells by active transport of K^+ from the external medium. Under salinity maintaining ionic ratio K^+/Na^+ of cyanobacteria in the cytosol and extrusion of Na^+ is very important. Transport of Na^+ is a passive process which is regulated by the negative potential difference of cell membrane and lower level of Na^+ concentration inside the cell that strongly favors Na^+ movement into the cells (Apsea and Blumwald, 2007). On the other hand, Na^+ removal from the cell is an active process mediated through Na^+/H^+antiporter (Blumwald et al., 2000). These are two key steps involved in the detoxification of Na^+ and osmotic adjustment of the cyanobacteria necessary for salinity tolerance.

Physiological Strategies of Cyanobacteria in Response to Salinity

Cyanobacteria maintain osmotic potential by import and export of cations from cells to the external environment (Hagemann, 2011). Two major physiological strategies are adopted by cyanobacteria for the tolerance in the stressful environments (Tripathi et al., 2017). The classical strategy is "salt-in strategy" frequently occurs in halophilic archaea, bacteria, and cyanobacteria (Thombre et al., 2016). Sequestration of Na^+ and K^+ into cytoplasm form external environments to maintain the osmotic equilibrium of cells (Maathuis et al., 2014). Although Na^+ is abundant in the extracellular environment, organisms accumulate intracellular K^+ and utilize ATP by the K^+ transporter (Assaha et al., 2017). The intracellular K^+ ion is higher than Na^+ and gradual expulsion of Na^+ using Na^+/H^+antiporter has been reported in cyanobacteria such as *Synechocystis* PCC 6714 and *Aphanothece halophytica* (Reed et al., 1985; Waditee et al., 2001). Accumulation of K^+ ion increased with increasing salinity in growth medium and accumulation of ion in the cytoplasm is, in fact, is a mechanism of osmoregulation of *Synechococcus* 6311 (Hagemann, 2011; Pade and Hagemann, 2014). Thus, K^+ and Na^+ ions play an important role in homeostasis of osmotic stress. Another strategy for the homeostasis of ionic composition of cytoplasm Na^+ and Cl^- efflux out into medium known as 'salt-out strategy' (Hagemann, 2011). Simultaneously, cells synthesized compatible solutes for adjustment of cellular osmotic potential (Kempf and Bremer, 1998). Compatible solute accumulation or synthesis widely occurs in bacteria, plants, fungi, and animals as an adaptive strategy to prevent deleterious effects of salinity (Hohmann, 1997; Burg et al., 1997; Tripathi et al., 2017).

Salinity Response of Cyanobacteria

General stress response in cyanobacteria are (1) water-deficit in cytosol and high solute concentrations in the cells (2) activation of P-ATPase activity and removal of Na^+ from cell to external medium (3) synthesis of osmolytes and over-expression of high affinity transporters expression of membrane proteins involved in ion transporter proteins, water channel proteins (Hagemann, 2011). Aquaporine regulates water transport in response to drought and salt stress (Afzal et al., 2016). Water channel proteins are preferably selected K^+ over Na^+ and regulate the integrity of the living cells. In this situation, cyanobacteria maintained homeostasis to enhanced uptake of K^+ and actively extrusion of Na^+ ions (Hasegawa, 2000; Checchetto et al., 2016).

Exact measurements of intracellular Na^+ in salt-treated cyanobacterial cells are difficult. The main reason is the localization of periplasm of inside the cells and another reason is Na^+ contamination in the medium. Cyanobacteria grown under hypersaline conditions show significantly higher Na^+ concentrations inside the cell compared to the external medium, suggesting that active transport of Na^+ out of the cells (Reed et al., 1984a). Reports are available on the correlation between cytoplasmic volume and total cell volume and intracellular volume of *Anabaena* strains is about 40% of the packed total cell volume (Apte and Thomas, 1986).

Antiporter (Na^+/H^+) for Ionic Homeostasis

Na^+/H^+ antiporter mechanism plays a vital role in ionic homeostasis, maintenance of cytoplasmic pH and turgor pressure (Conde et al., 2011). Export of Na^+ is mainly driven by the differences in a proton gradient. Blumwald et al. (1984) reported that exchange of Na^+ with H^+ in salt-treated cells of *Synechococcus* 6301. Alkaline conditions, however, activated primary active transport of Na^+ (Ritchie, 1992a). Padan and Schuldiner, (1993) reported that export of Na^+ by Na^+/H^+ antiporter in *E. coli*. Kaneko et al., (1996) reported that in the cyanobacterium *Synechocystis* 6803 six antiporter genes (Na^+/H+) were activated in response to salt stress. Billini et al., (2008) observed multiple Na^+/ H^+ genes in cyanobacteria. The occurrence of multiple gene families of antiporter (Na^+/H^+) indicates the survival strategies of cyanobacteria under salinity. Antiporters have also been studied in the halotolerant cyanobacterium *Aphanothece halophytica* and multiple Na^+/H^+ antiporter genes were reported (Wutipraditkul et al., 2005). Na^+/H^+ antiporter gene (*ApNhaP*) were cloned in the freshwater cyanobacterium *Synechocystis* 6803 from *A. halophytica* and transformed restored salt tolerance and capable of surviving at alkaline pH and hypersaline environment (Wutipraditkul et al., 2005).

ATP-Dependent K^+ Transport

K^+ plays a significant role in salt tolerance, osmotic potential, regulation of pH, enzymes activity, membrane energetics and gene expression (Wood, 1999; Ballal et al., 2007; Barragan et al., 2012). Unlike Na^+, exact estimation of intracellular K^+ is difficult.

Therefore, K^+ concentrations are represented in terms of total cell volume. NMR and radioactive isotopes techniques have been used for detection of K^+ ion and their concentration in the range of 100-200 mM in cyanobacterial (Ballal et al., 2007). The reduction of K^+ in salt-stressed cells and massive accumulation of osmolytes is essential for the survival of cyanobacteria (Hassan et al., 2016). Under salt stress, intracellular accumulation of trehalose and reduction of K^+ was studied in *E. coli* (Dinnbier et al., 1988). In animal cells, active transport of Na^+/K^+ ATPase acts as exchanger and osmoregulation. Inhibitions of Na^+/K^+ ATPase by Ouabain also indicate a close similarity of ATPase of *Nostoc* strain M-14 to the animal cells (Hangman et al., 2011).

REGULATION OF K^+ TRANSPORT THROUGH *KDP* SYSTEM

Cyanobacterial ATP-dependent K^+ transport is made up of the three *Kdp* ABC subunits (Checchetto et al., 2016). The *KdpA* subunit is the K^+ permease, KdpB subunit is a P-type ATPase and *KdpC* is responsible for assembly of the transport system (Hangman et al., 2011). Mutant analysis of *kdp* and *ktr* revealed distinct roles of K^+ transporters in the *Synechocystis* sp. PCC 6803 (Nanatani et al., 2014). *Anabaena* 7120 possesses two operons which encode all structural *kdp* genes designated *kdp*1 and *kdp*2 (Ballal and Apte, 2008). However, the unicellular freshwater cyanobacterial strains consist of single operon *kdp*2 types (Ballal et al., 2007). *Kdp* gene cluster is completely absent in marine *Prochlorococcus* and *Synocoechccus* species. Inducible *kdp* operon was reported when K^+ is limiting factor under salt stress in *E. coli* (Wood, 1999). *Kdp* system is important in *E. coli* for immediate uptake of K^+ during salt stress. The KdpD protein is, in fact, a histidine kinase which is present in cytoplasmic membrane and assumed that since internal K^+ or salinity stress signal and regulation of *kdp* operon is performed via the *Kdp* DE two-component system (Wood, 1999). Ballal et al., (2007) reported that *kdp* operons were up-regulated at a low level of K^+ in cyanobacteria. These results indicated that the activation of *Kdp* system is not necessary to salt stress but important for maintaining K^+ concentration in the medium. Kuo et al., (2005) observed that three classes of proteins are involved in K^+ transporters in *E. coil*. Follmann et al., (2009) studied K^+ channel proteins and its transport in the *Corynebacterium glutamicum*. Genome sequences analysis of *Anabaena* 7120 and *Synechocystis* 6803 revealed that two-gene operon in upstream in the *kdp2* gene which encodes two-component systems (Rre and Hik) protein (Ballal and Apte, 2008). Another K^+ transport system (*Ktr*) is responsible for the accumulation of K^+ in salt-stressed of cyanobacteria (Matsuda et al., 2004). *Trk* system of *E. coli* has closed resemblance of Ktr system of cyanobacteria and it consists of three subunits; KtrB function as permease, and two small subunits *KtrA* and *KtrE* (Matsuda et al., 2004). Deletion of *ktr* genes of salt tolerant strain *Synechocystis* 6803 made it sensitive. *kdpA* mutant could grow at high salt concentrations with efficient K^+ transport whereas Na^+

transport was unaffected. *Ktr* system is the main transporter than *Kdp* system in cyanobacteria. However, the Ktr system of cyanobacteria has a higher affinity for K^+ and *Km* value was around 60 mM and capable to catalyze K^+ transport rather than low K^+ concentration environment of cyanobacteria (Matsuda et al., 2004). On the basis of *Ktr* functions, K^+ uptake is dependent on Na^+ concentration; i.e., high K^+ transport rates were observed for Na^+ in environments (Matsuda et al., 2004; Checchetto V et al., 2016). Therefore, it was assumed that the transporter might exchange K^+ with Na^+ or need the Na^+ gradient for transport (Tashkin et al., 2017). However, *E. coli* mutants clearly revealed that the *Ktr* system is not able to transport Na^+ most probably and KtrB subunit induces conformational change upon Na^+ binding and subsequently stimulates K^+ transport (Matsuda et al., 2004). Activation by the Na^+ binding, K^+ transport activity is rapidly enhanced upon salt stress. Obviously, the Na^+ concentration of the external and internal of cells directly regulates the transport activity. On the other hand, the expression levels of the *Ktr* system do remain constant in the acclimation cyanobacterial cells (Marin et al., 2004). The driving force for K^+ transport by *Ktr* is probably due to the membrane potential that was recently supported by the bioenergetics study with inhibitors affecting either the membrane potential or proton gradient (Matsuda and Uozumi, 2006).

COMPATIBLE SOLUTE SYNTHESIS IN CYANOBACTERIA

Osmolytes are low-molecular-weight organic compounds, synthesized in a molar range in response to abiotic stress such as salinity, drought, and temperature (Yancey et al., 1982; Ventosa et al., 1998). These molecules are synthesized in archaea, bacteria, cyanobacteria, fungus, and plants (Brown, 1976; Roberts, 2005; Tripathi et al., 2017). Compatible solutes directing the remaining free water into the vicinity of macromolecules surface and create hydration shell to prevent denaturation and establishing turgor pressure of salt stress cells (Pade and Hagemann, 2014). Bremer and Kremer, (2000) reported that mechanism of compatible solutes can be explained on the basis of water exclusion principles. According to this model protective action of compatible solutes, indirect interacts with macromolecules such as proteins and/or a membrane protein. Biochemically osmolytes are diverse in nature and include amino acid and derivatives (proline and glycine betaine) nucleic acid derivatives (ectoine), and polyols (trehalose and mannitol) (Tripathi et al., 2017). Chen and Murata, (2002) reported that compatible solutes are able to scavenge the reactive oxygen species (ROS) which are byproducts of ionic and hyperosmotic stresses leading to cell death. The free and dissolved state of compatible solutes can be quantified and characterized by the techniques of HPLC and NMR (Norton et al., 1982; Nagata et al., 1996; Kalsoom et al., 2016). Among the cyanobacteria, the correlation was found between the salt tolerance and synthesis of compatible solutes (Klähn and Hagemann, 2011). Cyanobacterial strains grown at a low level of salinity (0.60 M) accumulates sucrose or trehalose as a major compatible solute. Moderately salt tolerant strains of cyanobacteria (1.7 M) synthesized glucosylglycerol (GG) or glucosylglycerate (GGA) and halophilic or halotolerant cyanobacterial grow up to in saturated level of salt concentrations synthesized ectoine and glycine betaine (GB) (Kurosova, 1982). Osmoregulation is one of the effective approaches which can be considered for the developing stress tolerant cyanobacteria (Singh et al., 2013). Transformation of gene *hetR* was reported by Chaurasia and Apte, (2011) which increased the functional heterocyst frequency of *Anabena* 7120. *Sirisattha* et al., (2012) transformed N-methyltransferases genes form *A. halophytica* to *Anabaena* 7120 to make it salt tolerant and capable to tolerate up to 120 mM NaCl. Singh et al., (2013) transformed N-methyltransferases genes encoded GB in freshwater *A. doliolum*. Transformed cells became halophilic and enhanced the N_2 fixing ability as compared to freshwater. Expression of compatible solutes in freshwater cyanobacteria is an eco-friendly approach for the sustainability and increasing soil nitrogen dynamics of paddy fields. Genetic engineering of cyanobacteria is emerging as a feasible approach for the development of salt-tolerant cyanobacteria.

CONCLUSION

Salt stress is a major constraint for agriculture productivity at the global level. Cyanobacteria are unique models to study the salinity stress in the photosynthetic microbes. Effective strategies are adopted by cyanobacteria under salinity which has been well characterized at physiological and molecular levels. However, much more remain to be explored. Cyanobacteria have developed efficient ionic mechanisms to regulate cellular homeostases such as ion transporters, synthesis/uptake of osmolytes and antioxidant enzymes level and regulation of several genes. Molecular studies have been carried out on the transporters and transformation of osmolyte specific genes. Transgenic approaches mainly manipulation of genes in freshwater N_2 fixing cyanobacteria also showed promising results. However, limited success has been obtained in cyanobacteria. Therefore, future research should be directed towards a better understanding of the physiological, biochemical and molecular mechanisms for ionic homeostasis to develop an effective strain of salt-tolerant cyanobacteria as bioinoculants.

ACKNOWLEDGMENTS

The author thankfully acknowledges the Department of Science and Technology (DST) for financial assistance.

Conflict of Interest

No conflict of interest among the authors.

References

Afzal, Z., Howton, T. C., Sun, Y., and Mukhtar, M. S. (2016). The Roles of aquaporins in plant stress responses zunaira. *J. Dev. Biol*, 4, 9.

Akbarimoghaddam, H., Galavi, M., Ghanbari, A., Panjehkeh, N. (2011). Salinity effects on seed germination and seedling growth of bread wheat cultivars. *Trakia J. Sci.* 9(1): 43-50.

Allakhverdiev, S. I., and Murata, N. (2008). Salt stress inhibits photosystems II and I in cyanobacteria. *Photosynth. Res,* 98: 529–539.

Almeida, D. M., Oliveira, M. M., and. Saibo, N. J. M. (2017). Regulation of Na^+ and K^+ homeostasis in plants: towards improved salt stress tolerance in crop plants. *Gent. Mol. Bio*, 40, 326-345 (2017).

Allen, M. B., Arnon, D. I. (1955). Studies on nitrogen-fixing blue-green algae, growth and nitrogen fixation by *Anabaena cylindrical* Lemm. *Plant. Physiol,* 30:366–372.

Apsea, M. P., and Blumwald, E. (2007). Na^+ transport in plants. *FEBS. Letters*, 581: 2247–225.

Apte, S. K., and Thomas, J. (1986). Membrane energetics and sodium transport in filaments nitrogen fixing cyanobacteria. *Euro. J. Biochem*, 154: 395-401.

Assaha, D. V. M, Ueda, A., Saneoka, H., Al-Yahyai, R., and Yaish, M. W. (2017). The Role of Na^+ and K^+ transporters in salt stress adaptation in glycophytes. *Front. Physiol,* 8:1-19.

Ballal, A., and Apte, S. K. (2008). Characterization of a response regulator protein that binds to *Anabaena sp*. strain. *Arch. Biochem. Biophys*, 474: 65–71.

Ballal, A., Basu, B., and Apte, S. K. (2007). The Kdp-ATPase system and its regulation. *J Bioscience,* 32: 559–568.

Barragan, V., Leidi, E. O., Andrés, Z., Rubio, L., De, Luca A., Fernandez, J. A., Cubero, B., and Pardo, J. M. (2012). Ion exchangers NHX1 and NHX2 mediate active potassium uptake into vacuoles to regulate cell turgor and stomatal function in *Arabidopsis*. *Plant. Cell.* 24:1127-1142.

Billini, M., Stamatakis, K., and Sophianopouou, V. (2008). Two membrane of a network of putative Na^+/H^+ antiporters are involved in salt and pH tolerance of freshwater cyanobacterium *Synechococcus elongatus*. *J. Bacteriol*, 190:6318-6329.

Blumwald, E., Ahron, G. S., and Apse, P. M. (2000). Sodium transport in plants. *Biochem. Biophys. Res. Commun.* 1465 (1-2) 140-151.

Blumwald, E., Wolosin, J. M., and Packer, L. (1984). Na^+/H^+ exchange in the cyanobacterium *Synechococcus* 6311. *Biochem. Biophys. Res. Commun*, 122: 452–459.

Bremer, E., and Krämer, R. (2000). Coping with osmotic challenges: osmoregulation through accumulation and release of compatible solutes. In: *Bacterial Stress Responses*. Storz and R. Hengge-Aronis (eds). ASM Press, pp. 79-97.

Brown, A. D. (1976). Microbial water stress. *Bacteriol. Rev*, 40:803–846.

Burg, M., Kwon, E., Kültz, D. (1997). Regulation of gene expression by hypertonicity. *Annu. Rev. Physiol*. 59:437–455.

Campbell, D., Hurry, V., Clarke, A. K., Gustafsson, P., Öquist, G., (1998). Chlorophyll fluorescence analysis of cyanobacterial photosynthesis and acclimation. *Microbiol. Mol. Biol. R,* 30: 667-680.

Castenholz, R. W. (2001). General characteristics of the cyanobacteria. In: *Bergey's manual of systematic bacteriology*, 2nd edition, 1. D. R. Boone and R. W. Castenholz. (eds) New York: Springer, Vol. 1, pp. 474–487.

Cavalier-Smith, (2002). Chloroplast Evolution: secondary symbiogenesis and multiple mosses. *Curr. Biol*, 12: 62–64.

Chaurasia, A. K., and Apte, S. K. (2011). Improved eco-friendly recombinant *Anabaena* sp. strain PCC7120 with enhanced nitrogen biofertilizer potential. *Appl. Environ. Microbiol*, 77 (2)395-399.

Checchetto, V., Segalla, A., Satro, Y., Bergantino, E., Szabo, I., and Uozumi, N. (2017). Involvement of potassium transport system in response of *Synechocystis* PCC 6803 cyanobacteria to external pH change, high intensity light stress and heavy metal stress. *Plant. Cell. Physiol*, 57(4):862-874.

Chen, T. H. H., and Murata, N. (2002). Enhancement of tolerance of abiotic stress by metabolic engineering of betaines and other compatible solutes. *Curr. Opin. Plant. Biol*, 5:250–257.

Checchetto, V., Segalla A, Sato Y, Bergantino E, Szabo I, and Uozumi N (2016) involvement of potassium transport systems in the response of *synechocystis* PCC 6803 cyanobacteria to external pH change, high-intensity light stress, and heavy metal stress. *Plant. Cell. Physiol*, 57(4): 862–877.

Conde, A., Chaves, M.M., and Geros, H. (2011). Membrane transport, sensing and signaling in plant adaptation to environmental stress. *Plant. Cell. Physiol*, 52(9): 1583–1602.

Das, K., and Roychoudhury, A. (2014). Reactive oxygen species (ROS) and response of antioxidants as ROS-scavengers during environmental stress in plants. *Front. Environ. Sci*, 53:1-13.

Dinnbier, U., Limpinsel, E., Schmid, R., and Bakker, E. P. (1988). Transient accumulation of potassium glutamate and its replacement by trehalose during adaptation of growing cells of *Escherichia coli* K-12 to elevated sodium chloride concentrations. *Arch. Microbiol*, 150: 348–357.

Dubey, A. K., and Rai, A. K. (1995). Application of algal biofertilizers (*Aulosira fertilissima* var. tenuis and *Anabaena doliolum* Bharadwaja) for sustained paddy cultivation in Northern India. *Israel. J. Plant. Sci.* 43:41–51.

Etana, P., Shimon, S. (1993). Na^+/H^+ antiporter, molecular devices that couple the Na^+ and H^+ circulation in cells. *J. Bioenerg. Biomembr*, 25, 6:647–669.

FAO, *High level expert forum-how to feed the world in 2050, economic and social development, food and agricultural organization of the United Nations, Rome, Italy, 2009.*

Fischer, B. B., Hideg, E., and Krieger-Liszkay, A. (2013). Production, detection, and signaling of singlet oxygen in photosynthetic organisms. *Antioxid. Redox. Sign*, 18:2154–216.

Flowers, T. J., and Colmer, T. D. (2008). Salinity tolerance in halophytes. *New. Phytol*, 179:945–963.

Follmann, M., Becker, M., Ochrombel, I., Ott, V., Krämer, R Marin, K (2009). Potassium Transport in *Corynebacterium glutamicum* is Facilitated by the Putative Channel Protein CglK, Which Is Essential for pH Homeostasis and Growth at Acidic pH. *J. Bacteriol,* 191. 9: 2944-2952.

Fufezan, C., Gross, C. M., Sjodin M, Rutherford, A. W., Krieger-Liszkay, A., Kirilovsky, D. (2007). Influence of the redox potential of the primary quinone electron acceptor on photoinhibition in photosystem II. *J. Biol. Chem*, 282: 12492–12502.

Fulda, S., Huckauf, J., Schoor, A., *and* Hagemann, M. (1999). Analysis of stress responses in the cyanobacterial strains *Synechococcus* sp. PCC 7942, *Synechocystis* sp. PCC 6803, and *Synechococcus* sp. PCC 7418*:* osmolyte accumulation and stress protein synthesis. *J. Plant. Physiol,* 154*:*240–249.

Glazer N. A. (1984). Phycobilisome a macromolecular complex optimized for light energy transfer. *Biochem. Biophy. Acta, Rev. Bioenerge* 768, (1):29-51.

Glick, R. B., Todorovic, B., Czarny, J., Cheng, Z., Duan, J., McConkey, B. (2007). Promotion of plant growth by ACC deaminase. *Cric. Rev. Plant. Sci.* 26 (5 6):227-242.

Golldack, D., Li, C., Mohan, H., and Probst, N. (2014). Tolerance to drought and salt stress in plants: unraveling the signaling networks. *Front. Plant. Sci*, 5, 10.

Gralla, J. D., and Vargas, D. R. (2005). Potassium glutamate as transcriptional inhibitor during bacterial osmoregulation. *EMBO. J,* 25: 1515–1521.

Gupta, B., and Huang, B. (2014). Mechanism of salinity tolerance in plants: physiological, biochemical and molecular characterization. *Int. J. Genomics*, 1-18.

Hagemann, M. (2011). Molecular biology of cyanobacterial salt acclimation. *FEMS. Microbiol. Rev*, 35: 87–123.

Hasegawa, P. M., Bressan, R. A., Jhu, J. K., and Bohnert, H. J. (2000). Plant cellular and molecular response to high salinity. *Ann. Rev. Plant. Physiol. Plant. Mol. Biol*, 51: 463-499.

Hohmann, S. (1997). Shaping up the response of yeast to osmotic stress. In, Hohmann S., Mager W. H. (eds). *Yeast stress responses*. Springer, Berlin Heidelberg, New York, pp. 101–145.

Hu, Y., Schmidhalter, U. (2002). Limitation of salt stress to plant growth. In: Hock, B., Elstner, C.F. (eds.), *Plant Toxicology*. Marcel Dekker Inc., New York, pp. 91–224.

Janet, M., Wooda, U., Erhard, Bremerb, Laszlo, N., Csonkac, Reinhard, K., Bert, P., Tiemen, van der Heidee., and Linda, T. (2001). Osmosensing and osmoregulatory compatible solute accumulation by bacteria. *Comparative. Biochem. Physiol*, 437-460

Kaneko, T., Sato, S., Kotani, H., Tanaka, A., Asamizu, E., Nakamura, Y., et al. (1996). Sequence analysis of the genome of the unicellular cyanobacterium *Synechocystis* sp. strain PCC 6803. II. Sequence determination of the entire genome and assignment of potential protein-coding regions. *DNA Res.* 3: 185-209.

Kempf, B., and Bremer, E. (1998). Uptake and synthesis of compatible solutes as microbial stress responses to high-osmolality environments. *Arch Microbiol* 170: 319-330.

Klähn, S., and Hagemann, M. (2011). Compatible solute biosynthesis in cyanobacteria. *Environ Microbiol* 13(3): 551-562.

Kronzucker, H. J., Devrim, M. C. L. N., Schulze, J. R., and Wong, B. D. T. (2013). Sodium as nutrient and toxicant. *Plant. Soil*. 369:1-23.

Lutgens, F. K., and Tarbuck, E. J. (2003). *Foundations of earth science*. Pearson, New Jersey

Maathuis, F. J. M., Ahmad, I., Patishtan, J. (2014). Regulation of Na^+ fluxes in plants. *Front. Plant. Sci*, (5)1-8.

Marin, K., Kanesaki, Y., Los, D.A., Murata, N., Suzuki, I., and Hagemann, M. (2004). Gene expression profiling reflects physiological processes in salt acclimation of *Synechocystis* sp. strain PCC 6803. *Plant. Physiol.*, 136: 3290–3300.

Matsuda, N., and Uozumi, N. (2006). Ktr-mediated potassium transport, a major pathway for potassium uptake, is coupled to a proton gradient across the membrane in *Synechocystis* sp. PCC 6803. *Biosci. Biotech. Bioch*, 70: 273–275.

Matsuda, N., Kobayashi, H., Katoh, H., Ogawa, T., Futatsugi, L., Nakamura, T., Bakker, E. P., and Uozumi, N. (2004). Na^+-dependent K^+ uptake Ktr system from the cyanobacterium *Synechocystis* sp. PCC 6803 and its role in the early phases of cell adaptation to hyperosmotic shock. *J. Biol. Chem*, 279: 54952–54962.

Mittler, R. E., and Blumwald. (2010). Genetic engineering for modern agriculture: challenges and perspectives. *Annu. Rev. Plant. Biol*, 61:443–462.

Nagata, S., Adachi, K., and Sano, H., (1996). NMR analyses of compatible solutes in a halotolerant *Brevibacteriums* p. *Microbiology* (1996), 142, 3355-3362.

Nanatani, K., T, Shijuku, Takano, Y., Zulkifli, L., Yamazaki, T., Tominaga, A., Souma, S., Onai, K., Morishita, M., Ishiura, M., Hagemann, M., Suzuki, I., Maruyama H., Arai, Fo, and Uozumi, N. (2015). Comparative Analysis of *kdp* and *ktr* Mutants Reveals Distinct Roles of the Potassium Transporters in the Model Cyanobacterium *Synechocystis* sp. Strain PCC 6803. *J. Bacteriol.* 197(4)676-687.

Nishiyama, Y., Allakhverdiev, S. I., and Murata, N. (2006). A new paradigm for the action of oxygen species in the photoinhibition of photosystem II. *Biochm. Biophy. Acta*, 1757: 742–749.

Nishiyama, Y., Allakhverdiev, S. I., Yamamoto, H., Hayashi, H., and Murata, N. (2004). Singlet oxygen inhibits the repair of photosystem II by suppressing the translation elongation of the D1 protein in *Synechocystis* sp. PCC 6803. *Biochemistry,* 43: 11321–11330.

Norton, R. S., Mackay, M. A., and Borowitzka, L. J. (1982). The physical state of osmoregulatory solutes in unicellular algae. A natural abundance carbon-13 nuclear-magnetic-resonance relaxation study. *Biochem J*, 202: 699–70.

Ohnishi, N., Allakhverdiev, S. I., Takahashi, S., Higashi, S., and Watanabe, M., Nishiyama, Y., and Murata, N. (2005). Two-step mechanism of photodamage to photosystem II: step 1 occurs at the oxygen evolving complex and step 2 occurs at photochemical reaction center. *Biochemistry,* 44: 8494–8499.

Olson, J. M. (2006). Photosynthesis in the *Archean* era. *Photosynth. Res*, 88 (2): 109–17.

Oren, A. (2000). Salts and brines. *The Ecology of Cyanobacteria,* Whitton B. A. and Potts, M., (eds), Kluwer, Academic Publishers, Dordrecht, Netherlands. 281–306.

Oren, A. (2007). Diversity of organic osmotic compounds and osmotic adaptation in cyanobacteria and algae. *Algae and Cyanobacteria in Extreme Environments.* Seckbach J ed. Springer, Dordrecht, Netherlands, pp. 639–655.

Padan, E and Schuldiner, S. (1993). Na^+/H^+antiporter molecular devices that couple the Na^+ and H^+ circulation in plant. *J. Bioenerg. Biomembran*, 25:647-669.

Pade, N., and Hagemann, M. (2014). Salt Acclimation of cyanobacteria and their application in Biotechnology, *Life*, 5: 25-49.

Patel, A., Sandhya, M., Pawar, R., Ghosh, P.K. (2005). Characterization of C of marine and freshwater habitat. *Protein. Expr. Purif,* 40(2): 248-255.

Patel, P. R., Kajal, S. S., Patel, V. R., Patel, V. J., and Khristi, S. M. (2010). Impact of salt stress on nutrient uptake and growth of cowpea. *Braz. J. Plant. Physiol*, 22:43-48.

Patterson, C. O., and Myers, J. (1973). Photosynthesis production of hydrogen peroxide by *Anacystis nidulans*. *Plant. Physiol*, 51: 104–109.

Rai, A. K., and Sharma, N. K. (2006). Phosphate metabolism in the cyanobacterium *Anabaena doliolum* under salt stress. *Curr. Microbiol*, 52:6–12.

Record, M. T. Jr., Courtenay, E. S., Cayley, D. S., and Guttman, H. J. (1998a). Responses of *E. coli* to osmotic stress: large changes in amounts of cytoplasmic solutes and water. *Trends Biochem. Sci*, 23:143–148.

Reed, R. H., J. A., Chudek, R., Foster and Stewart, W. D. P. (1984a) Osmotic adjustment in cyanobacteria from hypersaline environments. *Archs. Microbiol*, *138:*333–337.

Reed, R. H., Richardson, D. L., and Stewart, W. D. P. (1985). Na^+ uptake and extrusion in the cyanobacterium *Synechocystis* PCC 6714 in response to hypersaline treatment: evidence for transient changes in membrane permeability. *Biochim. Biophys. Acta. 814*, 347–355.

Ritchie, R. J. (1992a). Sodium transport and the origin of the membrane potential in the cyanobacterium *Synechococcus* R-2 (*Ana-cystisnidulans*) PCC 7942. *J. Plant. Physiol*, 139: 320–330.

Roberts M. F. (2005). Organic compatible solutes of halotolerant and halophilic microorganisms. *Saline. Syst.* 1: 1-30.

Sharma, N. K., Tiwari, S. P., Tripathi, K., Rai, A. K. (2011). Sustainability and cyanobacteria (blue-green algae): facts and challenges. *J. Appl. Phycol*, 23:1059–1081.

Singh, M., Sharma, N. K., Prasad, S., Yadav, S. S., Narayan, G., and Rai A. K. (2013). The freshwater cyanobacterium *Anabaena doliolum* transformed with *ApGSMT-DMT* exhibited enhanced salt tolerance and protection to nitrogenase activity but became halophilic. *Microbiology*, 159:641–648.

Singh, N. K, Sonani, R. R., Rastogi, R. P., and Madamwar, D. (2015). The phycobilisomes: an early requisite for efficient photosynthesis in cyanobacteria. *EXCLI. J*, 14: 268–289.

Sirisattha, W., R., Singh, M., Kageyama, H., Sittipol, D., Rai, A. K. and Takabe, T. (2012) *Anabaena* sp. PCC7120 transformed with glycine methylation genes from *Aphanothece halophytica* synthesized glycine betaine showing increased tolerance to salt. *Arch. Microbiol,* 194, 909-914.

Soo, R. M., Skennerton, C. T., Sekiguchi, Y., Imelfort, M., Paech, S. J., Dennis, P. G., Steen, J. A., Parks, D. H., Tyson, G, W., Hugenholtz, P. (2014). An expanded genomic representation of the phylum Cyanobacteria. *Genome. Biol. Evol*, 6:1031-1045.

Sudhir, P. and Murthy, S. D. S. (2004). Effects of salt stress on basic processes of photosynthesis. *Photosynthetica,* 42:481-486.

Tester, M., and Davenport. (2003). Na^+ tolerance and Na^+ transport in higher plants. *Ann. Bot*, 91: 503-527.

Thajuddin N. G. and Subramanian (2005). Cyanobacterial biodiversity and potential applications in biotechnology. *Curr. Sci*, 89: 1-10.

Tripathi, K., Abraham, G., Rai, V., Dhar, D. W., Vishakarma, R., Yadav, R. K. (2017) *Compatible solute in relation to microbes and their biological significance* (editors

Ram Prasad, Narendra Kumar Microbes and Sustainable Agriculture) I. K. Publication. New Delhi.

Tripathi, K., Sharma, N. K., Kageyama, H., Takabe, T., and Rai, A. K. (2013). Physiological, biochemical and molecular studies of halophilic cyanobacterium *Aphanothece halophytica* to Pi-deficiency. *Euro. J. Phycol*, 48(4):461-473.

Tripathi, K., Sharma, N. K., Rai, V., and Rai, A. K. (2013). Low cellular p-quota and poor metabolic adaptations of the freshwater cyano-bacterium *Anabaena fertilissima* Rao during Pi-limitation. *Antonie Leeuw*, 103:277-291.

Thombre, R. S., Shinde, V. D., Oke, R. S., Dhar, S. K., and Shouche, Y. S. (2016). Biology and survival of extremely halophilic archaeon *Haloarcula marismortui* RR12 isolated from Mumbai salterns, India in response to salinity stress. *Sci. Rep.* DOI: 10.1038/srep25642.

Ventosa, A., Nieto, J. J., and Oren, A. (1998). Biology of moderately halophilic aerobic bacteria *Microbiol. Mol. Biol. Rev*, 62: 504-544.

Vicente, J.B., Gomes, C. M., Wasserfallen, A., and Teixeira, M. (2002). Module fusion in an A-type flavoprotein from the cyanobacterium *Synechocystis* condenses a multiple component pathway in a single polypeptide chain. *Biochem. Biophys. Res. Commun.* 294: 82-87.

Waditee R, Hibino T, Tanaka Y, Nakamura T, Incharoensakdi A, Takabe T (2001) Halotolerant cyanobacterium *Aphanothece halophytica* contains an Na^+ /H^+ antiporter, homologous to eukaryotic ones, with novel ion specificity affected by C terminal tail. *J. Biol. Chem*, 276:36931-36938.

Watanabae, M., Semchonokc, D. A., Webber-Birungic, M. T, Ehirad, S., Kondoa, K., Narikawaa, Rei, Ohmorid, M., Boekemac, E. J., and Ikeuchia, M. (2014). Attachment of phycobilisomes in an antenna–photosystem I super complex of cyanobacteria. *Proc. Natl. Acad. Sc,* 111 (7) 2512–2517.

Waditee, R., Bhuiyan, M. N. H., Rai, Vet, al. (2005). Genes for direct methylation of glycine provide high levels of glycine betaine and abiotic-stress tolerance in *Synechococcus* and *Arabidopsis. P. Natl. Acad. Sci, USA* 102: 1318-1323.

Waditee, R., Hibino, T., Nakamura, T., Incharoensakdi, A., and Takabe, T. (2002). Overexpression of a Na^+/H^+ antiporter confers salt tolerance on a freshwater cyanobacterium, making it capable of growth in sea water. *P. Natl. Acad. Sci*, 99: 4109-4114.

Wang, W., Vinocur, B., and Altman, A. (2003). Plant responses to drought, salinity and extreme temperatures: towards genetic engineering for stress tolerance. *Planta*, 218:1-14.

Woese, C. R. O., Kandler, M. L. and Wheeler (1990). Towards a natural system of organisms: Proposal for the domains *Archaea*, *Bacteria* and *Eucarya*. *Proc. Natl. Acad. Sci,* US 87:4576-4579.

Wood, J. M. (1999). Osmosensing by bacteria: signal and membrane based sensor. *Microbiol Mol. Biol*. 63:230-262.

Wutipraditkul, N., Waditee, R., Incharoensakdi, A., Hibino, T., Tanaka, Y., Nakamura, T., Shikata, M., Takabe, T., and Takabe, T. (2005). Halotolerant cyanobacterium *Aphanothece halophytica* contains NapA-type Na^1/H^1 antiporters with novel ion specificity that are involved in salt tolerance at alkaline pH. *Appl. Environ. Microbiol*, 71: 4176–4184.

Yamaguchi, T, E., and Blumwald. (2005). Developing salt-tolerant crop plants: challenges and opportunities. *Trends. Plant. Sci*, 10:615-620.

Zhu, J. K. (2003). Regulation of ion homeostasis under salt stress. *Curr. Opin. Plant. Biol,* 6: 441-445.

Zunaira, A. T. C., Howton, Y. S., and Shahid, M. M. (2016). The roles of aquaporins in plant stress responses. *J. Dev. Biol*. 4, 9.

About the Editors

Keshawanand Tripathi obtained a doctoral degree in Biotechnology from Banaras Hindu University, Varanasi (India), on the topic of "Phosphate acquisition, utilizations growth and metabolic activities of freshwater (Anabaena fertilissima) and halophilic Aphanothece halophytic cyanobacterium under P-limitation and salt stress". He has published ten research papers in peer reviewed journals, including international and national research paper. Dr Tripathi received DS Kothari Post-doctoral fellowship and DST-young scientist project on the transformation of salt tolerant genes in freshwater salt sensitive cyanobacterium. He has five year research experience in the field of salinity and nutritional stress of cyanobacteria. Currently he is working on the salinity stress of cyanobacteria at Centre for Conservation and Utilization of Blue Green Algae, ICAR-Indian Agriculture Research Institute, New Delhi.

Narendra Kumar is Assistant Professor, Department of Biotechnology, IMS Engineering College, Ghaziabad, Uttar Pradesh. He obtained his PhD (Microbiology) from Jiwaji University, Gwalior, Madhya Pradesh and MSc in Applied Microbiology, College of Life Sciences, CHRI, Gwalior. He has more than twenty publications and a book to his credit, which include research papers and book chapters at international and national levels. Dr Narendra has fourteen years of research and teaching experience. He has also served as Head of the Department and Coordinator- Internal Quality Assurance Cell, Department of Biotechnology, IMS Engineering College, Ghaziabad, Uttar Pradesh. He has also served as Member of Board of Studies and RDCs of State Technical and other Universities.

Gerard Abraham obtained his doctoral degree in Botany from Banaras Hindu University, Varanasi. At present he is working as Principal Scientist at Centre for Conservation and Utilization of Blue Green Algae, ICAR-Indian Agriculture Research Institute, New Delhi. Dr. Abraham works on the stress physiology and biochemistry of cyanobacteria and Azolla. He has teaching and research experience spanning a period of more than twenty years and guided doctoral and post doctoral research work. Dr. Abraham has handled research projects from different funding agencies and published several research papers in journals of national and international repute. Currently he is working on the proteomics of salinity stress in Azolla and cyanobacteria.

INDEX

B

C

D

E

F

G

H

I

K

L

M

N

O

P

T

U

V

W

X

Y

Z

β

γ

Environmental Biotechnologies for Bioremediation of Contaminated Lands and Soil by Microbes, Plants and Earthworms

Authors: Rajiv K. Sinha and Dalsukh Valani (Griffith University, Brisbane, Australia), Shweta Sinha (University of Rajasthan, Jaipur, India)

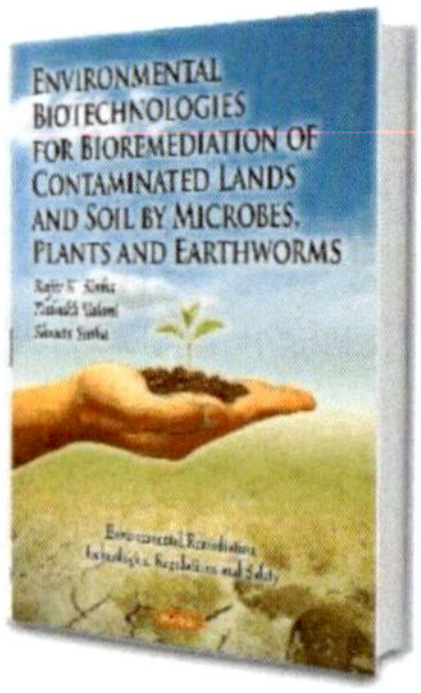

Series: Environmental Remediation Technologies, Regulations and Safety

Book Description: Bioremediation is a soft bioengineering technique to clean up contaminated lands and soils using microbes, plants and earthworms. This book addresses these issues regarding the benefits of microbes, plants and earthworms in bioremediation.

Softcover ISBN: 978-1-61668-148-7
Retail Price: $69

Soil Microbes and Environmental Health

Editor: Mohammad Miransari (Department of Book&Article, AbtinBerkeh Ltd. Company Isfahan, Iran)

Series: Environmental Science, Engineering and Technology

Book Description: In the present book, a team of researchers and scientists have contributed their research, ideas and thoughts regarding the microbial methods by which the environment can be kept clean and healthy.

Hardcover ISBN: 978-1-61209-647-6
Retail Price: $175

Microalgae and Man

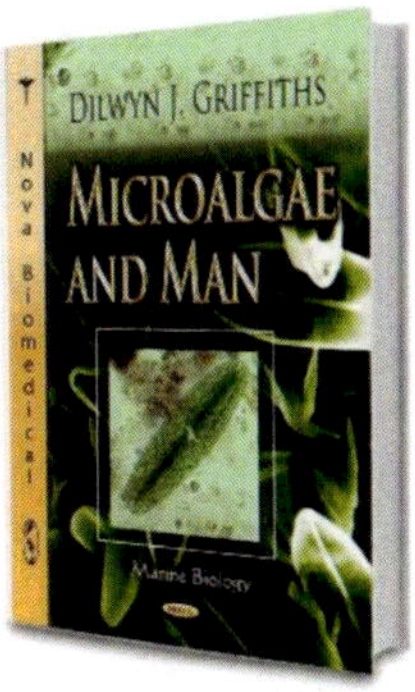

Author: Dilwyn J. Griffiths (School of Biological Sciences, James Cook University, Queensland, Australia)

Series: Marine Biology

Book Description: This book surveys our current understanding of those aspects of the biology of microalgae which constitute the basis of the range of practical applications now under consideration for their potential contribution to human health and well-being.

Hardcover ISBN: 978-1-62808-215-9
Retail Price: $179

Microbes in Soil and Their Agricultural Prospects

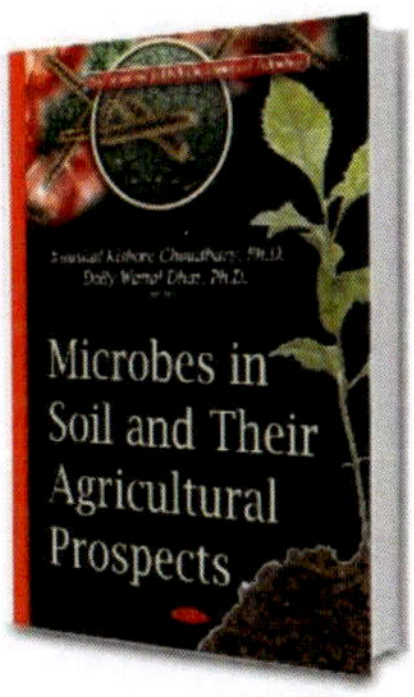

Editors: Kaushal Kishore Choudhary, Ph.D. and Dolly Wattal Dhar, Ph.D. (Dr Jagannath Mishra College, Affiliated to Babasaheb Bhimrao Ambedkar Bihar University, Muzaffarpur, Bihar, India and others)

Series: Air, Water and Soil Pollution Science and Technology

Book Description: The book *Microbes in Soil and Their Agricultural Prospects* is a collection of advantageous, informative, simulative and holistic viewpoints presenting basic and applied aspects of microbial functioning in soil. This book covers physiological, biochemical and molecular mechanisms of microbiomes pertaining to the production of available nitrogen (nitrogen fixation), phosphorus (P mobilization) and plant-growth promoting hormones for adaptation in agricultural soil.

Hardcover ISBN: 978-1-63482-824-6
Retail Price: $325

Microalgae and Other Phototrophic Bacteria: Culture, Processing, Recovery and New Products

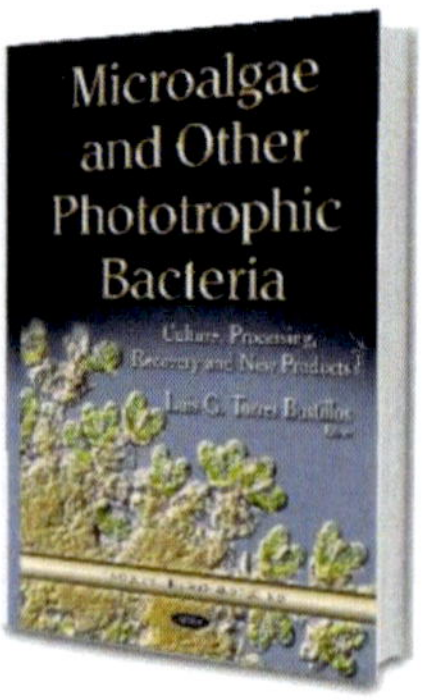

Editor: Luis G. Torres Bustillos (Bioprocess Department, UPIBI-Instituto Politecnico Nacional, Calle Acueducto, Col. Barrio La Laguna-Ticoman, Del Gustavo Madero, Mexico)

Series: Plant Science Research and Practices

Book Description: Microalgae are primary producers of oxygen and exhibit an enormous potential for biotechnological applications. Every chapter in this book has been written by an expert in the field, covering both the state of the art and the specific contributions of each group of authors in the worldwide literature.

Hardcover ISBN: 978-1-63482-078-3
Retail Price: $255

Plants and Microbes in an Ever-Changing Environment

Editor: Satya Shila Singh (Assistant Professor, Department of Botany, Guru Ghasidas Vishwavidyalaya, Bilaspur, Chhattisgarh, India)

Series: Plant Science Research and Practices

Book Description: In brief, present-day plants and microbes have obtained the resistance power from their precursors, which they have developed during the course of evolution. Hence, this book entails the consequences of environmental changes in plants and microbes, and the strategies adapted by them for survival.

Hardcover ISBN: 978-1-53610-918-4
Retail Price: $310